Electric Vehicle Design

Scrivener Publishing
100 Cummings Center, Suite 541J
Beverly, MA 01915-6106

Publishers at Scrivener
Martin Scrivener (martin@scrivenerpublishing.com)
Phillip Carmical (pcarmical@scrivenerpublishing.com)

Electric Vehicle Design

Design, Simulation and Applications

Edited by
Krishan Arora
Suman Lata Tripathi
and
Himanshu Sharma

Scrivener
Publishing

WILEY

This edition first published 2024 by John Wiley & Sons, Inc., 111 River Street, Hoboken, NJ 07030, USA
and Scrivener Publishing LLC, 100 Cummings Center, Suite 541J, Beverly, MA 01915, USA

For more information about Scrivener publications please visit www.scrivenerpublishing.com.

Wiley Global Headquarters
111 River Street, Hoboken, NJ 07030, USA

For details of our global editorial offices, customer services, and more information about Wiley products visit us at www.wiley.com.

Library of Congress Cataloging-in-Publication Data

ISBN 9781394204373

Front cover images supplied by Pixabay.com
Cover design by Russell Richardson

Set in size of 11pt and Minion Pro by Manila Typesetting Company, Makati, Philippines

Printed in the USA

10 9 8 7 6 5 4 3 2 1

Contents

Preface

Welcome to the exciting world of Electric Vehicle Design! In recent years, the automotive industry has witnessed a profound shift towards sustainable transportation, driven by the pressing need to mitigate environmental concerns and reduce dependency on fossil fuels. At the forefront of this revolution are electric vehicles (EVs), poised to redefine the way we move and reshape the landscape of transportation.

This book, *Electric Vehicle Design: Design, Simulation and Applications*, serves as a comprehensive guide to understanding the intricacies of designing and Engineering Electric Vehicles. Whether you're an automotive enthusiast, a student delving into the realm of transportation engineering, or a seasoned professional seeking to expand your knowledge, this book is designed to cater to your curiosity and provide valuable insights into the principles, technologies, and challenges inherent in EV design.

Within these pages, you will embark on a journey through the fundamental concepts of Electric Vehicle architecture, propulsion systems, energy storage solutions, and advanced technologies shaping the future of mobility. From the principles of electric drivetrains to the integration of cutting-edge battery technologies, each chapter is meticulously crafted to offer a blend of theoretical knowledge and practical applications.

Furthermore, this book goes beyond the confines of traditional Engineering texts by exploring the interdisciplinary nature of Electric Vehicle Design. You will discover the pivotal role of software and electronics in enhancing vehicle performance, the importance of aerodynamics in maximizing efficiency, and the crucial considerations in ensuring safety and regulatory compliance.

Moreover, "Electric Vehicles Design" aims to inspire innovation and foster a deeper understanding of the holistic approach required to develop sustainable transportation solutions. Through case studies, design challenges, and real-world examples, readers will be encouraged to think critically, problem-solve creatively, and contribute to the ongoing evolution of electric mobility.

As we stand on the cusp of a transformative era in transportation, the insights gleaned from this book will empower readers to navigate the complexities of electric vehicle design with confidence and vision. Whether your goal is to pioneer ground breaking technologies, shape policy decisions, or simply gain a deeper appreciation for the future of mobility, this book serves as your indispensable companion.

I extend my heartfelt gratitude to all the contributors, researchers, and pioneers whose dedication and innovation continue to drive the advancement of electric vehicles. It is my hope that this book will serve as a catalyst for further exploration, collaboration, and progress in the quest for sustainable transportation solutions.

So, without further ado, let us embark on this electrifying journey into the world of Electric Vehicle Design.

Happy reading!

1

Development of Braking Systems in Fuel Cell Electric Vehicles

Harpreet Singh Bedi*, Shakti Raj Chopra and Krishan Arora

School of Electronics and Electrical Engineering, LPU, Punjab, India

Abstract

Researchers are developing environmentally friendly, zero-emission cars to address concerns about greenhouse gas emissions. One promising solution is fuel cell vehicles, which use hydrogen as the primary energy source. These vehicles generate electricity by combining hydrogen with oxygen from the air, powering electric motors that drive the wheels. Additionally, they incorporate electric batteries to store excess energy during braking and aid during driving. The paper discusses the implementation of regenerative and anti-lock braking systems and models of vehicle components using MATLAB/Simulink software. The goal is to analyze the vehicle's performance under real working conditions using ADVISOR GUI models and demonstrate the benefits of fuel cell technology and advanced braking systems.

***Keywords*:** Vehicle model, braking systems, advisor

1.1 Introduction

Automotive engineers are concerned that burning fossil fuels in cars will lead to increased air pollution, increased carbon levels in the environment, and global warming. For this reason, engineers adopt electric vehicles. These vehicles are environmentally friendly. Battery electric vehicles are the most common vehicles on the market, but these vehicles require external charging. As a result, they are limited and require charging infrastructure to travel several kilometers. Recognizing these limitations of

**Corresponding author*: harpreet.17377@lpu.co.in

Krishan Arora, Suman Lata Tripathi and Himanshu Sharma (eds.) Electric Vehicle Design: Design, Simulation and Applications, (1–18) © 2024 Scrivener Publishing LLC

EVs, engineers introduced the fuel cell electric vehicle (FCEV). In this car, hydrogen was generated for locomotion and charging the onboard battery. Hydrogen is a clean fuel that does not pollute the environment. In a hydrogen society, FCEVs are not just eco-cars. The FCEVs with hydrogen fuel cells require stable performance, robust durability, and powerful performance. Therefore, automakers could test new technologies. Considering that the hydrogen required for fuel cell vehicles can be stored indefinitely and the manufacturing process is environmentally friendly, it will also help boost the growth of related industries such as power, steel, chemicals, and new materials needed for infrastructure. Such drive systems are not only used in urban traffic.

In this paper, we have also discussed the braking system used in this vehicle. The braking system discussed here is a regenerative braking system, which is also known as a kinetic energy recovery system which converts the kinetic energy of a moving object into potential or stored energy during the slowing down of the vehicle, which improves fuel economy. Another type of braking system discussed in this article is the anti-lock braking system (ABS). It is a security system used in ground vehicles such as cars, buses, motorcycles, and trucks. The ABS keeps the clutch in touch with the ground surface and gives a driver greater control of the vehicle, which prevents wheel lockup when braking. It is most impressive when traveling long distances of hundreds of miles. Fuel cell electric vehicles excel even with short refuelling times of just a few minutes. The fuel cell electric drive consists of various components that Bosch introduces into series production through appropriate research and development.

1.2 Historical Background of Fuel Cell

Keeping in mind the need to satisfy the energy requirements of the world, researchers have found an alternative source to fulfill the energy requirements. In the 1800s, English scientist William Nicholson described the process of generating electricity by separating hydrogen from water through the process named electrolysis.

$$2H_2O_{(l)} \rightarrow 2H_{2(g)} + O_{2(g)}$$

Various researchers have further led to the development of hydrogen fuel cells. The first fuel cell was invented by Sir William Robert Grove in mid-1838 and was used commercially for over a hundred years. Francis Thomas

Bacon is the inventor of [1] hydrogen-oxygen in 1932. In 1959, a team led by Harry Gehrig built a 15-kW tractor. Alkaline fuel cells are also known as Bacon fuel cells. It has been one of the most advanced technologies since its invention. This technology was used by NASA on the Apollo space missions in the late 1990s to generate electricity for satellites and spacecraft [1]. In today's terms of fuel cell technology, these gadgets were used as a primary or secondary power source in many facilities, including industrial and commercial centers and dwellings.

1.3 ADVISOR

This paper was written at the National Sustainable Energy Lab, and it refers to one of the US Department of Energy's (DOE) Advanced Vehicle Simulation (ADVISOR) applications built in the MATLAB/Simulink environment. ADVISOR provides the automotive engineering community with easy-to-use, customized, supported, and resilient analysis tools for a wide variety of vehicle models as shown in Figure 1.1 [2]. This tool is primarily used to measure fuel efficiency, vehicle performance, and emissions for vehicles using alternative technologies. This tool is great at calculating the relative change to be expected from technology usage compared to a reference situation [2].

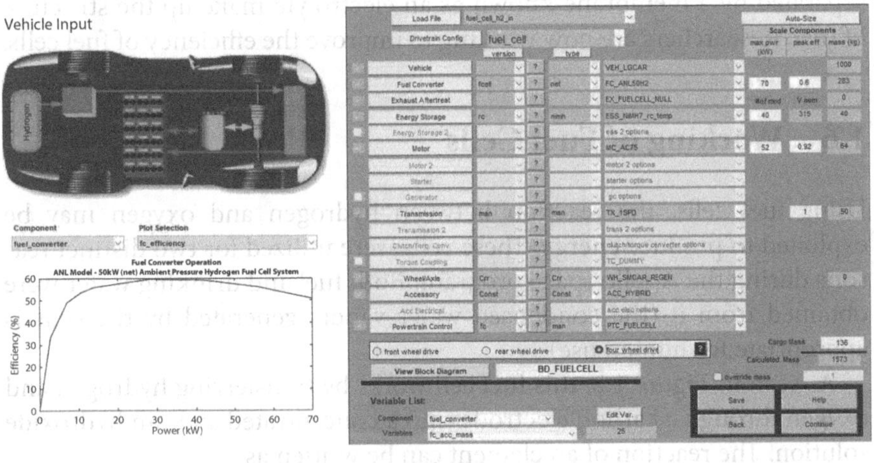

Figure 1.1 Advanced vehicle simulation interface.

1.4 Why Hydrogen is Preferred

Hydrogen is not an energy source, but it is an energy carrier. This means that energy can be stored or it can be released in a working form. As an energy conveyor, hydrogen has several edges. It can be created utilizing a variety of domestic energy sources, including nuclear power, solar energy, wind energy, biomass energy, and fossil fuels like natural gas and coal [3]. We are not reliant on any one energy source or external energy source because of the diversity of our energy sources [4]. Emissions of greenhouse gases are produced when hydrogen is produced using nuclear, renewable, and fossil fuel–based systems that also sequester carbon. All areas of the economy, including transportation, energy, business, and structures, can be powered by hydrogen. But for this, fuel cells are necessary [3].

1.5 What is a Fuel Cell?

A fuel cell is an electrochemical cell that uses two redox reactions to dynamically convert the chemical energy of a fuel (primarily hydrogen) and an oxidant (primarily oxygen) into electricity [3]. Fuel cells differ from most batteries in that they require a constant supply of fuel and oxygen (usually from the air) to sustain the chemical reaction, whereas chemical energy is typically derived from substances already present in the cell in batteries. As long as fuel and oxygen are available, fuel cells can keep producing electricity [3]. Two electrodes—anode and cathode—that are separated by a membrane known as an electrolyte make up the structure of fuel. Researchers are now working to improve the efficiency of fuel cells.

1.6 Working of Fuel Cells

Using fuel cells, the reaction between hydrogen and oxygen may be exploited to produce energy. These cells were utilized for two distinct reasons during the Apollo space program. Both fuel and drinking water were obtained from it (the condensed water vapors generated by the cells is appropriate for human use).

As seen in Figure 1.2, this fuel cell works by transferring hydrogen and oxygen through a carbon electrode into a concentrated sodium hydroxide solution. The reaction of an element can be written as

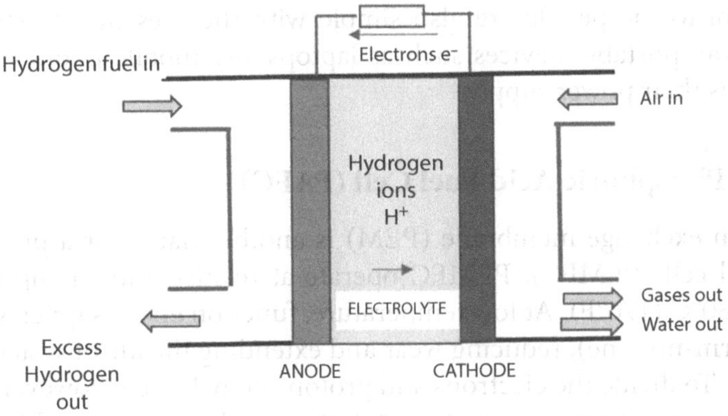

Figure 1.2 Fuel cell.

$$\textbf{Cathode Reaction} \quad : \quad O_2 + 2H_2O + 4e^- \rightarrow 4OH^-$$

$$\textbf{Anode Reaction} \quad : \quad 2H_2 + 4OH^- \rightarrow 4H_2O + 4e^-$$

$$\textbf{Net Cell Reaction} \quad : \quad 2H_2 + O_2 \rightarrow 2H_2O$$

1.7 Types of Fuel Cells

Fuel cells are one of the energy sources that could help address the coming energy crisis and power the world of the future. The fuel's chemical energy is converted into electricity through an electrochemical process [4]. They are like batteries in action. However, fuel cells can be refilled when chemical reactants run low. The types of fuel cells are as follows.

1.7.1 Direct Methanol Fuel Cell (DMFC)

Many fuel cells use hydrogen, which may be created from modern high hydrogen fuels like ethanol, methanol, and hydrocarbon fuels, either directly into the fuel cell system or within the fuel cell system. Nevertheless, DMFCs need real methanol, often mixed with water, and delivered directly to the fuel cell anode.

Unlike some fuel cell systems, DMFCs have fewer complications regarding the fuel storehouse because the energy density of methanol is significantly greater than hydrogen. However, it is lower than diesel and gasoline [4]. Because it is a fluid, like gasoline, the transportation and supply of methanol to the people are also simple with the present infrastructure. In general, portable devices such as laptops and mobile phones can use DMFC as their power supply.

1.7.2 Phosphoric Acid Fuel Cell (PAFC)

A proton exchange membrane (PEM) is another name for a phosphoric acid fuel cell (PEMFC). PEMFC operate at relatively low temperatures, around 80°C (176°F). At low temperature, function enables quick start-up (low warm-up time), reducing wear and extending the life of system components. To divide the electrons and protons of hydrogen, however, necessitates the use of precious metal catalysts, mostly platinum, which raises the system's cost and makes a second reactor necessary because platinum catalysts are extremely vulnerable to carbon monoxide poisoning. This is because when hydrocarbon fuels produce hydrogen, the carbon monoxide content of the fuel gas must be reduced. These reactors also require additional costs.

PAFC offers high power and small volume and weight advantages over other fuel cells. PEM fuel cells use porous carbon electrodes and hard polymers with platinum or platinum impurities as an electrolyte, which requires only hydrogen, oxygen, and water for work. Typically, PEMFCs are supplied with pure hydrogen from storage and reformers.

The PEM fuel cell is primarily used in transportation and some stationary applications. It is especially applicable for automotive work such as buses, cars, and heavy trucks. Compared to fuel-fired power plants, which run at 33% capacity, PAFCs are more efficient. Also, given their similar volume and weight, PAFCs are less durable than conventional fuel cells.

This process typically results in large and heavy fuel cells. PAFCs are also expensive. They require a much more expensive platinum catalyst than other options, which increases costs significantly.

1.7.3 Alkaline Fuel Cell (AFC)

AFCs are among the first technologies developed for the use of electricity and water in spacecraft and are the first of their type to be widely used by the US space industry. Potassium hydroxide exists as a solvent in these fuel cells using various non-metals as catalysts at the anode and cathode. Recently, a new AFC using a polymer membrane as the electrolyte has also been invented. This fuel cell is closely related to the standard PEM fuel cell but is non-acidic. They use an alkaline layer. AFCs are highly efficient due to the speed with which the electrochemical components interact within the cell. Space operation efficiency also exceeded 60%.

A significant drawback of this fuel cell is its vulnerability to carbon dioxide (CO_2) contamination. Carbonate production can result in extremely low CO_2 concentrations, which can have a considerable influence on the strength and quality of cells. Alkaline cells with a liquid electrolyte can limit the regeneration impact of carbonate production in the electrolyte by using the recirculation mode [4]. However, recirculation mode prevents the shunt current from flowing.

Other challenges also exist in liquid electrolyte installations, including weatherability, increased oxidation, and managing pressure drops. This is a concern because liquid electrolyte AFCs are more vulnerable to CO_2 poisoning than alkaline membrane fuel cells (AMFCs). However, the performance and durability of AMFCs are still limited by CO_2 emissions compared to PEMFCs. The AMFC has now been evaluated for watt- to kilowatt-scale implementations. Issues that hinder AMFCs include carbon sensitivity, membrane potential and durability, elevated temperature, water management, electrical density, and anodic electrocatalysis.

1.7.4 Solid Oxide Fuel Cell (SOFC)

An electrolyte constructed of solid, non-porous ceramic is found inside the SOFC. When converting fuel, SOFC efficiency is about 60%. In an arrangement that collects and uses waste heat (cogeneration) for the operation, combined fuel utilization rates can be as high as 85%. SOFC can operate at extremely hot temperatures up to 1,000°C (1,830°F). The extreme thermal process also eliminates the need for precious metal catalysts, reducing costs. Additionally, SOFCs can convert the fuel internally, allowing for a wider range of fuel usage and eliminating the need for system equipment [5, 6].

The best form of sulfur-tolerant fuel cells is SOFCs, which can resist higher sulfur concentrations than many other fuel cells. As a fuel, carbon monoxide is also employed. These properties allow SOFCs to use natural gas and coal-based biogas. However, operating at hot temperatures also has disadvantages. The result is a slow start and the need for enhanced thermal protection to keep system components warm and protected. Researchers are currently exploring the possibility of producing low-temperature SOFCs below 700°C with fewer durability issues and greater economic efficiency. Nevertheless, low-temperature SOFCs are not inferior in power to high-temperature systems. Stack materials are also being studied to work in this low-temperature region.

1.7.5 Molten Carbonate Fuel Cell (MCFC)

The MCFC is being developed for use in power, economy, and defense applications. Energy sources and coal-fired power plants are also considered. Like SOFCs, MCFCs are fuel cells that can function at hot temperatures by employing an electrolyte that blends molten carbonate floating in a porous lithium aluminum oxide ceramic model with a chemically inert substance [5]. Non-metals can be used as reactors in electrochemical cells at hot temperatures of 650°C (approximately 1,200°F) to reduce costs.

Another aspect where MCFCs offer significant cost efficiencies over PAFCs is increased productivity. The MCFCs can achieve efficiencies of around 65% when connected to turbines; they are significantly higher than the 37%–42% efficiency of PAFCs. The total fuel efficiency of collected and utilized heat losses is over 85%. The MCFC also does not require another reformer to convert natural gas and biogas to hydrogen.

1.8 Block Diagram of Vehicle on MATLAB/Simulink

The block diagram in Figure 1.3 displays an electric vehicle designed in MATLAB/Simulink with the help of the ADVISOR tool. In this block diagram, each component in this model is built separately, adding a subset model which is embedded in a fuel cell vehicle.

1.9 Braking System in Vehicle

A brake is a mechanical tool that reduces speed by sapping energy from a moving system. It is used to slow down or halt moving wheels, axles, or

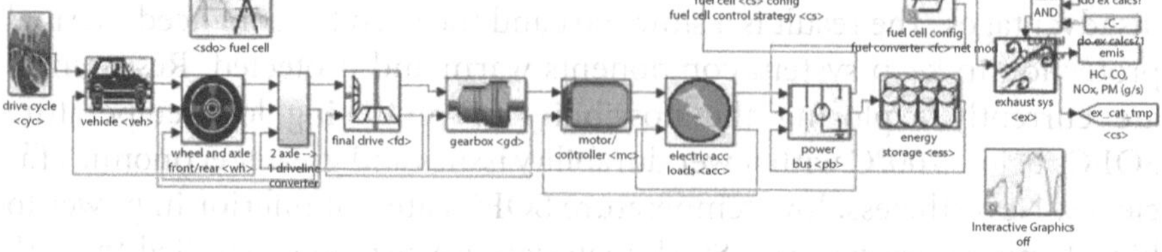

Figure 1.3 Block diagram of vehicle.

other moving parts, usually by friction. Two distinct kinds of braking systems are covered in this paper:

i. Regenerative braking system
ii. Anti-lock braking system

1.10 Regenerative Braking System

Regenerative braking is a special technique used in electric cars to recover kinetic energy lost during braking or deceleration. This energy is retrieved because of the vehicle's motion. Electric car regenerative braking systems can increase the vehicle's range; however, experience indicates that these gains are not necessarily brought about by inefficiencies in the energy conversion process. Regenerative braking systems' economic feasibility is frequently restricted by increased bulk and expense. Most electric cars come with a hybrid hydraulic regenerative braking system. When regenerative braking torque is inadequate to produce the same rate of deceleration as a normal vehicle, hydraulic braking torque is applied [7].

Due to the traffic management systems that are in place around cities and towns, an urban driving cycle has a substantial acceleration value and deceleration interval. As a result, a significant amount of energy is wasted during deceleration. The "wasted" energy can instead be collected with the aid of a regenerative braking system and utilized to move the vehicle. Using the Renault Zoe EV as an example, if a driver travels at a speed of 120 km/h (33.36 m/s), his kinetic energy value will be 0.25 kWh. His vehicle mass is approximately 1,600 kg. This energy is fully lost in existing internal combustion engine vehicle (ICEV) when they are close to a halt. The Renault Zoe's data from many driving cycles indicate that it has an energy efficiency of 0.2 kWh/km. A complete stop would waste energy equivalent to 1.25 km of travel, according to a study of vehicle energy efficiency and kinetic energy lost by braking. Losses in this range correspond to complete battery cycles, starting at 100% capacity and ending with a depleted battery.

The battery is the power source that fuels the engine of an electric car. By giving the vehicle's wheels the power to turn, the engine is depleted. The engine can turn the other way. The engine functions as a generator while it is rotating in reverse. The alternator transforms kinetic energy into electrical energy when the vehicle slows down to charge the battery.

Because of friction between the brake pads and the wheels, when a typical vehicle brakes, the energy is lost as heat energy. Electric cars' range can be extended through regenerative braking. Nonetheless, it is estimated that

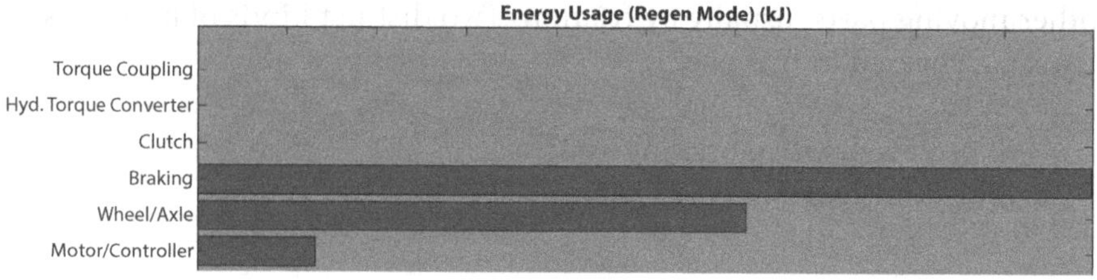

Figure 1.4 Energy usage during regenerative mode.

this energy may be captured with an efficiency of between 16% and 70% (Boretti, 2013) as shown in Figure 1.4. The cause of this significant efficiency disparity is related to the driver's driving technique (gradual or hard braking). Efficiency is also significantly impacted by both system temperature and ambient air temperature.

The efficiency of a regenerative braking system is determined in part by the driver's driving style. As contrast to slower braking with a regenerative braking system, the EV has friction brake pads, allowing for quicker braking. The effectiveness of the regeneration system is altered when friction braking is used. This trait supports the requirement for consumer education to instruct EV owners on how to maximize the effectiveness of the vehicle's energy recovery technology in order to prolong battery life [8].

Regenerative braking may be used by ICEVs to cut pollution in addition to enhancing the fuel efficiency of electric cars (Clarke, Muneer & Cullinane, 2010). This shorter charge/discharge time is cheaper and more energy dense than the flywheel method. This device also contains fewer hazardous substances than other storage devices that harm the environment. Ultracapacitors are a better alternative to short-lived electric batteries due to their associated very high efficiency (McCluer & Christin, 2008). It has been reported that the coulombic efficiency of ultracapacitors can reach efficiencies greater than 99% (Maxwell, 2010).

1.11 Anti-Lock Braking System (ABS)

Wheel locking during braking is avoided with the ABSs. While braking regularly, the driver is in charge; nevertheless, when braking abruptly or on slippery roads when the wheels are about to lock up, the anti-lock brakes engage. To return the wheel speed to the slip level range required for the best braking performance, the ABS regulates the brake line pressure independently of pedal force. Full-wheel lock is prohibited while braking by the ABS [9, 10]. Simply said, as the wheel does not lock even when the brake pedal is completely pushed, skidding is avoided during emergency braking. It facilitates driving, particularly on wet or slick roads brought on by rain, snow, or mud. The electronic control unit (ECU), wheel speed sensor, and hydraulic modulator make up the ABS's brain. Due to its closed-circuit design, ABS employs a feedback control system to modify the braking force in response to the rotational speed and deceleration of the controlled wheel to prevent locking. Modern cars include the ABSs to stop wheels from locking up and skidding after using the brakes. It is a built-in safety feature, and a controller is offered to change the required torque to keep the right slip ratio [9, 10]. The slip ratio is displayed along with the speed of the vehicle and wheel rotation. The automatic system adheres to the same threshold and cadence braking principles that experienced drivers did in preceding braking system versions. Driving is easy for the driver due to its quick response time. In dry, level terrain, ABS typically enhances vehicle control and reduces stopping distance, but on loose ground, such as gravel or snow-covered highways, ABS can dramatically lengthen stopping distance.

1.11.1 Components of ABS

1.11.1.1 Wheel Speed Sensor

Every (ABS) wheel speed sensor monitors the wheel speed in relation to the wheel it is paired with. The sensor is made up of a tone wheel, a twist

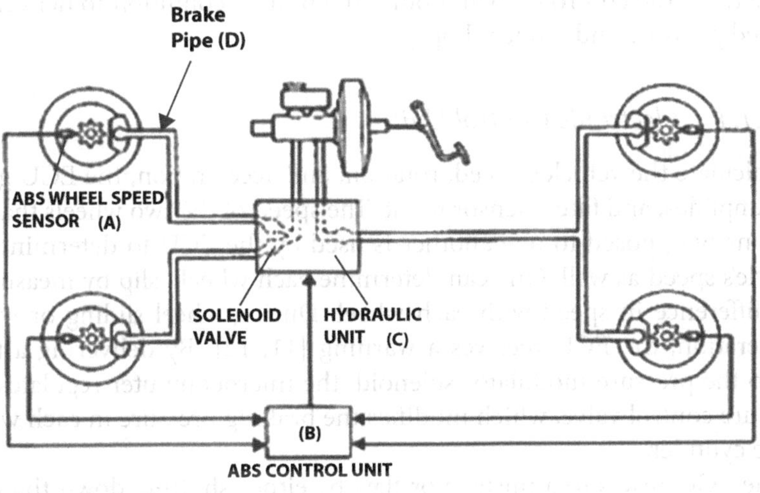

Figure 1.5 Anti-lock braking system.

wheel, and an eternal magnet. As each tooth of the tone wheel, which also spins around the wheel, travels in front of Paul Peace, the magnetic flux produced by the permanent magnets varies. The varying magnetic flux causes voltages to be generated at a frequency proportionate to the speed of the wheel. It also includes a ring with teeth, as shown in Figure 1.5. The interaction between the ring and the magnet results in an electric field that generates a sign. The ABS controller converts this signal into a digital signal and receives it. The controller then selects the correct speed for each wheel. The management of the car's drift depends on the ABS sensor.

1.11.1.2 Valves

Because valves control pressure and impose a limit on it, the ABS strongly depends on them. When the brake pedal is pressed hard, the valves limit the constraint on the brakes. A stuck valve is incapable of opening, closing, or shifting positions. The system will not be able to modify or regulate the pressure applied to the brakes if the valve is broken. There are three positions for the valve: block, release, and open.

1.11.1.3 Pumps

Once the valve releases the hydraulic brake's pressure, the pump oversees reestablishing it. The controller displays a valve open indicator when it notices wheel slip. The pump restores the desired level of pressure in the brake system as soon as the valve releases the pressure created by the occupant. Then, the controller will modify the pump's condition to deliver the needed pressure and reduce slippage.

1.11.1.4 Electronic Control Unit

To calculate the vehicle's speed, rotation, and acceleration, the ECU gathers, amplifies, and filters sensor input. The speed of the two wheels that are diagonally opposed to one another is used by the ECU to determine the vehicle's speed as well. One can determine each wheel's slip by measuring the difference in speed with each wheel. During wheel sliding or wheel acceleration, the ECU receives a warning [11, 12]. By delivering a trigger to the pressure modulator solenoid, the microcomputer regulates the pressure control valve, which modifies the braking pressure in each wheel brake cylinder.

The ECU reacts to a mistake or flaw by either shutting down the ABS completely or turning off the troublesome part of the system.

1.11.2 Types of the ABS Model

1. One-Channel ABS
Vehicles with a rear wheel ABS and pickup vehicles frequently have this configuration, also shown in Figure 1.6. The rear axle houses a one-speed sensor and a single valve to manage both rear wheels. This has a lot in common with the back of a three-channel arrangement. Both of the back wheels must lock up while being monitored for ABS to kick in. In this configuration, there is also a potential that one of the back wheels will lock, which would make the brakes less effective [12]. This system is easily recognized because normally only one brake line connects to both rear wheels through a T-fitting.

2. Three-Channel ABS
The two front wheels each have a valve and a speed sensor, whereas the two back wheels each have a valve and a sensor. This type of system is often set up in monster trucks with four-wheel ABS. The rear axle houses the speed sensor for the back wheels [11]. This method gives each of the front wheels a separate control to produce the most braking power. The ABS will not activate on the back wheels, which are controlled collectively, until both of them begin to lock up. When accelerating, if one of the rear wheels locks up, then the brake's stopping power is reduced.

3. Four-Channel ABS
The speed sensors on all four wheels and the individual valves for each wheel make it a more desirable type. With this configuration, the controller keeps track of each tire separately to ensure that it is applying the maximum amount of braking force.

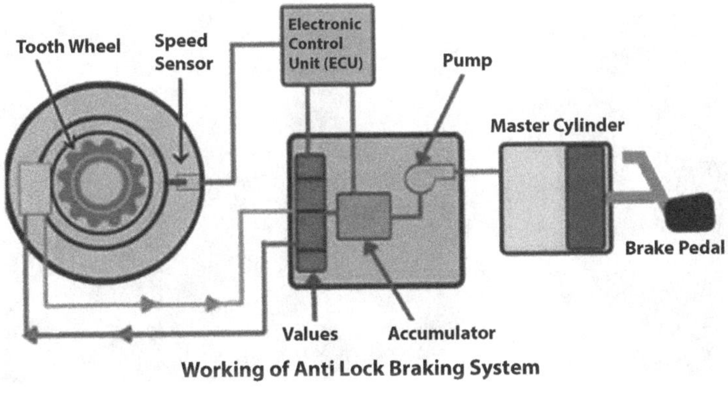

Figure 1.6 Working of the ABS model.

1.11.3 Anti-Lock Braking System Plot

In Figure 1.7, it is visible that the wheel slip of the vehicle is at a constant value, i.e., 0.2–0.3 and does not go to a value of 1 (that is wheel lock up) at a rapid rate which would cause the tire of the vehicle to slip.

Comparing the plots of Figures 1.8 and 1.9 of wheel speed and vehicle speed, it is clear from the wheel slip graph that the wheel speed does not lock up and remains lower than the vehicle speed, and the vehicle speed approaches zero within 15 s.

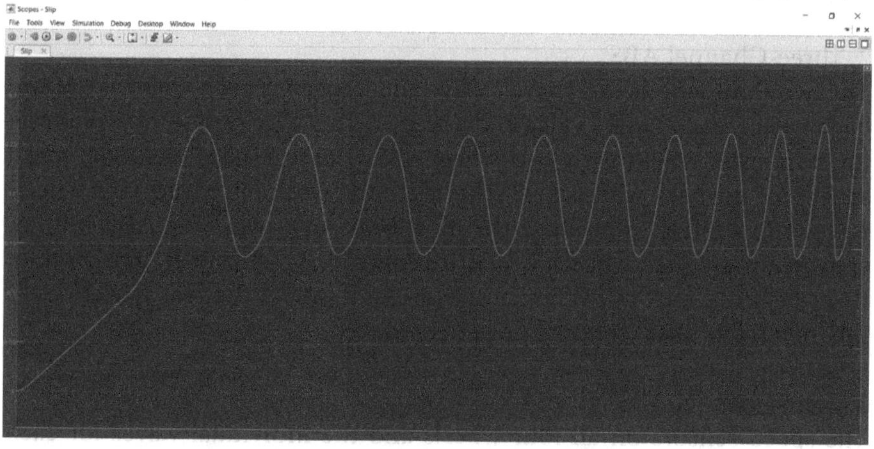

Figure 1.7 Wheel slip.

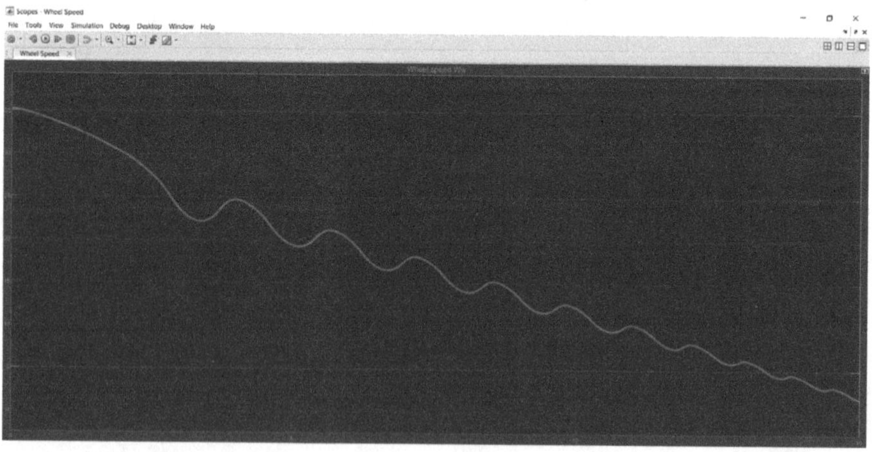

Figure 1.8 Wheel speed.

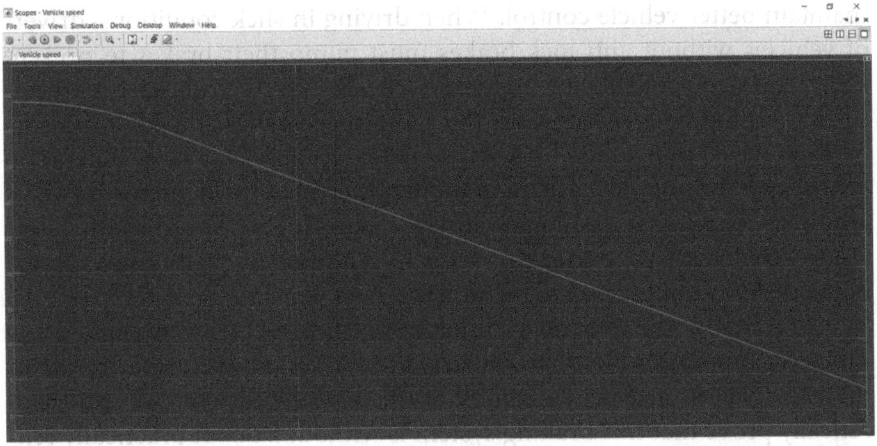

Figure 1.9 Vehicle speed.

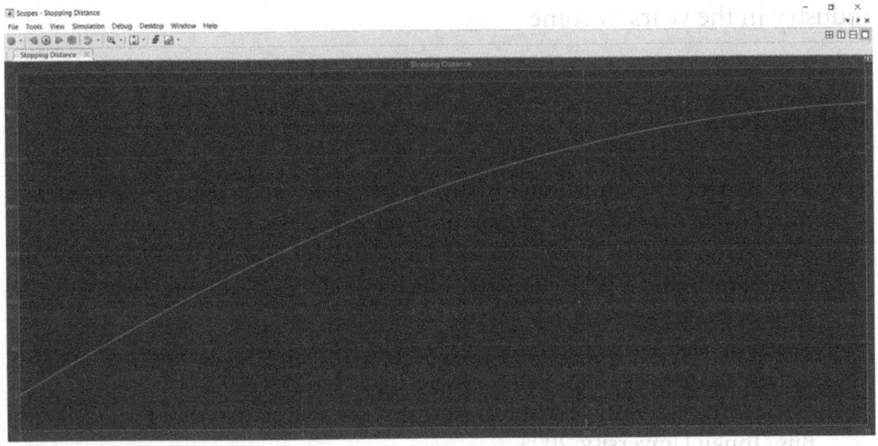

Figure 1.10 Stopping distance.

In Figure 1.10, the vehicle comes to a stop at about 14 s and skidding at a lesser distance of 300 ft, 3 s and about 100 ft less than without the use of ABS.

1.12 Conclusion

As car technology develops, the braking system becomes more complex. When sudden braking may be necessary, anti-lock brakes help drivers maintain better vehicle control. When driving in slick conditions, drivers of vehicles without anti-lock brakes must pump their brakes to prevent spinning uncontrollably due to locked wheels. The ABS adjusts the wheels so that all four wheels move in the same speed range thanks to sensors on each wheel that adjust brake pressure as needed. Our next-generation vehicles can be not only faster and more powerful but also safer with ABSs. Anti-lock braking technologies surely ensure higher speeds with fewer risks. Problems that frequently arise in automobiles with conventional brake systems can be resolved with ABS.

Additionally, the practical simulation on MATLAB/Simulink using multiple component blocks to construct the ABS's model further confirms the effectiveness and dependability of this braking system. We can theoretically prove that this braking system is without a doubt practically reliable for the safe operation of our vehicles at somewhat greater speeds, as mentioned in the paper and shown by the numerous graph plots. The ABS has increased vehicle dependability and will be helpful to the automobile industry in the years to come.

References

[1] Yin Tingting, Simulation Analysis of Fuel Cell Vehicle Powertrain, *Automobile Applied Technology*, vol. 3, pp. 3-6, 2019.

[2] Zhao Yane, Zhang Jianwu, Parameter Matching and Optimization for the Powertrain of a FWD Fuel Cell Electric Vehicle, *Automot. Eng.*, vol. 29, issue 5, pp. 409-414, 2007.

[3] Yang Yu, Xiao Rongge, Strategy to Develop New Energy in Fuel Cell Vehicle Industry, *Automot. Eng.*, vol. 12, issue 2, pp. 16-20, 2010.

[4] Chen Mingwei, Basic Technical Research for Powertrain of a Fuel Cell City Bus, Tongji University, 2005.

[5] Wang Ping, Huang Xiaofeng, Parameter Matching and Optimization of Fuel Cell Vehicle Hybrid Power System, *Shanghai Auto*, issue 3, pp. 7-11, 2010.

[6] Arora, K.; Kumar, A.; Kamboj, V.K.; Prashar, D.; Shrestha, B.; Joshi, G.P. Impact of Renewable Energy Sources into Multi Area Multi-Source Load Frequency Control of Interrelated Power System. *Mathematics*, 2021, 9, 186.

[7] Manikanta, D. Venkata, Mohinder Bassi, and Harpreet Bedi. Enhanced topology-based 9 level Hbridge inverter. *2017 IEEE 8th Annual Ubiquitous Computing, Electronics and Mobile Communication Conference (UEMCON)*. IEEE, 2017

[8] Arora, K.; Kumar, A.; Kamboj, V.K.; Prashar, D.; Jha, S.; Shrestha, B.; Joshi, G.P. Optimization Methodologies and Testing on Standard Benchmark

Functions of Load Frequency Control for Interconnected Multi Area Power System in Smart Grids. *Mathematics*, 2020, *8*, 980.

[9] Sharma, Vinod Kumar, *et al.* Imperative Role of Photovoltaic and Concentrating Solar Power Technologies towards Renewable Energy Generation. *Int. J. Photoenergy*, 2022 (2022).

[10] Verma P., Bedi H., and Sharma R. K., 2016, Integrated utilization of solar energy in a zero energy building (ZEB): an approach towards sustainable development, International Science Press, Phagwara, India.

[11] Shukla, Vaibhav, *et al.* A Survey on the Performance of the Various MPPT Techniques of Standalone PV Generation System.

[12] Bedi, Harpreet Singh. Utilization of a Wireless Network Performing CSMA/CA with Random Backoff Algorithm. *J. Phys. Conf. Ser.*, Vol. 2327. No. 1. IOP Publishing, 2022.

2

Design and Applications of Fuel Cells

Pradeep Singh[1*], Krishan Arora[1†] and Umesh C. Rathore[2‡]

[1]SEEE, LPU-Jalandhar, India
[2]ABVGIET-Pragtinagar, Shimla, HP, India

Abstract
The reliance on fossil fuels in transportation can lead to economic costs, such as increased fuel prices and the cost of environmental damages. These fossil fuels are finite and eventually run out. Such a drawback highlights the need for a transition to alternative fuels and more sustainable transportation systems, such as electric vehicles. There are some devices like fuel cells which convert chemical energy from a fuel into electricity through an electrochemical reaction. They are considered a promising technology for clean and efficient energy generation, offering several advantages over traditional power generation methods. Fuel cells can operate with a variety of fuels, including hydrogen, natural gas, and biofuels and can produce electricity with high efficiency and low emissions. The design of fuel cells involves optimizing factors such as materials, catalysts, and system configuration to achieve high performance, durability, and cost-effectiveness. Fuel cells have found applications in a wide range of fields, including stationary power generation, portable devices, transportation and backup power, aerospace, marine, and remote and off-grid areas among others. With ongoing research and development, fuel cells are expected to lead to new, innovative uses and a significant role in the transition to a more sustainable energy future. This chapter includes the design and applications of various types of fuel cells for electric vehicles.

***Keywords*:** Fuel cells, electric vehicles, types, chemical energy, stationary power generation, battery charging systems, fossil fuels

**Corresponding author*: dadwalsingh5226@gmail.com
†*Corresponding author*: krishan.12252@lpu.co.in
‡*Corresponding author*: rathore7umesh@gmail.com

Krishan Arora, Suman Lata Tripathi and Himanshu Sharma (eds.) Electric Vehicle Design: Design, Simulation and Applications, (19–56) © 2024 Scrivener Publishing LLC

2.1 Introduction

There is a growing consensus among experts and policymakers that vehicles that run on gasoline and diesel will eventually be phased out in favor of more sustainable and efficient alternatives. Some of the main reasons for this include climate change in which fossil fuels, including gasoline and diesel, are significant sources of the greenhouse gas (GHG) emissions, which causes climatic changes. The widespread adoption of alternative transportation technologies, such as electric vehicles (EVs), is considered as a crucial step in lowering pollutants and limiting the effects of climatic changes. Finite resources which include fossil fuels will eventually run out, and their extraction and use can lead to environmental degradation and other negative impacts. EVs offer a sustainable alternative that can help reduce our dependence on finite resources. Air pollution includes traditional vehicles powered by internal combustion (IC) engines which release harmful pollutants into the air, leading to respiratory problems and contributing to air pollution. EVs produce significantly lower emissions and can help improve air quality. They can also improve energy security. The transportation sector's dependence on foreign oil sources can lead to political instability and conflicts over access to resources, posing a threat to energy security. The widespread adoption of EVs could help reduce this dependence. Economic benefits like widespread adoption of EVs could create new jobs and economic opportunities and reduce the costs associated with environmental damages from traditional vehicles. These challenges highlight the need for a transition to more sustainable and efficient forms of transportation, and EVs are seen as an important part of this transition. However, the widespread adoption of EVs will require significant investments in charging infrastructure, battery technology, and other supporting systems. The need for electric vehicles (EVs) stems from several critical challenges facing the transportation sector, including the following:

i. Climate Change

The usage of EVs widely might contribute to lowering pollutants and reducing the effects of climatic change, as the transport industry contributes significantly to GHG releases.

ii. Dependence on Finite Resources

The transportation sector is heavily dependent on fossil fuels, which are finite resources that will eventually run out. EVs offer a sustainable alternative to traditional vehicles that rely on gasoline and diesel.

iii. Air Pollution

Traditional vehicles powered by IC engines release harmful pollutants into the air, leading to respiratory problems and contributing to air pollution. EVs produce significantly lower emissions and can help improve air quality.

iv. Energy Security

The transportation sector's dependence on foreign oil sources can lead to political instability and conflicts over access to resources, posing a threat to energy security. The widespread adoption of EVs could help reduce this dependence.

v. Economic Benefits

The widespread adoption of EVs could create new jobs and economic opportunities and reduce the costs associated with environmental damages from traditional vehicles.

2.2 Types of Electric Vehicles

These challenges highlight the need for a transition to more sustainable and efficient forms of transportation, and EVs are seen as an important part of this transition. However, the widespread adoption of EVs will require significant investments in charging infrastructure, battery technology, and other supporting systems. Electric vehicles can be provided by variety of fuels. There are different types of EVs available, which are shown in Figure 2.1. Each type of vehicle offers different benefits and trade-offs, and the choice of fuel will depend on factors such as driving range, refueling time, and cost.

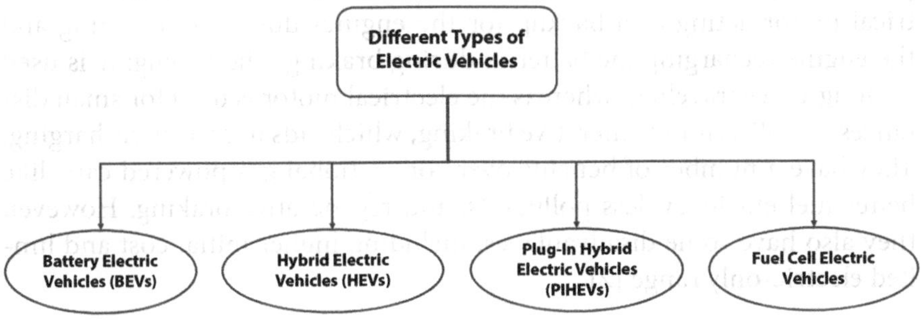

Figure 2.1 Different types of electric vehicles.

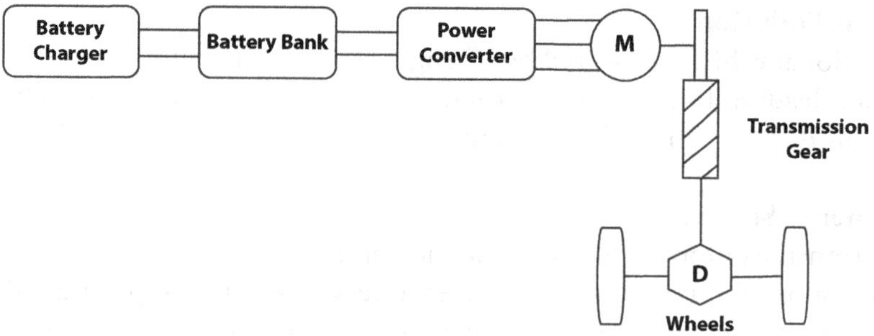

Figure 2.2 Battery electric vehicles (BEVs).

2.2.1 Battery Electric Vehicles (BEVs)

These EVs are powered exclusively by energy that is saved in batteries that are charged by outside power sources. Battery-operated EVs are a subset of EVs, powered exclusively by the energy saved in battery, as shown in Figure 2.2. In comparison to conventional gas-powered cars, they have a number of benefits, such as lower running costs, lower carbon emission, great economy, and a peaceful and relaxing ride. They have, however, significant drawbacks, such as a short range, a lengthy charging period, and greater capital costs [1]. In contrast, there are PHEVs, which feature a tiny battery which may be recharged from an independent supply, and HEVs, which combine the IC engine with an electrical motor.

2.2.2 Hybrid Electric Vehicles (HEVs)

An HEV is a form of EV that integrates an electrical motor, a battery source, and a traditional IC engine as shown in Figure 2.3. Propulsion is provided by both the IC engine and the electrical motor, with the electrical motor acting as a backup for the engines during accelerating and the engine recharging the batteries during braking. The IC engine is used for long-drive traveling, whereas the electrical motor is used for small distances as well as for regenerative braking, which aids in battery recharging. They have a number of benefits over conventional gas-powered cars, like better fuel efficiency, less pollutants, and regenerative braking. However, they also have some disadvantages, including higher initial cost and limited electric-only range [2].

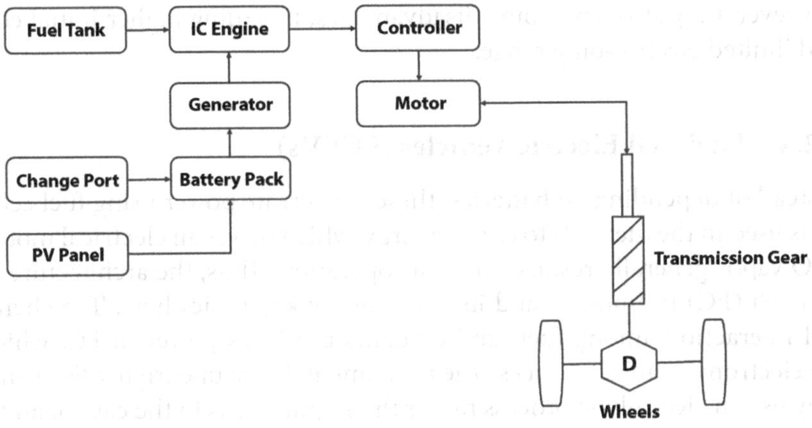

Figure 2.3 Hybrid electric vehicles.

2.2.3 Plug-In Hybrid Electric Vehicles (PHEVs)

These cars resemble hybrid electric vehicles (HEVs), except that they feature a bigger battery which may be recharged by an outside power supply. HEVs that can be recharged via an additional resource of power, such as a recharging station or a household socket, are known as PHEVs, as shown in Figure 2.4. Unlike conventional HEVs, which rely on regenerative braking to charge their battery, PHEVs can be charged to provide extended electric-only range, allowing them to run on electricity for longer periods of time. They have a number of benefits over conventional gas-powered cars, including better fuel efficiency, less pollution, and easy charging [3].

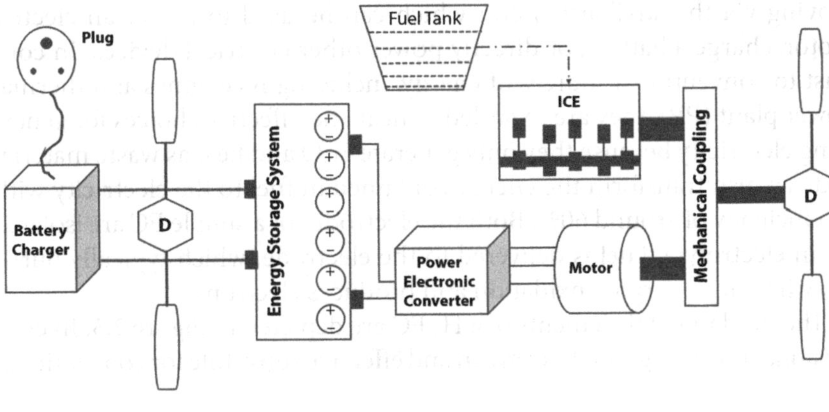

Figure 2.4 Plug-in hybrid electric vehicles (PHEVs).

However, they also have some disadvantages, including higher initial cost and limited electric-only range.

2.2.4 Fuel Cell Electric Vehicles (FCEVs)

Instead of depending on batteries, these cars create power using fuel cells. H_2 is used in the fuel cells to create energy, which drives an electrical motor. H_2O vapor generally results from this operation. Thus, the architecture of fuel cell (FC) technology and its uses are the key topics here. The chemical interaction among fuels and oxidants produces power in FC, which are electromechanical devices. The fundamental idea underlying FC is that they use an electrolytic process rather than ignition, as in the case of an IC engine, to transform the energy held in fuels into electric power [4]. The anode electrode of an FC receives H_2 which is delivered and then oxidized to liberate electrons. The oxidizer (often O_2) and oxidizing agent interact with electrons in the cathode terminal to create H_2O and heat. These electrons are then sent there via an external circuitry. Electrons moving across external circuit produce an electricity in the form of current that can be utilized directly to deliver power for recharging of batteries or run an electrical motor or other devices. The electrochemical reaction in a fuel cell can be represented by the following equation:

$$Fuel + Oxidant \rightarrow Electricity + Water + Heat$$

The oxidizer (often O_2) and oxidizing agent interact with electrons in the cathode electrode to create heat and H_2O. These electrons can then be sent there via an auxiliary circuit. Electrical current is produced by electrons moving via the auxiliary circuit which can be used to power an electric motor, charge a battery, or directly power other electrical devices. In contrast to conventional sources of energy, including IC engines and thermal power plant (PP), they are regarded as neat and effective choices for generating electricity because they only generate H_2O and heat as waste material and can also transform the energy contained in fuel to the electricity with an efficiency of around 60%. Both the electrodes of a simple FC are isolated by an electrolyte. Fuel is delivered to the electrode, which typically burns H_2, where it undergoes oxidation and produces electrons.

The fundamental elements of a H_2 FC are depicted in Figure 2.5. In conclusion, FCs are regarded as a green and effective substitute for conventional

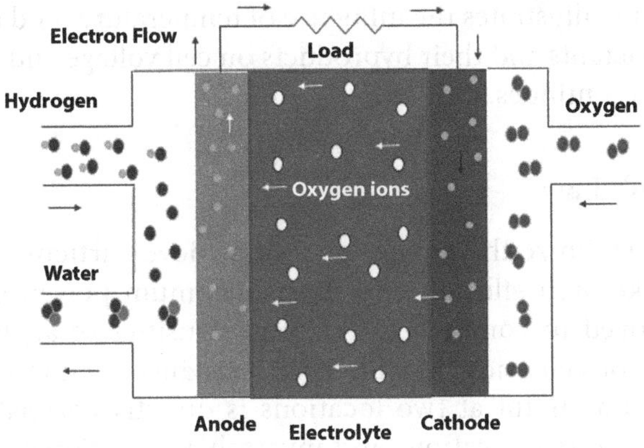

Figure 2.5 Schematic of fuel cell.

power generating techniques because they produce power through an electrolytic process involving a fuel as well as an oxidizer [5].

2.3 Design Equations of Fuel Cells

Numerous calculations are employed in the construction of an FC to determine different characteristics, such as cell potential, current flow rate, and generated power output.

2.3.1 Nernst Equation

The cell potential or voltage of an FC can be determined on the basis of concentration of hydrogen ions and oxygen ions at the electrodes [6]. The equation given by Nernst is shown in Equation 2.1

$$E = E_0 - \left(\frac{RT}{nF}\right) ln\left(\frac{\Delta_r}{\Delta_p}\right) \tag{2.1}$$

where E is the cell voltage, E_0 is the standard potential of electrode, R is the universal constant for gas, T is the temperature in kelvin, n is the number of electrons transferred, F is the Faraday's constant, Δ_r is the partial pressure of reactant(s), and Δ_p is the partial pressure of product (s).

This equation illustrates the influence of temperatures and relative pressure of the reactants and their byproducts on cell voltage and how it varies as the process continues.

2.3.2 Ohm's Law

In order to maximize the FC design and achieve particular operational objectives, like high efficacy as well as maximum power, Ohm's equation is employed to compute the current densities of an FC based on the potential of cell and their internal resistance. Ohm's law says that the current in a circuit at two locations is directly correlated with the voltages between two locations and inversely proportionate to the resistance between both, which describes the connection between *I*, *V*, and *R*. Moreover, an FC's internal *r* may be computed according to the density of current and voltage of cell, which is helpful for assessing the effectiveness of FC and identifying its weak points [7]. The equation for Ohm's law is given by Equation 2.2

$$I = \frac{V}{R} \tag{2.2}$$

where I is the density of current, V is the voltage of cell, and R is the internal resistance of cell.

2.3.3 Power Output

The density of *I* as well as the *V* of an FC may be used to determine its energy output. The formula is provided by Equation 2.3

$$P = IV \tag{2.3}$$

where P is the power output, I is the current density, and V is the cell voltage.

The quantity of electricity produced by an FC over a specific time duration is measured by the cell's power output. Engineers can determine the anticipated maximum power of FC based on the *V* and *I* density of cell by utilizing the total power equation. The *V* and *I* density in an FC are both influenced by the operating environment and the architecture of the cell. The total power of an FC may be increased by maximizing the structure and operational parameters. In order to evaluate various FC designs and identify potential areas for development, the total power of an FC may also be utilized as a statistical parameter.

2.3.4 Efficiency Equation

The efficiency of an FC can be determined by using Equation 2.4

$$\eta = \frac{P_{fc}}{P_{fc} + P_{H_2} + P_w} \tag{2.4}$$

where η denotes efficiency, P_{fc} is the power output of fuel cell, P_{H_2} is the rate of hydrogen consumption, and P_w is the rate of water consumption.

The degree to which the energy included in the H_2 fuel is turned into electric power is the efficacy of FC. This efficacy equation takes into account the rate at which H_2 is consumed, power output of FC and the rate at which water is created.

By optimizing the structure and operating conditions of an FC, engineers can increase the efficiency and maximize the amount of energy that is converted into electricity. Additionally, the efficacy of an FC can be used as a performance metric to compare different fuel cell designs and to determine areas for improvement.

2.3.5 Kinetic Current Density Equation

The kinetic current density in an FC can be derived from the equation given by Butler–Volmer, which expresses the link between the reaction rate of an electrochemical reaction and the over-potential. The Butler–Volmer equation is given in Equation 2.5

$$j = j_0 \frac{\{\exp[\alpha F(V_{cell} - E_0)] - \exp[-\alpha F(V_{cell} - E_0)]\}}{RT} \tag{2.5}$$

where j is the reaction rate, j_0 is the exchange current density, α is the transfer coefficient, V_{cell} is the cell voltage, E_0 is the standard electrode potential, R is the gas constant, T is the Temperature, and F is the Faraday's constant.

By rearranging the Butler–Volmer equation given in Equation 2.6 and taking the natural logarithm of both sides, we can obtain the equation for the kinetic current density in a fuel cell:

$$\ln\left(\frac{j}{j_0}\right)=\frac{\alpha F(V_{cell}-E_0)}{RT} \tag{2.6}$$

Rearranging this equation further which is shown by Equation 2.7 and solving for j, we get the final form of the kinetic current density equation of a fuel cell:

$$j=j_0\left\{1-\frac{\exp[-\alpha(V_{cell}-E_0)]}{RT}\right\} \tag{2.7}$$

Equation 2.7 describes the link between the process rate of the electrochemical reaction in the FC and cell voltage, temperatures, and other parameters.

2.3.6 Over-Potential Equation

The over-potential equation is given by Equation 2.8:

$$\eta=V_{cell}-E_0 \tag{2.8}$$

where η is the over-potential, V_{cell} is the cell voltage, and E_0 is the standard electrode potential.

The over-potential can be influenced by various factors such as the reactant transport rate, reaction kinetics, and the electrical resistance of the cell components. By minimizing the over-potential, engineers can improve the efficiency and performance of an FC. Additionally, the over-potential can be used to calculate the activation energy and other parameters in the reaction kinetics using methods such as the Tafel equation. These parameters are important in understanding the behavior of the fuel cell and can be employed in its design and simulation.

2.3.7 Heat Generation Equation

Heat generation in an FC is an important aspect of FC design and its effectiveness, as it affects the efficiency and reliability of the cell. The heat generated in a fuel cell can come from various sources, such as the exothermic reactions taking place within the cell, the resistance of electrical conductors, and the transfer of heat between the cell components. The heat generation equation in a fuel cell can be described by the energy balance equation, which states that the energy generated in the FC is equivalent to the energy lost to the surroundings [8]. The energy balance equation is given by Equation 2.9:

$$Q = \Delta H_{rxn} + Q_{elec} + Q_{cond} + Q_{conv} \tag{2.9}$$

where Q is the heat generated in the cell, ΔH_{rxn} is the heat generated from the exothermic reactions in the cell, Q_{elec} is the heat generated from the electrical resistance of the conductors in the cell, Q_{cond} is the heat generated from the transfer of heat between the cell components, and Q_{conv} is the heat generated from the transfer of heat to the environment through convention.

By understanding the sources of heat generation in a fuel cell and their relative contributions, engineers can design fuel cells with improved efficiency and performance. Additionally, the heat generation equation can be used to evaluate the impact of various parameters on the overall heat generation in the FC, such as its temperature of operation, reaction kinetics, and cell geometry.

These equations are used to design and optimize an FC for numerous applications, including fixed power generation, portable electricity, and also vehicles. By manipulating various parameters, such as the concentration of hydrogen ions and oxygen ions, temperature, and internal resistance, the productivity of an FC can be enhanced [9].

2.4 Designing of Fuel Cells

The following steps are involved in the designing of fuel cells:

i. Selection of the Type of FC

Choosing the right kind of FC for the specified purpose and operating parameters is the initial step in building an FC. Proton exchange membrane fuel cells (PEMFCs), direct methanol fuel cells (DMFCs), phosphoric acid fuel cells (PAFCs), alkaline fuel cells (AFCs), and solid oxide fuel cells (SOFCs) are the most prevalent forms of FCs.

ii. Electrode and Membrane Selection

The electrodes and membrane must be selected on the basis of the type of FC and the operating conditions. The electrodes must be able to efficiently catalyze the reaction of FC and the membrane must also be able to transport ions between the electrodes.

iii. Design of the Fuel Cell Stack

The stack is the core component of an FC, consisting of a multiple number of individual FCs connected in sequence to produce the required output voltage. The design of stack takes into consideration the number of factors such as the number of cells, the size of cells, and the material used for separation of plates.

iv. Balance of Plant Design

The balance of plant design includes the components that support the fuel cell stack, such as the fuel supply system, air supply system, cooling system, and electrical components. The design of these components must be optimized for the stack of FC and the intended application.

v. System Integration and Testing

Finally, the FC system must be integrated and examined to make sure that it is tested to ensure that it meets the required operational standards. This might entail optimizing the stack of FC and balancing the plant design based on the results of testing, as well as integrating the fuel cell system into the larger application, such as a vehicle or a stationary power system. It is essential to remember that designing an FC system involves a multidisciplinary approach, incorporating expertise in materials science, electrochemistry, thermodynamics, and electrical engineering.

2.5 Types of Fuel Cells

FCs come in a variety of forms, each having special qualities and benefits. Some common types are listed in Figure 2.6.

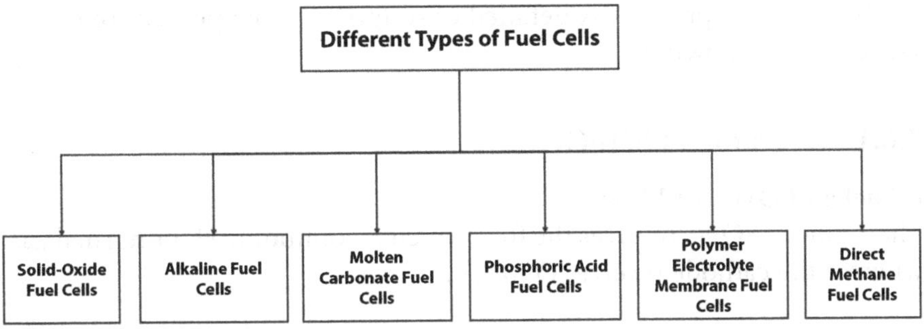

Figure 2.6 Different types of FCs.

2.6 Solid Oxide FCs (SOFCs)

SOFCs are a type of FC that move electrons in between the anode and cathode of a cell using a hard electrolyte, commonly composed of ceramic materials, as shown in Figure 2.7. Some FC designs, such as PEMFCs, employ liquids or polymeric electrolytes. The basic operation of an SOFC involves the oxidation of a fuel, such as H_2 or normal gas, at anode, which generates electrons and ions with positive charge, whereas the ion is carried via the hard solution to the cathode where they interact with O_2 to form H_2O and the charges flow via an outer electrical circuit to produce electricity [10]. For the function of SOFCs, chemical energy that is contained in fuel must be converted into electricity via an

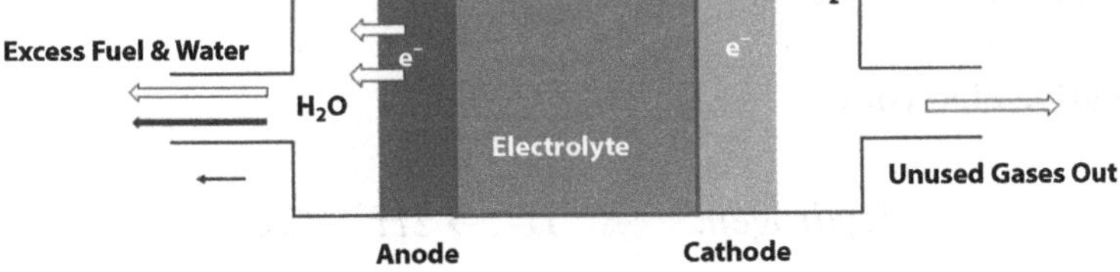

Figure 2.7 Solid oxide fuel cell.

electrochemical process. A detailed description of the procedure is provided in the following.

2.6.1 Working of SOFCs

i. Fuel and Oxidant Flow

The anode of SOFC receives the fuel, which is commonly H_2 or normal gas, whereas the cathode receives O_2 or air.

ii. Oxidation of Fuel

The fuel is oxidized at the anode, resulting in positive charge ions and electrons. The electrons go via an outer circuit to produce energy and the ions get transferred via the hard electrolyte, which is commonly formed of ceramic materials, to the cathode.

iii. Reduction of Oxidant

As O_2 or air and ions combine to generate H_2O at the cathode, electrons are liberated and travel via the outer circuits back to anode, completing the circuit.

iv. Electrochemical Reaction

The overall reaction is an electrochemical reaction, with the ions and electrons participating in separate reactions at the anode and cathode, respectively. This process generates electrical energy, with the amount of energy generated dependent on the fuel, oxidant, and operating conditions.

v. Heat Generation

The reaction of fuel and oxidant generates heat, which can be harnessed to provide additional energy through integrated heat and power (IHP) applications. The main reactions that occur in an SOFC involve the oxidation of a fuel, such as H_2 or normal gas, at the anode and the decreasing an oxidant, such as O_2 or air, at the cathode [11]. Here are the reactions that occur in an SOFC.

At negative electrode

$$\textit{Hydrogen:} \quad H_2 \rightarrow 2H^+ + 2e^-$$

$$\textit{Methane:} \quad CH_4 + H_2O \rightarrow CO_2 + 4H^+ + 4e^-$$

At positive electrode [12]

$$\textit{Oxygen:} \qquad O_2 + 4e^- \rightarrow 2O_2^-$$

The overall reaction can be expressed as

$$\textit{Hydrogen:} \qquad H_2 + O_2 \rightarrow 2H_2O$$

$$\textit{Methane:} \qquad CH_4 + 2H_2O \rightarrow CO_2 + 4H_2$$

The electrochemical reactions that occur in the SOFC generate electrical energy through the flow of ions and e^- between the anode and cathode [13]. The amount of electrical energy generated is dependent on the fuel and oxidant used, as well as the operating conditions of the cell.

2.6.2 Advantages of SOFCs

i. High Efficiency
SOFCs have a high energy transformation efficiency, typically in the range of 50%–60%, which is significantly higher than other kinds of FCs.

ii. Broad Fuel Compatibility
SOFCs can operate on a variety of fuels, including H_2, normal gas, and biofuels, making them a flexible and diverse resource of energy.

iii. Long Life
SOFCs have a long life span, typically in the range of 20–30 years, making them a durable and reliable power source.

iv. High-Temperature Operation
SOFCs can operate at high temperatures, which can help to improve their performance and efficiency.

v. Low Emissions
SOFCs produce relatively low emissions compared to traditional fossil fuel–based power source.

2.6.3 Disadvantages of SOFCs

i. High Cost
SOFCs might not be appropriate for the applications that require significant power output because of their low power density.

ii. Complex Manufacturing Process
The manufacturing process for SOFCs is complex and requires specialized equipment and materials, which can make them difficult to produce on a large scale.

iii. High Operating Temperature
The high operating temperature of SOFCs can increase the risk of degradation and shorten the life span of the cell.

iv. Durability Concerns
SOFCs are subject to degradation over time due to the high temperatures and harsh operating conditions, which can impact their performance and life span.

v. Limited Power Density
SOFCs have a limited power density, which means that they may not be suitable for applications where high power output is required.

2.6.4 Applications of SOFCs

SOFCs have large number of applications due to excellent energy conversion efficacy and the possibility to operate on different types of fuels, including H_2, normal gas, and biofuels. The following are some of the main applications of SOFCs.

i. Stationary Power Generation
SOFCs can be used as a source of electricity in isolated or off-grid areas, such as telecommunications towers and military bases, where the fuel can be transported or stored on site.

ii. Integrated Heat and Power (IHP) Systems
SOFCs can also be used for integrated heat and power (IHP) applications, where the heat produced by a cell during its functioning can be exploited for home heating or for heating water.

iii. Micro-Integrated Heat and Power (mIHP) Systems
Small-scale mIHP setups can be utilized for households and small enterprises which can incorporate SOFCs to produce both heat energy and power with a single module.

iv. Transport
SOFCs can be used as a range extender for electric vehicles, providing additional power when the battery is depleted, or as a primary power source for vehicles that operate on hydrogen fuel.

v. Backup Power
SOFCs can be utilized as alternative energy resources in important sectors, such as hospitals and data centers, where reliability is essential [14].

2.7 Alkaline Fuel Cells (AFCs)

AFCs function via electrochemically transforming the chemicals energy contained in fuels into electricity. Two electrodes and a medium, i.e., electrolyte, are the essential parts of AFC, as shown in Figure 2.8. The electrolyte, which is commonly made up of a basic mixture of potassium hydroxide (KOH), separates the anode electrode from the cathode electrode [15]. AFCs are a specific kind of FC that employ nickel (Ni) or iron (Fe) electrodes and a basic electrolytic solution that is commonly composed of KOH. Nevertheless, compared to various kinds of FCs, such as PEMFCs or SOFCs, these FCs are less often employed in large-scale electricity production.

2.7.1 Working of AFCs

The working of an AFC involves the following steps.

i. Fuel and Air Flow
H_2 gas is provided as the fuel for the anode part of the FC, and O_2 is provided to the cathode part.

ii. Electrochemical Reaction
The H_2 gas is divided into protons as well as e^- particles at the anode. When the e^- go via an outer circuitry and create an electric power, the protons are then transported via the electrolytes to the cathode part.

iii. Oxygen Reduction
At the electrode, i.e., cathode, the protons and e^- react with O_2 to form H_2O and release heat in this process.

iv. Voltage Generation
The voltage imbalance among both the electrodes, i.e., anode and cathode, generates a voltage, which is used to drive an electrical load.

In AFCs, the base solution serves as the electrolytic solution; typically, it is KOH. The alkaline environment provides a high proton conductivity, which enables efficient transport of ions and enhances the reaction kinetics. AFCs have a relatively low operating temperature and are often used in portable power applications, such as spacecraft and handheld devices. The overall chemical reaction equation for an alkaline fuel cell (AFC) is given as follows.

At negative electrode (fuel side)

$$Hydrogen\,(fuel) + OH^- \rightarrow H_2O + e^-$$

At positive electrode (oxygen side)

$$O_2 + 2H_2O + 2e^- \rightarrow 4OH^-$$

Overall reaction

$$Hydrogen + O_2 \rightarrow H_2O$$

H_2 serves as the main fuel, and O_2 serves as the oxidizer in this equation. H_2 is oxidized at the negative electrode, i.e., anode, to produce hydrogen (H^+) ions and electrons (e^-). In the positive electrode, i.e., cathode, when they interact with O_2 and e^- to form H_2O, the (H^+ spread via the electrolytes). spread via the electrolytes). An electrical energy was produced by the movement of electrons via an outer circuitry at the negative electrode. The overall reaction is exothermic, meaning that the overall energy is released in the kind of hot air, and the efficacy of the AFC depends on the degree to which this energy can be harnessed to produce electricity. AFCs are highly efficient, with potential efficiencies of up to 80%. These too are suitable for employment in large-power plants because of their low cost and ability to

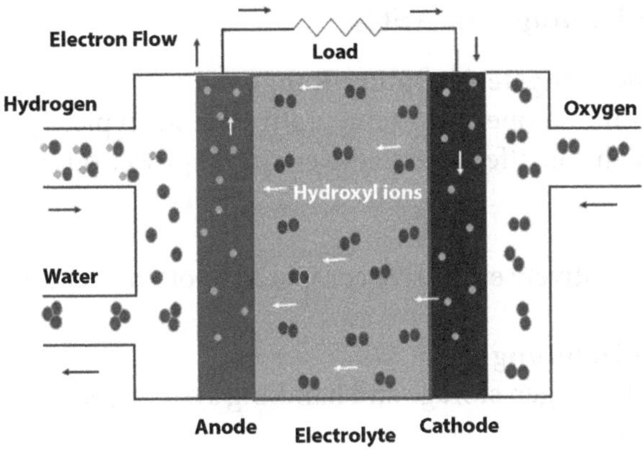

Figure 2.8 Alkaline fuel cell (AFC).

function at extreme temperatures. They are also capable of using a variety of fuels, including hydrogen, methanol, and ethanol, which makes them flexible and adaptable to changing energy needs.

2.7.2 Advantages of AFCs

i. High Efficiency
AFCs have high energy conversion efficiency, typically around 60%.

ii. High Power Density
AFCs have a high power density, meaning that they can generate a large amount of electrical power per unit of volume.

iii. Simple Design
AFCs have a simple design, with few components, making them relatively easy to manufacture and maintain.

iv. Robustness
AFCs are robust and have a long life span, with a low rate of degradation over time.

v. Hydrogen Availability
AFCs can run on hydrogen, which is a readily available fuel, making them attractive for large-scale power generation.

2.7.3 Disadvantages of AFCs

i. Limited Operating Temperature Range
AFCs have a limited operating temperature range, typically between 60°C and 80°C, which can affect their efficiency and power output.

ii. High Cost
AFCs are still relatively expensive compared to other fuel cell technologies.

iii. Hydrogen Handling
AFCs require hydrogen storage and handling systems, which can be expensive and complex.

iv. Poor Durability
The effectiveness and durability of AFCs can indeed be impacted by contaminants inside the fuel as well as the oxidant, which might be susceptible to them.

v. Limited Fuel Options
AFCs are limited to hydrogen as a fuel source, making them less flexible compared to other fuel cell technologies.

2.7.4 Applications of AFCs

i. Stationary Power Generation
AFCs can be used for stationary power generation, providing electrical power to homes, buildings, and remote locations.

ii. Portable Devices
Handheld electronics like notebooks, mobile phones, and hospital instruments can all be powered by AFCs.

iii. Spacecraft
AFCs have been used to power spacecraft, providing a reliable and long-lasting power source for missions.

iv. Transportation
AFCs can be used to power vehicles, such as electric cars and buses, as an alternative to traditional IC engines.

v. Backup Power
AFCs can be used as backup power sources in case of power outages or emergencies.

vi. Remote Monitoring
AFCs can be used to power remote monitoring systems, such as those used in environmental monitoring and weather monitoring.

vii. Military Applications
AFCs can be used in military applications, providing a reliable and mobile power source for military in the battlefield.

2.8 Molten Carbonate Fuel Cell (MCFC)

MCFC is indeed a form of FC that generates energy using an electrochemical process between fuel (often normal gas or biofuels) and air using a liquid limestone as the electrolyte, as shown in Figure 2.9. Excellent power transfer efficacy and high stability are two benefits of MCFCs that operate them at a temperature of about 650°C–700°C. These offer the potential to be employed in major-scale energy generation in integrated heat and power (IHP) plants and are mostly used in portable power generating applications [16].

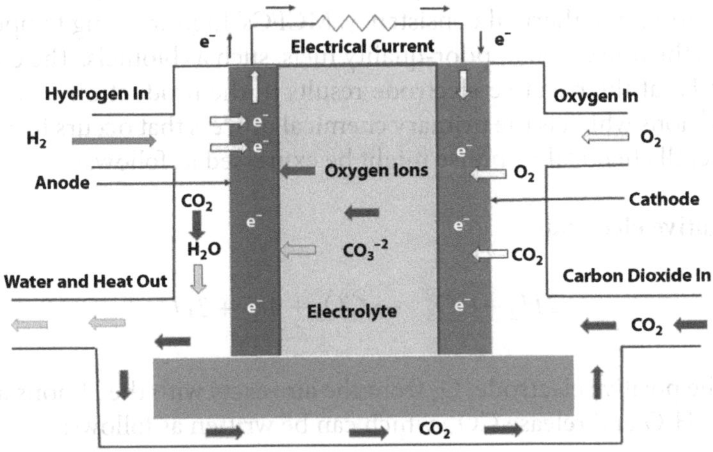

Figure 2.9 Molten carbonate fuel cell (MCFC).

2.8.1 Working of MCFC

i. Fuel and Air Flow

Normal gas or biofuel is supplied as the fuel to the negative electrode, i.e., anode, of the FC, and air is provided to the positive electrode, i.e., cathode.

ii. Electrochemical Reaction

The fuel (H_2) combines using the carbonate ions of the solution at negative electrode to produce CO_2, e^-, and H^+ ions. With the use of an outer circuitry, electrons are transported to the positive electrode and generate an electric power.

iii. Oxidation Reduction

At the positive electrode, O_2 from the open air interacts with the H^+ ions and e^-, producing H_2O and releases CO_2.

iv. Heat Generation

The reaction at the positive terminal produces heat, which needs to be eliminated to retain the temperature of the FC within the operable range [17].

v. Voltage Generation

The voltage variation between the negative and positive electrode generates a voltage, which is used to drive an electrical load.

The carbonate solution is kept molten in the case of MCFCs, which improves the reaction rate and permits effective flow of ions [18]. In addition to providing thermal consistency, MCFCs' high working temperature permits the utilization of poor-quality fuels, such as biofuels. The combustion of H_2 at the negative electrode results in the production of e^-, CO_2, and H^+ ions which is the primary chemical process that occurs in MCFCs. The overall chemical response might be expressed as follows:

At negative electrode

$$2H_2 + CO_3^{2-} \rightarrow CO_2 + 4e^- + 2H^+$$

At the positive electrode, O_2 from the air reacts with the H_2 ions and e^- to form H_2O and release CO_2, which can be written as follows:

At positive electrode

$$O_2 + 2H^+ + 2e^- \rightarrow H_2O$$

Heat is produced by reaction at positive electrode, and it must be dissipated in order to keep the temperature of the FC stable. The voltage variation between positive and negative electrode generates a voltage, which is used to drive an electrical load. In the MCFC, the carbonate electrolyte is maintained in a molten state, which allows for efficient transfer of ions and enhances the kinetics response. The high operating temperature of MCFCs also provides thermal stability and allows for the use of low-quality fuels, such as biogas.

2.8.2 Advantages of MCFCs

i. High Energy Conversion Efficiency
MCFCs have a high energy conversion efficiency, which results in low fuel consumption and low emissions of GHGs.

ii. Long-Term Stability
MCFCs are ideally suitable for usage in large-scale power production applications because they have a long working lifetime and are reliable under a variety of working circumstances.

iii. High-Temperature Operation
The high operating temperature of MCFCs provides thermal stability and allows for the use of low-quality fuels, such as biogas.

iv. High-Temperature Heat Production
MCFCs generate high-temperature heat as a byproduct of the reaction, which can be used for integrated heat and power (IHP) applications.

2.8.3 Disadvantages of MCFCs

i. High Cost
The high cost of materials and fabrication makes MCFCs relatively expensive against the other kinds of FCs.

ii. Complex Design
MCFCs have a complex design, which makes them difficult to fabricate and maintain.

iii. Need of High-Temperature Components
MCFCs require high-temperature components, such as interconnects and heat exchangers, which can be difficult to manufacture and maintain.

iv. Corrosion and Degradation
MCFCs are susceptible to corrosion and degradation due to the high-temperature and corrosive environment, which can lead to reduced performance over time.

2.8.4 Applications of MCFCs

i. Stationary Power Generation
MCFCs are well-suited for large-scale electricity generation in PP. They are very energy-efficient and can run on a number of fuels, namely, bioenergy and normal gas, which results in low emissions of GHGs

ii. Integrated Heat and Power (IHP)
MCFCs generate high-temperature heat as a byproduct of the reaction, which can be employed for IHP services, such as room heating and H_2O heating.

iii. Industrial Processes
MCFCs can be used to power industrial processes that require high-temperature heat, such as melting and refining processes.

iv. Heavy-Duty Vehicles
MCFCs can be employed to operate heavy-duty automobiles, such as buses and trucks, which require high-temperature heat for the engine and cab.

v. Remote or Off-Grid Locations
MCFCs can be used to provide power in secluded or off-grid areas, such as remote villages and islands, where grid connectivity is not available.

Overall, MCFCs are a promising technology for large-scale electricity generation, but their high cost and complexity limit their use in smaller-scale applications.

2.9 Phosphoric Acid Fuel Cells (PAFCs)

PAFC is also a category of FC that use H_3PO_4 acid as an electrolytic medium as shown in Figure 2.10. They run at quite high temperatures, i.e., in between 150°C to 200°C, and are mainly used in stationary applications, such as for distributed power generation and in cogeneration systems. PAFCs are known for their high efficiency, durability, and long life span, making them well-suited for these types of applications. These seem to be less suited for mobile purposes than other kinds of FCs due to their high working temperatures, related cost, and service requirements [19].

2.9.1 Working of PAFCs

i. H_2 is injected into the negative electrode compartment of this FC, whereas O_2 (usually from air) is injected into the positive electrode compartment.
ii. At the negative electrode, the H_2 molecules are broken into two parts, i.e., H^+ protons and e^- electrons, which are separated by the electrolytic solution.
iii. The H^+ protons pass through the electrolytic solution and migrate to the positive electrode; on the other hand, the e^- electrons travel through an outer circuitry to produce electrical energy.
iv. At the positive electrode, the O_2 molecules and protons combine to form H_2O, which is released as a byproduct.
v. Phosphoric acid acts as the electrolyte, facilitating the movement of protons through the cell and preventing the mixing of the hydrogen and oxygen gases.

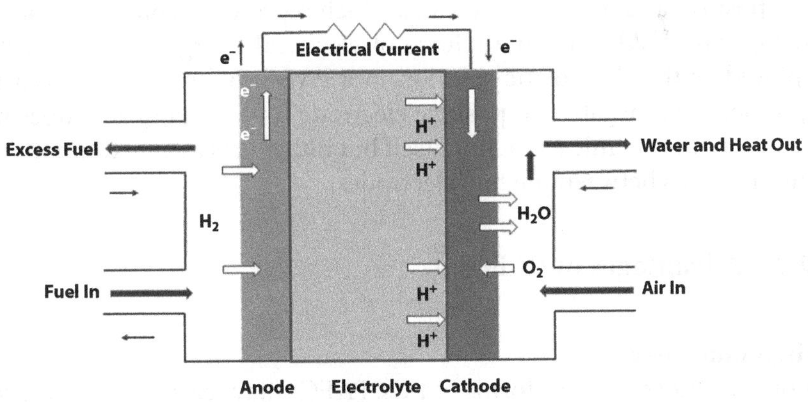

Figure 2.10 Phosphoric acid fuel cell (PAFC).

vi. The cell works at a temperature of about 150°C to 200°C, which is necessary to keep the phosphoric acid in a liquid state and facilitate the electrochemical reactions.
vii. Multiple number of small cells could be connected in series to form a fuel cell stack, which could generate higher voltages and power levels as needed.

Generally, the PAFC generates electricity efficiently by transforming the molecular energy in H_2 fuel, with H_2O as the only waste. This process is highly efficient, and the fuel cells can operate continuously for long periods of time, making them well-suited for stationary power generation applications. The chemical reaction that occurs in a PAFC is described as follows:

Chemical reaction at negative electrode

$$2H_2(Hydrogen) \rightarrow 4H^+(Protons) + 4e^-(Electrons)$$

Chemical reaction at positive electrode

$$O_2(Oxygen) + 4H^+(Protons) + 4e^-(Electrons) \rightarrow 2H_2O\ (Water)$$

Overall chemical reaction

$$2H_2 + O_2 \rightarrow 2H_2O$$

Overall reaction is the combination of the two half-chemical reactions at the positive and negative electrode, which represents, conversion of H_2 and O_2 into H_2O, with the release of electrical energy. H_3PO_4 acid is employed as the electrolytic solution in a PAFC to facilitate the transfer of protons from negative to positive electrode [20]. H_3PO_4 acid does not take part in the chemical reaction itself but rather acts as a medium for the transfer of ions between the two electrodes.

2.9.2 Advantages of PAFCs

i. High Efficiency
Among the most effective forms of FCs, PAFCs may reach efficiency levels of near to 50%.

ii. Durability
PAFCs have a long life span (up to 40,000 h) and can operate continuously for extended periods of time.

iii. Low Emissions
PAFCs produce water and carbon dioxide as byproducts and have lower emissions of other pollutants compared to traditional combustion-based power systems.

iv. Flexibility
PAFCs offer higher fuel supply flexibility because they can run on a wide range of fuels, namely, H_2, normal gas, and methanol.

v. Reliability
PAFCs are well-suited for stand-alone uses, likewise for distributed power production and cogeneration systems, because they offer high efficacy and reliability.

2.9.3 Disadvantages of PAFCs

i. High Cost
PAFCs are relatively expensive to produce energy, because the catalyst and material used for this is very expensive, which can make them less economically viable compared to other power generation technologies.

ii. Maintenance
Owing to the acidic properties of the electrolyte, i.e., H_3PO_4 as well as the extreme working temperature required, PAFCs necessitate continuous servicing and supervision.

iii. Low Power Density
In comparison with conventional types of FCs, PAFCs possess a very low density that may also restrict their usage in some fields.

iv. Slow Start-Up Time
PAFCs can take several hours to reach full operating capacity, which can be a disadvantage in applications where quick response times are required.

v. Limited Mobility
Due to their high operating temperature, PAFCs are less suitable for mobile applications compared to some other varieties of FCs.

2.9.4 Applications of PAFCs

PAFCs have several applications, primarily in stationary power generation and cogeneration systems. Some of the specific applications of PAFCs include the following.

i. Backup Power

PAFCs can be used for backup power in case of a power outage or other emergency situation, providing reliable and efficient power in remote or off-grid locations.

ii. Distributed Power Generation

PAFCs are well-suited for distributed power generation systems, where power is generated locally and distributed to nearby consumers. PAFCs can serve the purpose of providing power and energy to buildings, hospitals, schools, and other facilities.

iii. Cogeneration

PAFCs can be employed in IHP systems, where the wasted heat from the FC is captured and used to supply hot water or room heating, in addition to electricity generation. This can significantly improve the system's total performance while decreasing pollutants and power costs [21].

iv. Military Applications

PAFCs have been used by the military for backup power and mobile power generation, due to their durability and reliability.

v. Space Applications

PAFCs have been used in space applications due to their high reliability and efficiency.

Overall, PAFCs are well-suited for applications that require high efficiency, durability, and reliability, particularly in stationary power generation and cogeneration systems. Although they are less well-suited for mobile applications, their unique features make them helpful techniques for various purposes [22].

2.10 Polymer Electrolyte Membrane Fuel Cell (PEMFC)

PEMFC is a kind of FC that works at reasonably low temperatures and employs polymer as an electrolyte membrane, as shown in Figure 2.11.

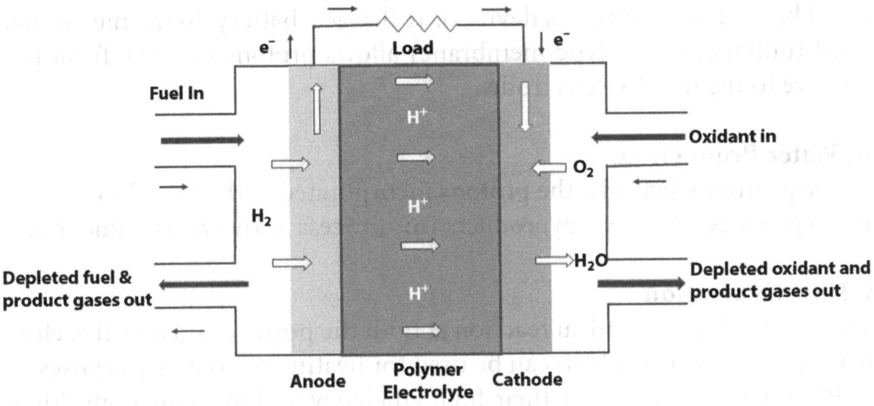

Figure 2.11 Polymer electrolyte membrane fuel cell.

The PEMFC was first developed in the 1960s and has since become one of the most widely researched and developed types of fuel cells [23]. In a PEMFC, O_2 is utilized as the oxidizer and H_2 is normally employed as main fuel. At the negative electrode, the fuel is oxidized, releasing electrons as well as protons. The polymeric electrolytic membrane carries the particles to the positive electrode, where they mix with O_2 and e^- to create H_2O. An electrical charge is created as the electrons go via an outer circuitry [24].

PEMFC functions by transforming chemical energy, usually hydrogen (H_2) stored inside the fuel into an electrical current.

2.10.1 Working of PEMFC

i. Fuel Delivery

H_2 is delivered to the positive electrode of the FC. The fuel may be stored in a high-pressure tank, or it may be generated on the site from the resources of fuel like normal gas or methanol.

ii. Oxygen Delivery

O_2 (usually from air) is delivered to negative electrode of the FC. The negative electrode is usually with the air through a diffusion layer, which helps to distribute the oxygen evenly across the surface of the cathode [25].

iii. Electrochemical Reaction

At the positive electrode, the H_2 molecules are broken into two parts, i.e., electrons as well as protons through a process called oxidation. The electrons flow through an outer circuitry, producing an electrical energy which could be utilized to supply a device or recharge a battery. In the meantime, PEM (polymer electrolytic membrane) allows protons to move from the positive to the negative electrode.

iv. Water Production

At the positive electrode, the protons incorporated with O_2 molecules and e^- to produce H_2O as a byproduct. This process is known as reduction.

v. Heat Generation

The electrical and chemical reaction at both the positive and negative electrode produces heat, which can be used for heating or cooling purposes.

PEMFCs are known for their high efficiency and low emissions. They function at comparatively low temperatures that lie between 50°C and 100°C, making them easier and safer to operate than some other types of fuel cells. PEMFCs are also fast-starting and responsive; they are therefore ideally designed for areas like automobiles or handheld power sources where short reaction times are necessary. The chemical reaction that occurs in a PEMFC can be summarized as follows:

At the negative electrode

$$H_2(Hydrogen) \rightarrow 2H^+(Protons) + 2e^-(Electrons)$$

At the positive electrode

$$\frac{1}{2}O_2(Oxygen) + 2H^+(Protons) + 2e^-(Electrons) \rightarrow H_2O\ (Water)$$

Overall reaction

$$H_2 + \frac{1}{2}O_2 \rightarrow H_2O$$

2.10.2 Advantages of PEMFCs

i. High Efficiency

PEMFCs offer a high energy conversion efficiency, with some designs achieving efficiencies of over 50%.

ii. Low Emissions
PEMFCs emit only water and heat as byproducts, making them a clean and environmentally friendly energy source.

iii. Quick Start-Up
PEMFCs can start up quickly, reaching their operating temperature in just a few minutes, making them well-suited for portable power applications.

iv. Low Operating Temperature
At extremely lower temperatures, PEMFCs function in between 50°C and 100°C, making them easier and safer to operate than some other types of fuel cells.

v. Modular Design
PEMFCs are modular and can be easily combined to produce the desired power output.

2.10.3 Advantages of PEMFCs

i. High Cost
PEMFCs are currently more expensive than some other types of FCs, mostly as a result of high expense of the positive and negative electrode platinum catalyst.

ii. Durability
PEMFCs are not as durable as some other kinds of FCs, and their lifetime might be limited by factors such as the degradation of the polymer electrolytic membrane [26].

iii. Sensitivity to Impurities
PEMFCs are sensitive to impurities in the fuel and oxidant, which can cause damage to the membrane and reduce the fuel cell's efficiency.

iv. Hydrogen Storage
PEMFCs require a source of hydrogen fuel, which must be stored in high-pressure tanks or produced on-site using a reformer. Hydrogen storage and production can be complex and expensive.

v. Water Management
PEMFCs require careful management of water levels, as too much or too little water can damage the polymer electrolytic membrane and reduce the FCs' efficiency.

2.10.4 Applications of PEMFCs

PEMFCs have a range of potential applications, including the following.

i. Transportation
PEMFCs can be used as an alternative to IC engines in vehicles, including cars, buses, and trucks. PEMFCs are quite suitable for use in EVs, as they offer high efficiency, quick start-up times, and low emissions.

ii. Stationary Power
PEMFCs can be used to provide electricity for stationary applications, including homes, businesses, and remote areas without access to the power grid. PEMFCs could be employed in integrated heat and power (IHP) systems to supply them both electricity and heat, increasing efficiency and reducing costs.

iii. Portable Power
PEMFCs could also be employed to deliver power for handheld devices, which includes laptops, mobile phones, and other electronics. PEMFCs are well-suited for portable power application as a result of their great efficiency, quick start-up times, and low emissions.

iv. Military Applications
PEMFCs are used by the military as a portable power source for soldiers in the field. PEMFCs can be used to power a range of equipment, including radios, GPS devices, and other electronics.

v. Backup Power
PEMFCs can also be employed to offer backup power in the event of a power failure. PEMFCs can be used to provide electricity to critical infrastructure, including hospitals, data centers, and other facilities.

Overall, PEMFCs offer a range of potential applications across multiple industries, and ongoing R&D are concentrated on improving durability, efficiency, and cost-effectiveness of PEMFCs to make them more widely available and practical for a variety of applications.

2.11 Direct Methanol Fuel Cells (DMFCs)

Methanol is used as the fuel in DMFCs, a kind of FC that runs at extremely low temperatures, usually between 60°C and 90°C, as shown

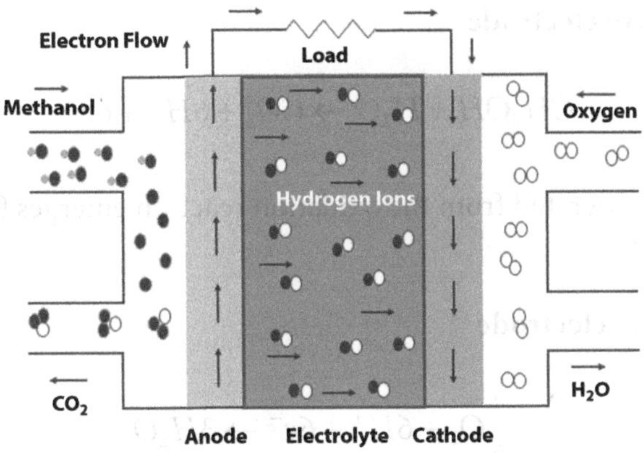

Figure 2.12 Direct methanol fuel cell (DMFC).

in Figure 2.12. Although O_2 is lowered at the positive electrode, oxidizing the methanol and producing H_2O as a waste, O_2 is lowered at the negative electrode, liberating electrons that travel via an outer circuitry and produce electricity [27]. A particular class of PEMFC called DMFCs burns methanol for fuel.

2.11.1 Working of DFMC

i. Fuel and oxidizer are supplied to the negative and positive electrode of the FC. The anode is coated with a catalyst, typically made of platinum or palladium, that makes it easier for methanol to be converted into CO_2 and protons. The positive electrode reduces the oxidizer that is usually O_2 to H_2O.
ii. The protons produced at the negative electrode crosses the PEM which divides the positive and negative electrode. The PEM keeps the fuel and oxidizer from combining, whereas it allows mainly protons to flow through [28].
iii. The electrons generated during the methanol oxidation reaction are not allowed to flow through the PEM, but instead travel through an outer circuitry, generating an electrical energy that could be supplied to an external device.
iv. Both protons and e^- combines with O_2 at the positive electrode, supplying H_2O as the only byproduct.

At the negative electrode

$$CH_3OH + H_2O \rightarrow CO_2 + 6H^+ + 6e^-$$

The CO_2 generated from the oxidation reaction emerges from negative electrode [29].

At the positive electrode

$$\frac{3}{2}O_2 + 6H^+ + 6e^- \rightarrow 3H_2O$$

This reaction shows the reduction of O_2 to form H_2O by consuming protons and electrons generated at the anode.

The overall chemical equation for the DMFC can be expressed as follows:

$$CH_3OH + \frac{1}{2}O_2 \rightarrow CO_2 + 2H_2O$$

This equation shows the full oxidation of methanol to CO_2 and H_2O. In the FC, this reaction is split into two, half-cell reactions:

2.11.2 Advantages of DMFCs

i. These cells are relatively simple in design and can operate at low temperatures, making them an attractive option for portable and mobile power applications.

ii. Methanol is an easily accessible fuel allowing for longer run times compared to other FC types.

iii. DMFCs are highly efficient at converting fuel to electrical power, with reported efficiencies up to 40%.

2.11.3 Disadvantages of DMFCs

i. Methanol crossover is a major issue in DMFCs, where methanol crosses the membrane and provides reaction at the positive electrode, reducing the efficiency of the cell.

ii. DMFCs have a low power density compared to other fuel cell types, limiting their use in high-power applications.

iii. Methanol has a lower density of energy compared with the traditional fuels, which restricts the practical usage of DMFCs to low-power applications.

iv. DMFCs require a high concentration of methanol (typically > 3 M) to achieve reasonable power output, which makes fuel storage and transportation more challenging.

Overall, DMFCs have the potential to be a cost-effective, efficient, and environmentally friendly source of power for portable and mobile applications. To enhance the efficiency and overcome the issues brought on by the charge transfer and poor energy output, more research is necessary.

2.11.4 Applications of DMFCs

i. Handheld Electronic Devices
DMFCs can supply small portable devices such as notebooks, smartphones, and cine cameras, providing longer run times than traditional batteries.

ii. Remote Sensors
DMFCs can provide a reliable and continuous power source for remote sensors, eliminating the need for frequent battery replacements.

iii. Automotive
DMFCs can power auxiliary systems in vehicles, such as air conditioning and heating, reducing the load on the main battery.

iv. Military
DMFCs can provide a reliable and efficient power source for military operations, where traditional power sources may not be available or practical.

v. Medical Devices
DMFCs can power medical devices, such as infusion pumps and ventilators, where reliability and portability are essential.

vi. Backup Power
DMFCs can provide backup power for critical systems, such as emergency lighting and communication equipment, during power outages.

Overall, the potential applications of DMFCs are broad and diverse, with a particular focus on portable and mobile power sources. However, further research and development are needed to address the challenges associated with methanol crossover and lower density of power and to increase efficiency and productivity of DMFCs for practical applications.

2.12 Parameters Affecting the Performance of FCs

i. Cell Potential

A high cell potential is desirable since that will allow the FC to generate more power. However, too high of a cell potential would mean the current density is too low due to the polarization curve of an FC, which leads to a sluggish reaction and losing more fuel to cover for the activation losses. So, a moderate cell potential is ideal if it is coupled with a decent current density, as that will give good power and good efficiency.

ii. Temperature

Among the most essential factors, the productivity enhancement at greater temperatures is primarily caused by expanded proton resistivity, better mass exchange of the reagents, and upgraded kinetics electrode for the oxide reducing reaction (ORR) and H_2 oxidizing reaction (HOR). Moreover, raising temperature can help electro-catalysts tolerate impurities better, especially those of the susceptible FCs such as the AFC and PEMFC. Yet, greater operational temperatures can limit the life span of an FC by causing membrane drying, a rise in the H_2 crossing speed, and the deterioration of parts like bipolar tiles.

iii. Number of moles of electrons transferred

The more electrons are transferred, the better the flow of current and thus more electricity produced.

iv. Partial Pressure

The open circuitry voltage will rise with larger relative pressures of the reagents in the FCs. However, pressure of both H_2 and O_2 decreases when temperature increases for PEMFCs (decreases exponentially after 80°C), which increases activation losses and reduces cell performance. So, a good balance between both partial pressure and temperature is needed, which will probably be around 80°C for PEMFCs [30].

References

[1] Hari Om Bansal and Praveen Goyal, "Enablers and Barriers of Electric Vehicles in India - A Review", *IEEE International Symposium on Sustainable Energy, Signal Processing and Cyber Security (ISSSC)*, IEEE, pp. 1-5, 2020.

[2] Behrooz Mashadi and David Crolla, *Vehicle Powertrain Systems*, Wiley, 2012.

[3] Claudel Mombeuil and, Hemantha Premakumara Diunugala. "Green brand awareness, green brand association, green perceived quality, and intention to purchase electric vehicles: The mediating effect of green trust", ResearchSquare Platform LLC, pp. 1-15, 2023.

[4] Leo J. M. J. Blomen and Michael N. Mugerwa, *Fuel Cell Systems*, Springer Science + Book Media, LLC, 2013.

[5] M.W. Ellis, M.R. Von Spakovsky and D.J. Nelson, "Fuel cell systems: efficient, flexible energy conversion for the 21st century", *Proceedings of the IEEE*, Vol. 89, Issue 12, December 2001.

[6] Kai Sundmacher, Achim Kienle, Hans Josef Pesch, Joachim F. Berndt and Gerhard Huppmann, "Molten Carbonate Fuel Cells", Wiley Online Library, 2007.

[7] Xiao-Zi Yuan, Chaojie Song, Haijiang Wang and, Jiujun Zhang, "Electrochemical Impedance Spectroscopy in PEM Fuel Cells", SpringerScience and Business Media LLC, pp. 1-37, 2010.

[8] Guangxu Zhang, Xuezhe Wei, Siqi Chen, Jiangong Zhu, Guangshuai Han, Xueyuan Wang, and Haifeng Dai, "Revealing the Impact of Fast Charge Cycling on the Thermal Safety of Lithium-Ion Batteries", *ACS Appl. Energy Mater.*, Vol. 5, Issue. 6, pp. 7056-7068, 2022.

[9] Xiaoyao Tan and Kang Li, *Inorganic Membrane Reactors: Fundamentals and Applications*, Wiley Online Library, 2014.

[10] Eklas Hossain, Slobodan Petrovic, *Renewable Energy Crash Course - A Concise Introduction*, Springer Publication, Chapter 9, Energy Storage, pp. 83-94, 2021.

[11] YannickGarsany, Karen E. Swyder Lyons, "Kinetic Activity in Electrochemical Cells", *Springer Handbook of Electrochemical Energy*, Chapter 14, pp. 423-445, January 2017.

[12] Abid Rabbani and Masoud Rokni, "Modeling and Analysis of Transport Processes and Efficiency of Combined SOFC and PEMFC Systems", *Energies in MDPI*, Vol. 7, Issue 9, 2014.

[13] Albert Boretti and Stefania Castelletto, "MXenes in Polymer Electrolyte Membrane Hydrogen Fuel and Electrolyzer Cells", *Ceram. Int.*, RMIT University, Vol. 48(23), pp. 34190-34198, 2022.

[14] Veenasri Vallem, Yasaman Sargolzaeiaval, Mehmet Ozturk, Ying-Chih Lai and Michael D. Dickey, "Energy Harvesting and Storage with Soft and Stretchable Materials", *Adv. Mater.*, Wiley, Vol. 33, 2004832, pp. 1-37, 2021.

[15] S. Kakaç, A. Pramuanjaroenkij and L. Vasiliev, *Mini-Micro Fuel Cells-Fundamentals and Applications*, Springer book series, NATO Science for Peace and Security Series C: Environmental Security, 2008.

[16] Christoph Straub, Armin Vetter and A. VonFelde, *Biogas Production and Energy Crops-Encyclopedia of Sustainable Science and Technology*, Springer Book, pp. 1097-1145, 2012.

[17] Arora, K.; Kumar, A.; Kamboj, V.K.; Prashar, D.; Jha, S.; Shrestha, B.; Joshi, G.P. Optimization Methodologies and Testing on Standard Benchmark Functions of Load Frequency Control for Interconnected Multi Area Power System in Smart Grids. *Mathematics* **2020**, *8*, 980.

[18] Cornelia Breitkopf and Karen Swider Lyons, *Springer Handbook of Electrochemical Energy*, 2017.

[19] Can O Colpan, Yagmur Nalbant, and Mustafa Ercelik, "Fundamentals of Fuel Cell Technologies", *Comprehensive Energy Systems*, Vol. 4, 2018.

[20] P Singh, K Arora, UC Rathore, Control Strategies for Improvement of Power Quality in Grid Connected Variable Speed WECS with DFIG – An Overview, *J. Phys. Conf. Ser.*, 2022.

[21] James Larminie, *Electric Vehicles Technology*, Chapter 5, "Fuel Cells", Wiley Publication, pp. 87-114, 2002.

[22] Boris I. Kharisov, Oxana V. Kharissova and H. V. Rasika Dias, *Nanomaterials for Environmental Protection*, Wiley Online Library, 2014.

[23] Daniel Soeder, *Energy Futures - The Story of Fossil Fuel, Greenhouse Gas, and Climate Change*, Springer Book, Chapter 8: "Zero Carbon Vehicles", pp. 185-204, 2022.

[24] P. Thounthong and, P. Sethakul, "Analysis of a Fuel Starvation Phenomenon of a PEM Fuel Cell", *IEEE Power Conversion Conference - Nagoya*, pp. 731-738, 2007

[25] Prof. Dr. Andreas Züttel, Dr. Andreas Borgschulte and Prof. Dr. Louis Schlapbach, *Hydrogen as a Future Energy Carrier*, ISBN: 9783527622894, 2008.

[26] C. Ozgur Colpan, Ibrahim Dincer and Feridun Hamdullahpur. "Reducing Greenhouse Gas Emissions from a Landfill Site Using Various Thermal Systems", SpringerScience and Business Media LLC, Chapter 8, pp. 161-178, 2010.

[27] Arora, K.; Kumar, A.; Kamboj, V.K.; Prashar, D.; Shrestha, B.; Joshi, G.P. Impact of Renewable Energy Sources into Multi Area Multi-Source Load Frequency Control of Interrelated Power System. *Mathematics* **2021**, *9*, 186.

[28] Jiujun Jhang and Hansan Liu, *Electrocatalysis of Direct Methanol Fuel Cells: From Fundamentals to Applications*, Wiley Online Library 2009.

[29] Ayhan Demirbas, *Green Energy and Technology, Bio Hydrogen, for Future Engine Fuel Demands*, Chapter: 7, "Fuel Cells", pp. 221-239, 2009.

[30] Maher A. R. and Sadiq Al-Baghdadi. "A simple mathematical model of performance for proton exchange membrane fuel cells", *Int. J. Sustainable Energy*, Vol. 26, No. 2, pp. 79-90, June 2007.

3

Smart Energy Management and Monitoring System for Electric Vehicles with IoT Integration

Challa Krishna Rao[1,2,*], Sarat Kumar Sahoo[2] and Franco Fernando Yanine[3]

[1]Department of Electrical and Electronics Engineering Aditya Institute of Technology and Management Tekkali, A.P., India
[2]Dept. of Electrical Engineering, Parala Maharaja Engineering College, Berhampur, Odisha, India
[3]School of Engineering of Universidad, Finis Terrae, Providencia, Santiago, Chile

Abstract
Urban areas are searching for solutions to this issue as environmental deterioration and air pollution levels rise. Reducing vehicle pollution is one of the most important approaches to managing air pollution in urban areas. The creative move in this area is represented by electric vehicles (EVs). Electric cars are high-tech devices that rely on a ton of data to work at their best. The usage of electric cars must increase because they are the most advanced form of transportation currently available. To increase battery life and safeguard the associated components that they power, electric vehicle batteries must accurately evaluate the level of charge because they can be harmed by overcharging or over-discharging. This article covers a low-cost, real-time, Internet of Things (IoT)–based, and user-friendly technology for controlling and monitoring the batteries in electric cars. It displays crucial details about the battery's health, such as battery capacity and how it is being charged and discharged right now. Real-time display and updating of these data are possible. The proposed system is implemented using an ESP32 microcontroller, the Blynk mobile app, and the Blynk IoT platform. Among the performance, measurements are those for tracking speed, acceleration, mileage, battery management, charging, issue alarms, and preventative maintenance systems. The monitoring of these electric cars is significantly impacted by IoT.

***Keywords*:** Battery charging, state of charge, BMS, electric vehicle, IoT

**Corresponding author*: krishnarao.challa@gmail.com

Krishan Arora, Suman Lata Tripathi and Himanshu Sharma (eds.) Electric Vehicle Design: Design, Simulation and Applications, (57–80) © 2024 Scrivener Publishing LLC

3.1 Introduction

As petrol prices rise, electric vehicles (EVs) are growing in popularity. Due to these circumstances, many manufacturers are researching gas-alternative fuel sources. Electrical energy sources may be better for the environment because they produce fewer pollutants. Additionally, EVs give great energy-saving and environmental-preservation benefits. Lithium-ion batteries are utilized in the majority of electric cars [1, 2]. An electric vehicle's battery pack is made up of lithium-ion cells, which are often put together into modules and connected in series to produce the required voltage. There are specific voltage and temperature requirements that must be satisfied for lithium-ion batteries to function safely. If the temperature or voltage drops below the set point, then the cell may degenerate. The cell may catch fire if these values exceed what is permitted. A typical lithium-ion battery is limited to [2.50; 4.20] V and [13.0; 60.0] °C in terms of voltage and temperature, respectively. The term "cell safety window" refers to the set of values that falls inside this range [3]. The driving range is influenced by some variables, including battery capacity, engine complexity, and battery age. Due to current research aimed at increasing the range of electric vehicles (EVs), monitoring and management systems for batteries must now be created and improved [4, 5]. A battery management system (BMS) is a component used in electric vehicles that keeps track of the battery pack's health and ensures its safe functioning. The BMS also makes an estimated state of charge (SoC) for the battery [3]. A battery's status of charge is an important factor in extending battery life and giving the user the knowledge to evaluate the battery's state of charge or discharge [6]. For a battery to be charged and discharged properly and hence increase its life span, one of the most important key enablers is the SoC [7, 8]. Moreover, the SoC shows how much of a battery's maximum potential capacity is still available for charging. The battery SoC displays how much battery power is still available to run an EV, just like a fuel gauge does in a car that runs on gasoline. Reliable and safe vehicle operation is ensured by a proper battery SoC calculation; this contributes to giving current data on the energy and battery capacity. Because they are situated on an axis of very high energy density, batteries, and Li-ionized batteries in particular, over the whole vehicle operating cycle, are perfect for high energy density activities [9, 10]. The coulomb counting technique is easy to use. Theoretically, coulomb counting necessitates observing the battery's input and output currents to measure charging. Because of this, capacity is typically expressed in ampere-hours. Several all-electric car battery types can be employed with the technique [11, 12].

Nowadays, a fascinating topic of research is the Internet of Things (IoT). The Internet of Things which is presently being created and is being used widely may be connected to any electrical item in our environment [13, 14]. It is a network of physically linked devices, embedded electronics, sensors, actuators, and other parts that facilitate data collecting and exchange [15, 16]. The aforementioned uses can be fulfilled by a wide variety of additional IoT devices now available. Yet, due to how simple they are to use and run, the most popular IoT devices are Arduino and Raspberry Pi [19, 20].

3.2 The Control of Electric Vehicles Using IoT

Let us examine each component of an IoT management system for electric vehicles and how they improve the operation of these vehicles.

3.2.1 Battery Management System

What the battery management system is responsible for includes monitoring and controlling battery activities. This includes the charging and discharging cycle. Ensuring that the vehicle is receiving the greatest possible energy supply minimizes the possibility of battery damage and ensures battery health. The monitoring circuit of the BMS keeps track of the crucial battery characteristics [22]. This contains the temperature, voltage, and current under both charging and discharging conditions. Based on the measurement, it calculates factors like power, state of health (SoH), and state of charge (SoC) to assure health. As it is possible to log data remotely for battery metrics and conditions [17, 18], the IoT is essential for monitoring and regulating. High-quality Li-ion battery packs are becoming more and more popular among EV manufacturers because of their extended life span and improved high energy density range. There are, however, disadvantages in addition to these advantages [23].

The data from the onboard sensors collected through IoT can assist in controlling the difficulties in cases where the battery fails. They can then be put via AI-based performance evaluation algorithms [24]. For various Li-ions, tests may be performed to evaluate the patterns of charging and discharging, both partially and fully. The information gathered throughout each phase is used to describe models. Before being deployed on a server, these are AI-integrated. The EV transmits important sensor data to the server, providing information on the performance and future steps [25].

3.2.2 Safe and Intelligent Driving

Real-time monitoring of the vehicles and their essential components is also made possible by the deployment of IoT. It helps with the technology's preventative maintenance, which increases its dependability in the eyes of consumers. The following features can be offered to customers via IoT devices included in EVs [26].

- It can measure the driver's absolute and relative performance factors, including acceleration and speed, and provide real-time advice to ensure greater performance.
- Real-time tracking, geo-fencing, and immobilization can all be used to stop theft. This lessens the need for insurance, and safety and security will be improved.
- It will monitor car performance data so that EV and battery makers may improve their products. The criteria include a range for each charge; the use of the vehicle; performance variances caused by location, weather, and age; and changes in the range for each charge over a defined period [21].

3.2.3 System for Fault Alert and Preventative Maintenance

An electric car is a high-tech device; thus, technological issues are inevitable. IoT-based fault alert systems assist in warning drivers of EV defects so they have time to take corrective action. Even if EVs are meticulously constructed, there may be times when some parts malfunction. AI algorithms and distant IoT data may be crucial in predicting this. This assists in forewarning EV users so they can address any potential difficulties or prevent a complete meltdown [27]. Because people will see it as trustworthy, this will result in a better client experience. Monitoring the remote performance while keeping an eye on the general temperature and moisture conditions across diverse geographies is crucial. This will provide the users with a sense of comfort and security in addition to speeding up the resolution of the problems [28].

3.2.4 Data from Telemetry

Data are gathered when linked to car sensors using IoT-based telemetric technology, and it may be quickly presented through widgets, sent as fast notifications, and used to create automated reporting [29]. The advantages of employing telematics for the distant monitoring of electric vehicles are described below.

3.2.4.1 *Battery Usage Information*

Users of electric vehicles with telematics may monitor real-time battery use statistics. This data analysis offers information that enhances battery life. To prevent battery failure, it also enables the monitoring of crucial variables including voltage, current, and temperature [30]. By recording and sending the data to a remote server using an IoT device, the battery setup may be customized so as to enhance the best charging techniques [31].

3.2.4.2 *Report on Charging*

The use of telematics in electric cars enables the creation of reports on all charging sessions that may have taken place over a vehicle's lifetime. The charging report details the amount of time spent charging, the location of the charging station, and the percentage of charge that each car received [32].

3.2.4.3 *Notify About Nearby Charging Stations*

Knowing the state of charge (SoC) might help a motorist to determine when and where to charge the electric car. Telematics-based monitoring systems for electric vehicles alert the user to low battery levels in the cars as well as the location of nearby charging outlets [33].

3.2.4.4 *Data on Driver Behavior*

Electric car remote monitoring systems with telematics focus on driver behavior data in addition to monitoring and analyzing data on electric vehicle performance [34]. Using IoT smartphone apps, telematics provides fleet managers with immediate input on changes in driver behavior to increase vehicle safety.

3.3 IoT Management Issues with Electric Vehicles

After going over the benefits of IoT for monitoring electric vehicles, we must now explore the drawbacks, as described below.

3.3.1 Internet Safety

A network is susceptible to hacker assaults and data theft because it transfers such a massive volume of data. So, it is necessary to reinforce and increase the security of the IoT networks that support the data from EV systems [35].

3.3.2 Higher Price

EV IoT solutions are extremely sophisticated and have expensive installation and maintenance expenses. Further R&D in this area may be done in the future to save expenses.

3.3.3 Considering the Challenges and Advantages

IoT is heavily utilized to track all-electric vehicles. Among the performance, there are measurements for tracking speed, acceleration, mileage, battery management, charging, issue alarms, and preventative maintenance systems. Its drawbacks, including cost and cybersecurity, must be carefully considered [36]. IoT is crucial to the success of electric vehicles as an innovative step toward better air pollution management. The entire globe is looking for new solutions to this issue as air pollution levels keep rising and the environment continues to deteriorate. Reducing vehicle emissions is one of the greatest strategies to manage air pollution. EVs are a revolutionary development in this area [37]. Moreover, using IoT may help businesses learn more about electric vehicles and enhance the customer experience. Electric cars are high-tech devices that gather a lot of data to provide consumers with the best performance possible. Tracking performance indicators including speed, acceleration, mileage, battery management, charging, trouble warnings, and system upkeep requires the Internet of Things [38].

3.4 Monitoring and Management Benefits of IoT

Let us look at some of the important IoT elements used in the monitoring and management system for electric cars and the various ways that they improve the overall operation of electric vehicles.

3.4.1 IoT and Battery Management Systems

The role of a battery management system (BMS) is to manage and observe battery operations. By certifying that the electric car is being powered by more effective energy, the BMS helps to monitor the battery's charging and discharging cycle, assures the battery's health, and reduces the danger of damage [33]. The vital battery statistics are tracked by the BMS monitoring circuit. This will include the temperature when charging and discharging as well as the ambient temperature. Moreover, it evaluates the battery's vital signals, such as power, charge level, and state of health, and uses the information to guarantee a long battery life. The Internet of Things (IoT), which makes it possible to log data remotely for battery characteristics, conditions, etc., is essential for regulating and monitoring the battery life span [35].

3.4.2 IoT for Safe and Intelligent Driving

Thanks to the Internet of Things, we can monitor real-time data on automobiles and the vital parts inside them. Customers are given a sense of the knowledge offered by the technology, boosting their confidence in it. Users of electric vehicles may receive the following information via IoT devices integrated with such vehicles [36]:

- It assists in keeping an eye on the driver's performance as well as important factors like vehicle speed, acceleration, and temperature and offers suggestions in real time to ensure improved performance. Real-time tracking, geo-fencing, and immobilization can help stop car theft. IoT improves safety and security and lessens the need for insurance as a result [40].
- IoT will track car performance information so that EV and battery manufacturers may enhance their wares. The information offered by IoT consists of the range for each charge; the battery life cycle; the usage of a vehicle; performance differences according to place, weather, and age; and shifts in the range for each charge over time [41].

3.4.3 Theft Prevention

Electric vehicles are currently more costly than conventional cars with gasoline or diesel engines. Due to the high cost of batteries, two-wheelers are more vulnerable to theft, and IoT might aid in more precise tracking. Of the overall cost of EVs, 40% goes toward the battery. So, it is imperative to keep the battery safe against theft. On-demand app companies employ electric vehicles for a variety of delivery services, including food delivery, freight delivery, bike sharing, and hyper-local deliveries [42]. Under these circumstances, IoT may be used to locate EVs in real time if they are stolen. These IoT gadgets may also sound an alarm and alert the owner of the theft.

3.4.4 Detection of Falling or Crashes

Most IoT devices come with built-in accelerometers that can measure vehicle acceleration and tilt. As a result, IoT can quickly identify abrupt changes in the vehicle acceleration and angle, alerting the owner to the possibility of a fall or accident. This is also a great way to alert people about serious incidents [43].

3.4.5 Battery Leasing Made Simple

One of the most crucial elements for battery swapping and leasing businesses is the performance and location tracking of batteries. From the battery switching stations, it is simple to follow and keep an eye on the battery shifting. IoT may play a significant role in tracking or monitoring batteries and in preventing theft and loss of the battery [44].

3.5 Predictive Maintenance System with Fault Alerts

An electric automobile will inevitably have technical problems because it is a high-tech device. It notifies drivers of vehicles of any defects or malfunctions via IoT fault warning systems, giving them time to take appropriate action [45]. With IoT, we can measure the car's charge cycle, miles driven, battery temperature, and other parameters. We can then utilize this information for routine type maintenance, battery life monitoring, and replacement of other worn-out elements of the vehicle. This might prevent harm to the car and extend the life of both the car's batteries and the car itself. As a result of the increased reliability, customers will have a better overall experience [46].

3.6 IoT Management and Monitoring Issues with Electric Vehicles

We have previously seen the benefits of IoT in managing and monitoring electric vehicles, but we now need to consider any potential drawbacks.

3.6.1 Threat from Cyber Attacks

The utilization of a large volume of data transferred across a network and the addition of so many devices to the Internet makes these data more susceptible to hacker assaults and data theft [47]. Hackers may get access to the system, take user information, and utilize it inappropriately. Thus, it is necessary to reinforce and make more hack-proof the IoT devices attached to the EV and gather the whole data.

3.6.2 Electric Car Prices are Quite High

Because the IoT systems employed in electric vehicles are extremely sophisticated and have substantial installation and operating expenses, the entire cost of an EV is increased. More R&D in this field has the potential to lower total expenses shortly [48].

3.6.3 Technological Difficulty

The overall design of EVs gets increasingly complex and difficult to handle in the IoT system because it is difficult to design, develop, maintain, and enable the comprehensive technology [49].

3.6.4 Connectivity and Reliance on Power

For effective operation and to give real-time data, IoT devices used in EVs need to be connected to the internet and have constant power. Both the gadget and everything connected to it will cease to function if either goes offline. So, it will also have an impact on EVs' overall performance [50].

3.6.5 Battery Management and Monitoring System

The recommended IoT-based solution for EV consumers is shown in Figure 3.1. A mobile application that provides monitoring and administration of the complete system may be used to materialize the

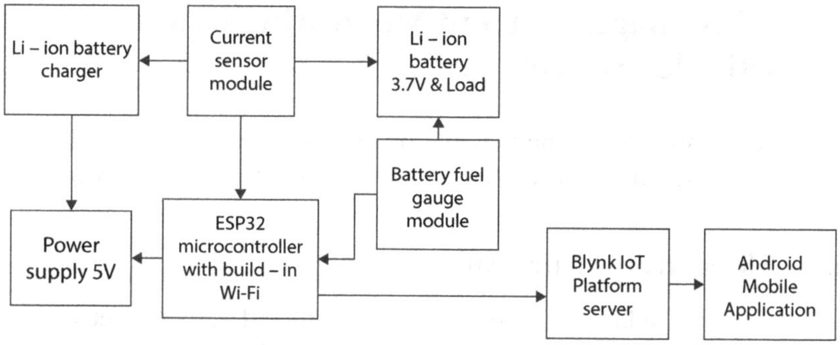

Figure 3.1 The battery management and monitoring system [1].

established system. Users may manage the charging process while also keeping track of the input current, output current, and a summary of their overall charging current for their electric car batteries [51]. Each of these parameters has real-time displays and updates. The suggested system consists of a Li-ion battery, a Li-ion battery charger, a fuel gauge sensor for lithium-ion batteries, two current sensors, an ESP32 microcontroller with integrated Wi-Fi, and two current sensors. A message that describes the total charging current is transmitted using a mobile telegram application throughout each charging procedure [52].

The fact that ESP32 has Bluetooth and Wi-Fi built-in is the main factor in choosing it. So, unlike on other Arduino boards, extra radio modules are not required. The ESP32 consists of simply one chip and one packaging. For the system to work, ESP32 must be linked to a Li-ion battery fuel gauge sensor combined with two current sensors [53]. After gathering and analyzing the sensor data, ESP32 uploads it over the Wi-Fi module to Blynk Server. Blynk's Android mobile application controls both the charging process and how all information is displayed and updated in real time [54].

3.6.6 Prototype of a Battery Charge Control and Monitoring System

Figure 3.2 depicts the design of the IoT-based battery charge control and monitoring system for electric vehicles. It has been incorporated into this prototype system using the ESP32 microcontroller. The system may be powered via a USB port on a computer or by any other source that can provide the 5-V DC it needs to operate. The Wi-Fi module that is included with the ESP32 is used to establish an internet connection [55]. Two current sensors are present in the ESP32, one of which measures the current

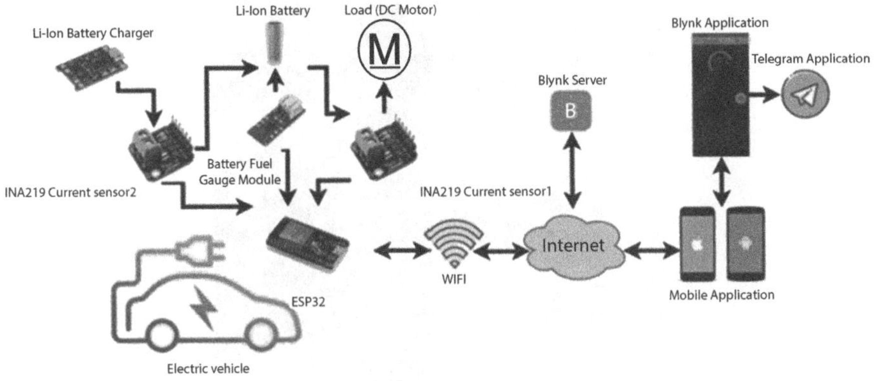

Figure 3.2 Battery charge control and monitoring circuit of the electric vehicle architecture [3].

entering the battery and the other which measures the current leaving the battery. The battery is charged via a coupled Li-ion battery charger, and a relay manages the charging procedure. Battery capacity or SoC is measured using a Li-ion battery fuel gauge sensor (state of charge) [56].

As demonstrated in Figure 3.3, the technology notifies the user when the battery needs to be charged so that it can be safeguarded against damage from over-discharge and have a longer life span. The charging procedure is finished when the battery is fully charged or when the off button is pushed. To stop the charging process, the Li-ion battery charger has a built-in protected circuit [57].

3.6.7 Scenario for the Battery Monitoring and Management System

ESP32 is used to process and gather sensor data, and two current sensors, a battery fuel gauge module, and other components are connected to the same serial port via the I2C protocol. Two bi-directional open-drain lines are all that are required for the transfer of SDA and SCL data. These two lines have both been boosted. Data are sent via serial data (SDA), whereas the clock signal is transmitted using serial clocks (SCL); the rate of each clock pulse on each bit of data sent across the SDA line is synchronized by the SCL line, from high to low [58]. Battery SoC was retrieved from MAX 17043 and the current in/out value was taken from INA219 via ESP32. The relay managed the charging activity by being linked between the battery and TP4054. Data transit between the ESP32 and A special authentication token provided while creating an account on the Blynk mobile application

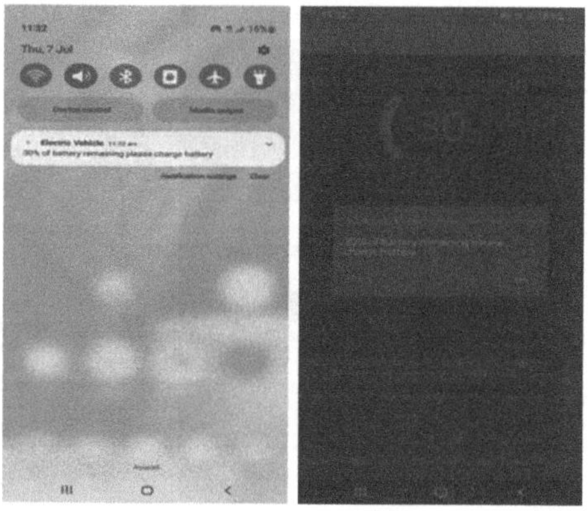

Figure 3.3 The system alert for low battery remaining [5].

is used to protect the Blynk server using devices running the mobile application; a digital dashboard was used to display the data obtained from the Blynk server [60].

3.7 Microcontroller

Figure 3.4 illustrates the ESP32, a powerful SoC microcontroller with built-in Wi-Fi 802.11 b/g/n, dual-mode Bluetooth 4.2, and peripherals (a). Because it contains two cores with independent clock speeds of up to 240 MHz, it is a superior substitute for the 8266 processor. Together with these improvements, there are now 4 MB of flash memory and 36 GPIO pins as opposed to 17 before [61].

The ESP32 has two essential components. Each CPU core may be changed independently. For data and instructions, 520.0 KB of on-chip SRAM is accessible. For instance, the controller has 8 MB of SPIPSRAM and 4 MB of external SPI flash according to the kind of board [22, 23].

3.7.1 DC Current Sensor

It offers digital voltage, current, and power data for precision-controlled systems to make educated decisions. Programmable registers enable flexible architecture for measurement precision and continuous versus triggered operation [39]. The device uses user-customizable conversion filtering and timing to monitor shunt voltage loss and bus supply voltage. Direct readouts of current in amps may be detected using a built-in multiplier and a programmable calibration value. Watts is the unit of measurement for

(a) Microcontroller [12]

(b) Current sensor [12]

(c) Relay [38]

(d) Li-Ion Battery Charging [38]

(e) Battery fuel gauge [38]

Figure 3.4 The different types of sensors used in the EVs.

power; a separate multiplication register is used. The I2C or SMBus compatible interface offers 16 programmable addresses. For the INA219, there are two courses: A and B. The INA219 gadget can identify shunts on buses with voltages between 0 and 26 V. Up to 1 mA of supply current can be sent to the device from a single 3.0- to 5.50-V power supply. The operating range of the INA219 is 40 to 125 C. Figure 3.5 describes the basic schematic diagram [24].

Figure 3.5 shows the breakout board's pin-out schematic. Vcc, Gnd, Scl, Sda, Vin, and Vin+ are the pinouts. The breakout board is powered by Vcc and Gnd, the I2C clock synchronization and data transmission are handled

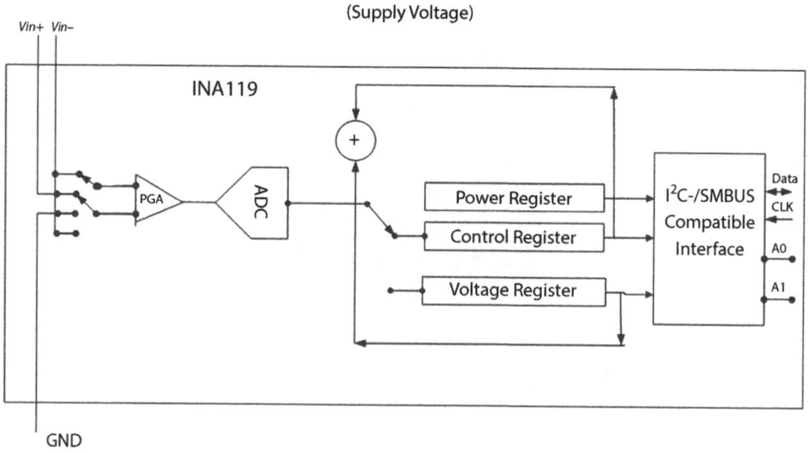

Figure 3.5 Functional block diagram current sensor [22].

by Scl and Soda, and the DUT (Device under Test) is powered by Vin and Vin+ through the INA219 sensing resistor [25].

Managing Charge-Discharge Relay Relays are used to handle the charging process. Figure 3.4 illustrates how relays work as electromechanical or electronic switches to open and close circuits (c). Relays easily join and reconnect connections that have been electrically disconnected between two circuits.

The relay diagram shows that when the relay is not activated, the contact is open. Even if the relay is not enabled, the link is severed by an normally closed (NC) contact. To modify the contacts' state in either scenario, an electrical current must be applied to them [58]. Relays are often used to switch lesser currents in a control circuit; they do not control power-consuming devices, except for small motors and low-ampere Solenoids. Because a low voltage may cause a relay's contacts to switch to a high voltage, relays can "control" higher voltages and amperes. By identifying electrical abnormalities including over current, undercurrent, overloads, and reverse currents, protective relays can prevent equipment damage. Relays are used to manage pilot lights, heating elements, starting coils, and audible alarms [26].

3.7.2 Fuel Gauge Module for Li-Ion Batteries

The ModelGauge method does not require extra current-sense resistors or battery relearn cycles, in contrast to conventional fuel gauges. Temperature adjustment is possible in the application with little micro-coulomb contact with the device. The cost and sourcing restrictions of the battery can be reduced by placing an integrated circuit on the system side to get data sets for capacity and measurement estimates, and an I2C interface is employed [11].

The battery's temporal impacts on impedance and chemical reactions are taken into account by the model. For MAX17043/MAX17044, the SoC calculation does not develop inaccuracies with time. This is an advantage over conventional coulomb counters, which have SoC drift as a result of current-sense offset and cell self-discharge. For diverse Li+ chemistry changes at various temperatures and ages, this model performs well. For the MAX17043/MAX17044 to work at its best, application-specific setup information is required [35].

3.8 IoT-Based Systems for Battery Management and Monitoring

The easiest IoT platform to use is Blynk, which has an iOS and Android application builder. Also, it features several libraries that allow the user to quickly develop amazing Internet of Things apps on any preferred hardware platform [28, 29]. A graphical user interface (GUI) is developed in a digital dashboard. From the widget box, widgets may be dragged and dropped. Every widget relies on points in some way to operate. A person that creates a Blynk account will receive 2,000 energy points. The energy balance falls when a widget is used [30, 31]. Users of the Blynk mobile application may monitor the battery capacity of electric cars, regulate the charging, and control the input/output battery currents absorbed by the load process by pushing a button on the digital application dashboard (start and stop). A message with the entire amount of current consumed in the charging operation will be delivered to the Telegram application after the battery charging procedure is over.

3.9 Design of Battery Charge Control and Monitoring System

Figure 3.6 shows the battery charge control system prototype. The configuration of the circuit connections is shown in Figure 3.7 [59]. The current sensor and microcontroller may both be connected to the same serial port on the microcontroller because of their common use of the Inter-Integrated

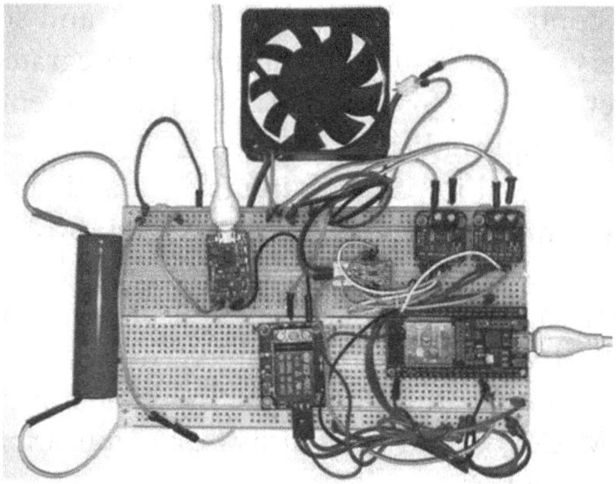

Figure 3.6 Design of battery monitoring and control circuit for practical verifications [1].

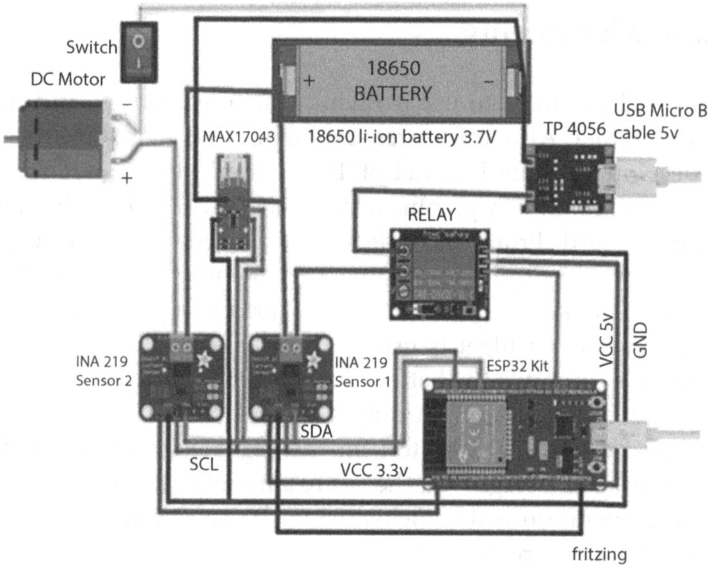

Figure 3.7 Battery control system circuit design [32].

Circuit Protocol, which enables one or more "controller" chips to interact with a large number of "peripheral" digital integrated circuits. As programmable addresses to pick between two INA219 devices, connect (A0, A1) to the solder board [45].

Pins for the serial clock line and the serial data of sensors are connected, and D22 and D21 are connected to MAX17043. To supply the sensor with the required power, the VCC and GND pins of the ESP32, as well as the VCC and GND pins of two INA219 devices, were connected to achieve a voltage of 3.30 V. For monitoring battery capacity, the MAX17043 is connected with its positive (+) pins to the battery terminals and its ground (GND) pin to the ESP32's GND pin [48]. A typically open pin, controlled by the application, is used by the relay in the circuit to regulate the charging activity. The button (push to charge) switches from off to on when the user clicks the application dashboard; it is also used to initiate charging [49].

3.10 Results and Discussion

After connecting all parts, a prototype system was created, put into practice, and tested, as seen in Figure 3.6. An ESP32 Wi-Fi module is used to link the ESP32 to the router, and the system is powered by a USB connection on a computer. The ESP32 gathers and processes all of EV's data before sending it to the Blynk server and displaying it on the Blynk mobile app. Updating and visualizing all data in real time, as shown in Figure 3.8, demonstrates the dashboard display for the user's mobile application for charging and discharging.

Every time the user presses the button to charge or discharge, the total charging current is shown, and the precise amount is subsequently communicated to the user's Telegram program.

As seen in Figure 3.8, throughout this process, the battery's capacity and voltage rise. The charging process comes to an end in one of two ways: either the (push to charge) button is depressed, or the TP4056 board

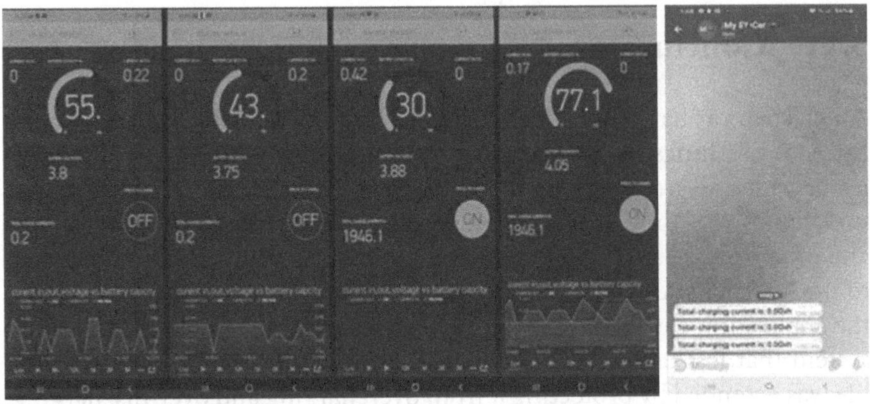

Figure 3.8 Message received by telegram application.

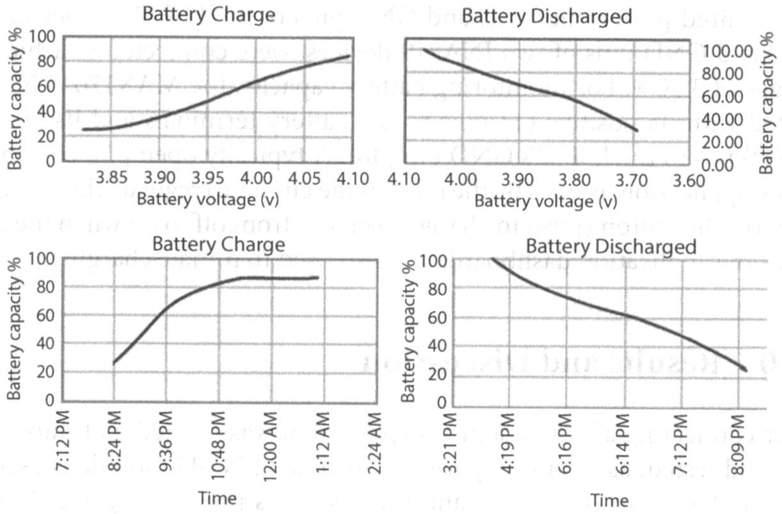

Figure 3.9 Voltage and time characteristics of charging and discharging of the battery.

protection circuit completely charges the battery. A message with a breakdown of the current used to charge the battery is delivered to the user's Telegram app each time the (ON and OFF) button is pressed to charge the device. This function allows you to know the charge date within a given time frame by using it as a bill after each charging procedure and keeping it as a booking. Figure 3.8 displays the message that was sent by telegraph (e). Figure 3.9 shows the length of time needed for the charging and discharging process using the average charging current (0.24 AH) and average discharging current.

3.11 Conclusions

An electric car battery management and monitoring system based on the Internet of Things is prototyped in this study. The ESP32 microcontroller provides all necessary data on a mobile application dashboard that is updated in real time, allowing users to regulate battery charge/discharge and monitor the battery SoC condition. This is accomplished by connecting to the internet and making use of its IoT features. A current out widget on the application dashboard is another feature of the suggested management system that lets users see any unusual current consumption. The battery life was extended by protecting it from overcharging and over-discharging.

In contrast to conventional coulomb-counter fuel gauges, whose accuracy drift is brought on by an offset error in the current-sense measurement, the method utilized to calculate battery SoC is better. Although the error in these systems is typically modest, it accumulates over time, cannot be fixed and requires repeated repairs. Corrections are frequently carried out at a preset SoC level, generally near full or empty. Some methods change the battery's voltage when it is at rest. To determine the precise SoC, these techniques use the battery voltage over a prolonged period of idleness. Both have the same flaw: if the rectified state is not consistently seen over time in the actual application, then the system fault is unbounded. In contrast to conventional coulomb-counter fuel gauges, whose accuracy drift is brought on by an offset error in the current-sense measurement, with simply simple cycles, the end-user accuracy cannot be understood. As MAX17043/MAX17044 are not dependent on the most recent information, they do not incur drift. We are all aware that IoT has the potential to completely change the EV sector and create a better environment. Electric vehicle owners and service providers now have access to more efficient and practical EVs thanks to IoT. Users will have instant access to all vehicle-related information, including maintenance due, charging locations nearby, battery condition, fault alerts, and many other things.

3.12 Future Scope of IoT in Electric Vehicles

In the approaching years, the Internet of Things (IoT) in the EV business will have a better future. The automotive IoT industry will have a global market value of $106.32 billion by the end of 2023, according to the Netscribes Market Research Report. Numerous governments all over the world intend to replace conventional automobiles powered by gasoline or diesel engines with electric ones. But, it is only feasible if the global car sector complies with all the regulations.

People now expect their vehicles to be a complete software platform with self-driving capabilities, easy control, and reduced operating costs. IoT technology may be used by electric vehicle producers to advance the production process and enhance the post-sale procedure. There will be several benefits once IoT technology is used globally for electric vehicles, including lower battery prices, safer and more secure driving, reduced manufacturing costs, easy maintenance, employment opportunities, and a reduction in air pollution. Therefore, it appears that IoT has a highly promising future in the automotive and electric vehicle industries.

References

1. Abd Wahab, M.H., Mohamad Anuar, I.N., Ambar, R., Baharum, A., Shanta, S., Sulaiman, M.S., *et al.* (2018) IoT-Based Battery Monitoring System for Electric Vehicle. *Int. J. Eng. Technol.*, 7, 505-510.
2. Rao, S.S. and Rangaswamy, D. (2021) Power Quality Mitigation and Transient Analysis in AC/DC Hybrid Microgrid for Electric Vehicle Charging. *Indones. J. Electr. Eng. Comput. Sci.*, 24, 1315-1322.
3. Arora, K.; Kumar, A.; Kamboj, V.K.; Prashar, D.; Shrestha, B.; Joshi, G.P. Impact of Renewable Energy Sources into Multi Area Multi-Source Load Frequency Control of Interrelated Power System. *Mathematics*, 2021, *9*, 186.
4. Dost, P., Spichartz, P. and Sourkounis, C. (2015) Charging Behaviour of Users Utilising Battery Electric Vehicles and Extended Range Electric Vehicles within the Scope of a Field Test. *2015 International Conference on Renewable Energy Research and Applications (ICRERA), Palermo, 22-25 November*, 2015, 1162-1167.
5. Hannan, M.A., Hoque, M.M., Hussain, A., Yusof, Y. and Ker, P.J. (2018) State-of-the-Art and Energy Management System of Lithium-Ion Batteries in Electric Vehicle Applications: Issues and Recommendations. *IEEE Access*, 6, 19362-19378.
6. Koko Friansa, Irsyad Nashirul Haq, Bening Maria Santi, Deddy Kurniadi, Edi Leksono, Brian Yuliarto, Development of Battery Monitoring System in Smart Microgrid Based on Internet of Things (IoT), *Procedia Eng.*, Vol. 170, pp. 482-487 (2017).
7. W. Iqbal, H. Abbas, M. Daneshmand, B. Rauf and Y. A. Bangash, An In-Depth Analysis of IoT Security Requirements, Challenges, and Their Countermeasures via Software-Defined Security, in: *IEEE Internet Things J.*, vol. 7, no. 10, pp. 10250-10276, Oct. 2020.
8. D. N. Tsyani, A. Kurniasari and C. Hudaya, Battery Monitoring System with LoRa Technology, *2018 3rd International Conference on Information Technology, Information System and Electrical Engineering (ICITISEE)*, 2018, pp. 125-129.
9. Alireza Zourmand, Chan Wai Hung, Andrew Lai Kun Hing and Mohammad AbdulRehman, Internet of Things (IoT) using LoRa Technology, *IEEE International Conference on Automatic Control and Intelligent Systems (I2CACIS 2019*), 29 June 2019, Selangor, Malaysia, pp 324 -330
10. Arora, K.; Kumar, A.; Kamboj, V.K.; Prashar, D.; Jha, S.; Shrestha, B.; Joshi, G.P. Optimization Methodologies and Testing on Standard Benchmark Functions of Load Frequency Control for Interconnected Multi Area Power System in Smart Grids. *Mathematics*, 2020, *8*, 980.
11. C. K. Rao, S. K. Sahoo and F. F. Yanine, Demand Response for Renewable Generation in an IoT based Intelligent Smart Energy Management System, *2021 Innovations in Power and Advanced Computing Technologies (i-PACT)*, Kuala Lumpur, Malaysia, 2021, pp. 1-7,

12. Ramadan, H.S., Becherif, M. and Claude, F. (2017) Energy Management Improvement of Hybrid Electric Vehicles via Combined GPS/Rule-Based Methodology. *IEEE Trans. Autom. Sci. Eng.*, 14, 586-597.
13. A. Lavric and V. Popa, Internet of Things and LoRa™ Low-Power Wide-Area Networks: A survey, *2017 International Symposium on Signals, Circuits and Systems (ISSCS)*, Iasi, Romania, 2017, pp. 1-5.
14. C. Choi, J. Jeong, I. Lee and W. Park, LoRa based renewable energy monitoring system with open IoT platform, *2018 International Conference on Electronics, Information, and Communication (ICEIC)*, 2018, pp. 1-2.
15. V. K, R. K. Nema and A. Ojha, Various Types of Wireless Battery Management System in Ev, *2020 IEEE International Students' Conference on Electrical, Electronics and Computer Science (SCEECS)*, 2020, pp. 1-5.
16. Gholizadeh, M. and Salmasi, F.R. (2014) Estimation of State of Charge, Unknown Nonlinearities, and State of Health of a Lithium-Ion Battery Based on a Comprehensive Unobservable Model. *IEEE Trans. Ind. Electron.*, 61, 1335-1344.
17. Sun, F. and Xiong, R. (2015) A Novel Dual-Scale Cell State-of-Charge Estimation Approach for Series-Connected Battery Pack Used in Electric Vehicles. *J. Power Sources*, 274, 582-594.
18. Jiya, I.N., Gurusinghe, N. and Gouws, R. (2019) Hybridisation of Battery, Supercapacitor and Hybrid Capacitor for Load Applications with High Crest Factors: A Case Study of Electric Vehicles. *Indones. J. Electr. Eng. Comput. Sci.*, 16, 614-622.
19. P Singh, K Arora, UC Rathore, Control Strategies for Improvement of Power Quality in Grid Connected Variable Speed WECS with DFIG – An Overview, *J. Phys. Conf. Ser.*, 2022.
20. Mahadik, Y. and Vadirajacharya, K. (2019) Battery Life Enhancement in a Hybrid Electrical Energy Storage System Using a Multi-Source Inverter. *World Electr. Veh. J.*, 2019, 10 (2), 17.
21. Rao, C. K., Sahoo, S. K., Balamurugan, M., & Yanine, F. F. (2021). Design of Smart Socket for Monitoring of IoT-Based Intelligent Smart Energy Management System. In: *Lect. Notes Electr. Eng.*, (pp. 503–518). Springer Singapore.
22. Rusimamto, P.W., Endryansyah, E., Anifah, L., Harimurti, R. and Anistyasari, Y. (2021) Implementation of Arduino Pro Mini and ESP32 Cam for Temperature Monitoring on Automatic Thermogun IoT-Based. *Indones. J. Electr. Eng. Comput. Sci.*, 23, 1366-1375.
23. Mohammed, E.A., Al-Allaf, A.F. and Altamer, B.R. (2020) IoT-Based Monitoring and Management Power Sub-Station of the University of Mosul. *IOP Conf. Ser.: Mater. Sci. Eng.*
24. Tehrani, Y.H. and Atarodi, S.M. (2019) Design & Implementation of a High Precision & High Dynamic Range Power Consumption Measurement System for Smart Energy IoT Applications. *Measurement*, 146, 458-466.

25. Nasser, A., Mohammed, E. and Ali, A. (2022) Smart Energy Management of Wind/PV/Battery Renewable Energy Sources Based on IoT. *Proceedings of 2nd International Multi-Disciplinary Conference Theme: Integrated Sciences and Technologies,* Sakarya, 7-9 September 2021, 355-370.
26. M. Bassoli, V. Bianchi, I. De Munari, and P. Ciampolini, An IoT approach for an AAL Wi-Fi-based monitoring system, *IEEE Trans. Instrum. Meas.*, vol. 66, no. 12, Dec. 2017.
27. A. Pötsch, A. Berger, and A. Springer, Efficient analysis of power consumption behaviour of embedded wireless IoT systems, in: *Proc. 2017 IEEE Int. Instrum. Meas. Technol. Conf.*, May 2017.
28. F. Mohammadi, G.-A. Nazri, and M. Saif, Modeling and simulation of hybrid electric vehicle using MATLAB/Simulink, in: *Proc. 2019 Int. Conf. Power Generation Syst. Renewable Energy Technol.*, Aug. 2019.
29. H. S. V. S. Kumar Nunna, S. Battula, S. Doolla, and D. Srinivasan, Energy management in smart distribution systems with vehicle-to-grid integrated microgrids, *IEEE Trans. Smart Grid*, vol. 8, no. 5, Sep. 2018.
30. M. Rizzi, A. Depari, P. Ferrari, A. Flammini, S. Rinaldi, and E. Sisinni, Synchronization uncertainty versus power efficiency in LoRaWAN networks, *IEEE Trans. Instrum. Meas.*, vol. 68, no. 4, Apr. 2019.
31. Rao, C. K., Sahoo, S. K., & Yanine, F. F. (2022). Forecasting Electric Power Generation in Photovoltaic Power Systems for Smart Energy Management. In: *2022 International Conference on Intelligent Controller and Computing for Smart Power (ICICCSP).*
32. Spinelli, G.M., Gottesman, Z.L. and Deenik, J. (2019) A Low-Cost Arduino-Based Datalogger with Cellular Modem and FTP Communication for Irrigation Water Use Monitoring to Enable Access to CropManage. *HardwareX*, 6, e00066.
33. Ambrož, M. (2017) Raspberry Pi as a Low-Cost Data Acquisition System for Human Powered Vehicles. *Measurement*, 100, 7-18.
34. Shinde, V.R., Tasgaonkar, P.P. and Garg, R.D. (2018) Environment Monitoring System through Internet of Things (IOT). *2018 International Conference on Information, Communication, Engineering and Technology (ICICET),* Pune, 29-31 August, 2018, 1-4.
35. Sung, W.-T. and Hsiao, S.-J. (2020) the Application of Thermal Comfort Control Based on Smart House System of IoT. Measurement.
36. Salih, T.A. and Noori, M.S. (2020) Using LoRa Technology to Monitor and Control Sensors in the Greenhouse. *IOP Conf. Ser.: Mater. Sci. Eng.*
37. Rao, C. K., Sahoo, S. K., Balamurugan, M., Satapathy, S. R., Patnaik, A., & Yanine, F. F. (2020). Applications of Sensors in Solar Energy Systems. In: *2020 International Conference on Renewable Energy Integration into Smart Grids: A Multidisciplinary Approach to Technology Modelling and Simulation (ICREISG).* IEEE.

38. Babiuch, M., Foltynek, P. and Smutny, P. (2019) Using the ESP32 Microcontroller for Data Processing. *2019 20th International Carpathian Control Conference*, Krakow- Wieliczka, 26-29 May 2019, 1-6.
39. Kurniawan, A. (2019) *Internet of Things Projects with ESP32: Build Exciting and Powerful IoT Projects Using the All-New Espressif ESP32*. Packt Publishing Ltd., Birmingham.
40. Texas Instruments (2011) INA219 Zerø-Drift, Bidirectional Current/Power Monitor with I2C TM Interface. No. September, 2011.
41. Lambert, J., Monahan, R. and Casey, K. (2021) Power Consumption Profiling of a Lightweight Development Board: Sensing with the INA219 and Teensy 4.0 Microcontrollers. *Electronics*, 10, Article No. 775.
42. Behera, S.K. (2019) Use of Ina219 Sensor for Locating. *Int. J. Res. Anal. Rev. (IJRAR)*, 6, 1-4.
43. Schwunk, S.; Armbruster, N.; Straub, S.; Kehl, J.; Vetter, M. Particle filter for state of charge and state of health estimation for lithium–iron phosphate batteries. *J. Power Sources* 2013, 239, 705–710.
44. Tulsyan, A.; Tsai, Y.; Gopaluni, R.B.; Braatz, R.D. State-of-charge estimation in lithium-ion batteries: A particle filter approach. *J. Power Sources*, 2016, 331, 208–223.
45. He, Y.; Liu, X.; Zhang, C.; Chen, Z. A new model for state-of-charge (SoC) estimation for high-power li-ion batteries. *Appl. Energy*, 2013, 101, 808–814.
46. Burgos-Mellado, C.; Orchard, M.E.; Kazerani, M.; Cárdenas, R.; Sáez, D. Particle-filtering-based estimation of maximum available power state in lithium-ion batteries. *Appl. Energy*, 2016, 161, 349–363.
47. Zhou, D.; Zhang, K.; Ravey, A.; Gao, F.; Miraoui, A. Online estimation of lithium polymer batteries state-of-charge using particle filter-based data fusion with multimodal's approach. *IEEE Trans. Ind. Appl.*, 2016, 52, 2582–2595.
48. Du, Q.; Han, Q.; Zhang, Y.; Liu, Z.; Tian, S.; Zhang, Z. Adopting combined strategies to make state of charge (SoC) estimation for practical use. *J. Renew. Sustain. Energy*, 2018, 10, 034102.
49. Xia, B.; Sun, Z.; Zhang, R.; Lao, Z. A cubature particle filter algorithm to estimate the state of the charge of lithium-ion batteries based on a second-order equivalent circuit model. *Energies*, 2017, 10, 457.
50. Li, B.; Peng, K.; Li, G. State-of-charge estimation for lithium-ion battery using the gauss-hermite particle filter technique. *J. Renew. Sustain. Energy*, 2018, 10, 014105.
51. Ye, M.; Guo, H.; Xiong, R.; Yu, Q. A double-scale and adaptive particle filter-based online parameter and state of charge estimation method for lithium-ion batteries. *Energy*, 2018, 144, 789–799.
52. Hu, X.; Sun, F.; Zou, Y.; Peng, H. Online estimation of an electric vehicle lithium-ion battery using recursive least squares with forgetting. In: *Proceedings of the American Control Conference (ACC)*, San Francisco, CA, USA, 29 June–1 July 2011.

53. Eddahech, A.; Briat, O.; Vinassa, J.-M. Adaptive voltage estimation for ev li-ion cell based on artificial neural networks state-of-charge meter. In: *Proceedings of the 2012 IEEE International Symposium on Industrial Electronics (ISIE)*, Hangzhou, China, 28–31 May 2012.
54. Lim, K.; Bastawrous, H.A.; Duong, V.-H.; See, K.W.; Zhang, P.; Dou, S.X. Fading kalman filter-based real-time state of charge estimation in lifepo4 battery-powered electric vehicles. *Appl. Energy*, 2016, 169, 40–48.
55. Lotfi, N.; Landers, R.G.; Li, J.; Park, J. Reduced-order electrochemical model-based SoC observer with output model uncertainty estimation. *IEEE Trans. Control Syst. Technol.*, 2017, 25, 1217–1230.
56. Safwat, I.M.; Li, W.; Wu, X. A novel methodology for estimating state-of-charge of li-ion batteries using advanced parameters estimation. *Energies*, 2017, 10, 1751.
57. Duong, V.-H.; Bastawrous, H.A.; See, K.W. Accurate approach to the temperature effect on state of charge estimation in the lifepo4 battery under dynamic load operation. *Appl. Energy*, 2017, 204, 560–571.
58. Xia, B.; Lao, Z.; Zhang, R.; Tian, Y.; Chen, G.; Sun, Z.; Wang, W.; Sun, W.; Lai, Y.; Wang, M. Online parameter identification and state of charge estimation of lithium-ion batteries based on forgetting factor recursive least squares and nonlinear kalman filter. *Energies*, 2018, 11, 3.
59. Shen, P.; Ouyang, M.; Lu, L.; Li, J.; Feng, X. The co-estimation of state of charge, state of health, and state of function for lithium-ion batteries in electric vehicles. *IEEE Trans. Veh. Technol.*, 2018, 67, 92–103.
60. Ali, M.; Kamran, M.; Kumar, P.; Nengroo, S.; Khan, M.; Hussain, A.; Kim, H.-J. An online data-driven model identification and adaptive state of charge estimation approach for lithium-ion-batteries using the lagrange multiplier method. *Energies*, 2018, 11, 2940.
61. Zhang, C.; Allafi, W.; Dinh, Q.; Ascencio, P.; Marco, J. Online estimation of battery equivalent circuit model parameters and state of charge using decoupled least squares technique. *Energy*, 2018, 142, 678–688.

4

A Review of Electric Vehicles: Technologies and Challenges

N. Suthanthira Vanitha[1]*, L. Manivannan[2], K. Radhika[1], A. Karthikeyan[1] and T. Meenakshi[3]

[1]*Muthayammal Engineering College, Rasipuram, India*
[2]*National Institute of Technology, Tiruchirappalli, India*
[3]*Jansons Institute of Technology Karumathampatti, Coimbatore, India*

Abstract

Electric vehicles (EVs) have shown unprecedented growth over recent decades around the world, and they are one of the future technologies that have entered India. Both EVs and hybrid EVs employ the most promising solutions to replace the conventional internal combustion (IC) engine–based vehicles. Many countries around the globe have implemented this technology, which contributes toward amelioration of environmental pollution. The Indian government has set a target of having 30% EVs on its roads by 2030. In India, a particular set of circumstances is conducive to a sustainable mobility paradigm, and this has created an opportunity for accelerated adoption of EVs over IC engine–based vehicles. The technological advancement and increased concern about controlling pollution has given EVs a tag of future mobility. The core components of an EV are EV battery and electric motor. This electric motor acts as a drive for an EV system. The drive of an EV is controlled by an electronic package controller named as power electronics converter. This converter operates between the battery and electric motor in an EV to control the speed and acceleration just as a carburetor does in IC engine–based vehicles. The converter plays a vital role in regulating the energy flow from the battery. This chapter primarily aims to review various types of electric motors and power electronic converters used in electric vehicles. Secondly, the popular types of battery in EVs and traditional battery charging approaches are discussed.

Keywords: EVPS, electric motor, power electronics–based controller (converter), battery

**Corresponding author*: varmans03@gmail.com

Krishan Arora, Suman Lata Tripathi and Himanshu Sharma (eds.) Electric Vehicle Design: Design, Simulation and Applications, (81–100) © 2024 Scrivener Publishing LLC

4.1 Introduction

India is the land of opportunity for any kind of new technology. The electric vehicle (EV) is one of the emerging technologies not only in India but also in the world. EVs have been a media focal point as they provide the most promising solutions to replace the conventional internal combustion (IC) engine–based vehicles. Recent years have seen a rapid development of EVs. An EV normally uses one or more electric motors rather than a traditional petrol or diesel engine for propulsion [1]. The electric motor is powered by rechargeable batteries that can be charged using household mains electricity through an EV charge point at home or at a more powerful EV charge station at work or in the street. Due to their sustainability and cheaper running costs, EVs are likely to be the way of the future. In fact, electric vehicles are nothing new as they were started way back in 1827.

4.2 Electric Motors

The core element of an electric vehicle, apart from electric vehicle batteries, which replaces the IC engines, is the electric motor. The electric motors used for automotive applications should have the characteristics such as high starting torque, high power density, and good efficiency. The various types of electric motors used in EVs are as follows:

A. DC series motor
B. Brushless DC (BLDC) motor
C. Permanent magnet synchronous motor (PMSM)
D. Three-phase AC induction motors
E. Switched reluctance motors (SRM)

4.2.1 DC Series Motor

The DC series motor is an excellent choice for traction applications because of its high initial torque capacity. In the early 1900s, it was the motor that was most frequently employed for traction applications. Its benefits include simple speed control and the ability to tolerate a dramatic rise in load. It is the optimum traction motor because of all these features. High maintenance required by brushes and commutators is the fundamental disadvantage of DC series motors. Indian railways employ these motors [2].

4.2.2 Brushless DC Motors

It resembles DC motors with permanent magnets in many ways. Because it does not include a commutator and brush setup, it is known as a brushless system. Due to the electronic commutation of this motor, BLDC motors require no maintenance. The traction properties of BLDC motors include high starting torque and high efficiency of 95%–98%. For high-power density design approaches, BLDC motors are appropriate [3]. Due to their traction properties, BLDC motors are the most popular motors for use in EV applications. BLDC motors further have two types, as described below.

4.2.2.1 Out-Runner-Type BLDC Motor

The stator is inside the motor, whereas the rotor is outside. Because the wheel is directly attached to the external rotor, it is also known as a hub motor. Such motors do not need an external gear system. In some circumstances, the motor itself comes equipped with planetary gears. This motor reduces the total size of the vehicle. The power output in the in-runner design is controlled by a limitation of the motor size. Manufacturers of electric cycles, like Hullikal, Tronx, Spero, and light-speed bicycles, preferred this motor [4]. Moreover, two-wheeler producers like 22 Motors and NDS Eco Motors use it. Figure 4.1 shows the BLDC hub motor used by the manufacturer of 22 Motors.

4.2.2.2 In-Runner-Type BLDC Motor

With this design, unlike normal motors, the rotor of the motor is inside and the stator is outside. The out-runner configuration is slightly bulkier

Figure 4.1 BLDC hub motor.

Figure 4.2 In-runner–type BLDC motor.

than the in-runner layout because these motors need an external transmission system to send power to the wheels. BLDC motors are used by many three-wheeler manufacturers, including Goenka Electric Motors, Speego Vehicles, Kinetic Green, and Volta Automotive. Manufacturers of low- and medium-performance scooters also employ BLDC motors for movement [5]. Figure 4.2 shows the in-runner–type BLDC motor used in the scooter.

These factors make the motor of choice for many EV applications. Permanent magnet high cost is the biggest disadvantage. Due to the temperature environment, permanent magnets have a shorter life span when the motor is overloaded over a certain point.

4.2.3 Permanent Magnet Synchronous Motor

The BLDC motor, which has permanent magnets on the rotor, and this motor are similar. These motors have traction properties including high power density and high efficiency, just like BLDC motors. The distinction is that BLDC has a trapezoidal back EMF and PMSM has a sinusoidal back EMF. Higher power ratings are possible with permanent magnet synchronous motors. The greatest option for high-performance applications like vehicles and buses is PMSM. Due to their higher efficiency than induction motors, PMSMs are giving them strong competition despite their expensive price [2, 3]. PMSM is also costlier than BLDC motors. PMSM motors are used by the majority of companies in their hybrid and EVs. For example, Toyota Prius, Chevrolet Bolt EV, Ford Focus Electric, Zero Motorcycles SR/S, Nissan Leaf, Honda Accord, and BMW i3 use PMSM motors for propulsion. Figure 4.3 shows the permanent magnet synchronous motor used in Toyota Prius in 2004.

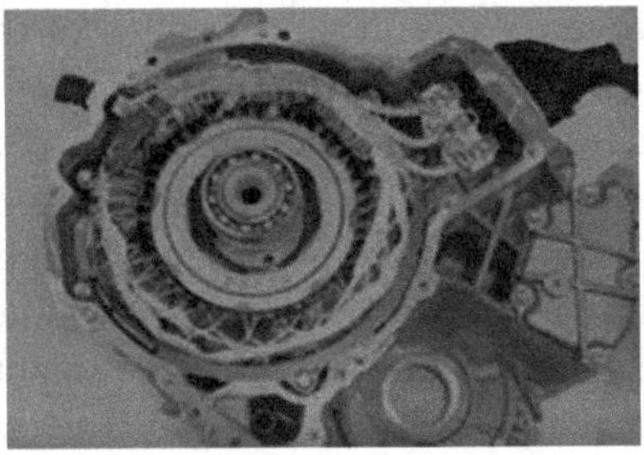

Figure 4.3 PMSM of Toyota Prius.

4.2.4 Three-Phase AC Induction Motors

In fixed voltage and frequency operation, induction motors do not have a high starting torque like DC series motors. But by utilizing different control techniques like field oriented control (FOC) or voltage per frequency ratio (v/f) approaches, this characteristic can be changed. These control techniques enable the motor to start with the greatest torque, which is excellent for traction applications, and they require minimum servicing. Squirrel cage induction motors have a long life span; up to 92%–95% efficiency can be achieved [3]. The drawback of an induction motor is that it

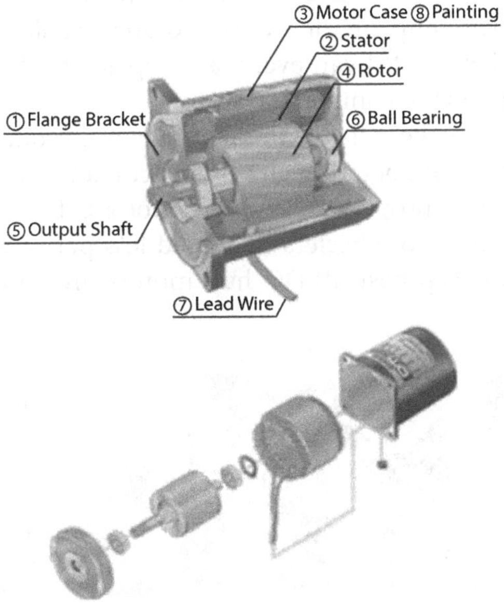

Figure 4.4 Oriental induction motor.

requires a complex inverter circuit, and control of the motor is difficult. Figure 4.4 depicts the internal structure of an oriental induction motor.

The best interpretation of the induction motor's superior performance compared to competitors is the Tesla Model S. By opting for induction motors, Tesla might have wanted to eliminate the dependency on permanent magnets. Mahindra Reva e2o's propulsion system is a three-phase induction motor. Induction motors will be used in automobiles and buses made by major automakers like TATA Motors. The two-wheeler company TVS Motors will introduce an electric scooter that is powered by an induction motor. Due to their low cost, induction motors are the favored option for EVs that focus on performance. The ability to tolerate harsh weather conditions is another benefit. The Indian Railways have begun switching over from DC motors to AC induction motors as a result of these benefits.

4.2.5 Switched Reluctance Motors

Switched reluctance motors (SRMs) are a category of variable reluctance motor with double saliency. SRMs are simple in construction and robust as shown in Figure 4.5. The SRM rotor is a piece of laminated steel that has neither permanent magnets nor windings attached to it. As a result, the rotor's inertia is reduced, aiding in rapid acceleration. The resilient nature of SRM makes it suited for high-speed application. High power density is another feature that SRM delivers for EVs. The SRM control complexity and expanding switching circuit are its two greatest shortcomings. It has some noise problems. SRM can eventually replace PMSM and induction motors once it enters the commercial sector.

It is essential to make a list of the performance requirements that the vehicle must meet, the operating circumstances, and the associated costs before choosing the proper electric car motors. For applications like go-kart vehicles and two-wheelers that need less performance (often less than 3 kW) at a cheap cost, BLDC hub motors are an excellent choice.

Figure 4.5 Switched reluctance motor.

Table 4.1 Comparison between DC and AC motor.

S. no.	DC motor	AC motor
1.	Multi-speed transmission	Single-speed transmission
2.	Heavy weight	Light weight for same power
3.	More expensive	Less expensive
4.	85% to 95% efficiency at full load	95% efficiency at full load
5.	Controller is simple and less expensive.	Controller is somehow complex and expensive.

For three-wheelers and two-wheelers, it is also good to choose BLDC motors with or without an external gear system. The best motor to employ for high-power applications would be a PMSM or an induction motor, which is used in two-wheelers, automobiles, buses, or trucks. One will have greater alternatives for motor types when using EVs once the synchronous reluctance motor and switching reluctance motor become as affordable as PMSM or induction motors [2, 3]. On the whole, EVs use either DC motor or AC motor based on the nature of the requirement. Table 4.1 shows a general comparison between DC and AC motor as per the report of Idaho National Laboratory.

4.3 Power Electronic Converters

Generally, voltage source inverters (VSIs) are widely used in variable speed electric motor drives, uninterrupted power systems, active power filters, and, more recently, in renewable energy conversion systems and EVs. In EVs, the electric motor is controlled by power electronics converters. Most of the EVs' configuration has two energy storage devices, namely, the main energy system (MES) and the rechargeable energy storage system (RESS). The MES normally provides the extended driving range, whereas the RESS provides good acceleration and regenerative braking capability. These energy storage devices vary with their output voltage with load or state of charge (SoC), and the high voltage of the DC link creates major challenges for vehicle designers when integrating energy storage devices with the drive of an EV.

DC-DC converters can normally be used to interface the functional elements in an EV by stepping up or chopping the voltage levels [4].

The power electronic converters' structure has to be reliable, lightweight, small volume, high efficient, low electromagnetic interference, and low voltage or current ripple. Moreover, as far as the electric vehicle is concerned, it is noticed that the use of a high-power DC-DC converter is necessary for providing the power supply system to it. The power of the DC-DC converter depends on the characteristics of the vehicle such as top speed, acceleration time, weight, maximum torque, and power profile (peak power and continuous power). The different configurations of DC-DC converter mentioned below have been discussed.

4.3.1 Bi-Directional DC-DC Converter

A bi-directional converter can move power in either direction, which is useful in applications requiring regenerative braking. By altering the duty cycle (the switch's ratio of on/off time), it is possible to regulate the amount of power flowing between the input and the output. Generally, this is done to manage the output voltage, the input current, and the output current or to maintain a steady power. Converters developed by transformers would offer separation between both the input and output. Bi-directional DC-DC converters can be grouped as described below.

4.3.1.1 Non-Isolated Converters

This type of converter is typically utilized when the voltage needs to be increased or decreased by a relatively modest factor (less than 4:1) and when there is no issue with there being no dielectric isolation between the input and output. In this non-isolated group, there are five basic converter types: the buck, boost, buck-boost, Cuk, and charge-pump converters. While voltage step-up is accomplished with a boost converter, voltage step-down is accomplished with a buck converter. Out of these converters, a DC-DC boost converter is widely used in EV application. But the DC-DC boost converter does not meet the criteria of electrical isolation. Furthermore, this architecture experiences strong current and voltage fluctuations due to the wide magnitude difference between the input and output [5]. This problem can be overcome by using isolated converters and interleaved multichannel DC-DC converter topology.

4.3.1.2 Isolated Converters

A high-frequency transformer is employed in these types of converters. An isolated converter is required in applications where the output must be fully separated from the input. In this category, converters are available in varieties, including half-bridge, full-bridge, fly-back, forward, and push-pull DC-DC converters. All of these converters have a high-voltage stepping-up or -down ratio and can be employed as bi-directional converters. Out of these isolated converters, the full-bridge isolated DC-DC converter is mostly used due to its capability of high power transmission.

i. Full-Bridge Isolated DC-DC Converter

A full-bridge DC-DC converter is the most frequently implemented circuit configuration for fuel-cell power conditioning when electrical isolation is required. Due to low switch voltage and current, the full-bridge DC-DC converter is appropriate for high power transmission. It exhibits minimal voltage and current ripples at the input and output. The full-bridge topology is a favorite for the zero-voltage switching–pulse-width modulation techniques. Figure 4.6 shows the topology of full-bridge isolated DC-DC converter.

The converter topology consists of power electronics switches with high switching speed which helps to reduce weight and volume of equipment; however, it causes some undesirable effects such as radio-frequency interference emission. Electromagnetic compatibility regulation fast semiconductor devices make it possible to have high-speed and high-frequency switching in power electronics converters. The main cause of EMI emissions is considered to be high dv/dt or di/dt by switching current power devices. Electrical equipment, particularly static power electronic converters, is being used in a wider variety of equipment. Electric networks are negatively impacted by power electronics converters, which are regarded as a significant source of electromagnetic interference. Some users, particularly those in the residential, business, and medical sectors, are extremely sensitive

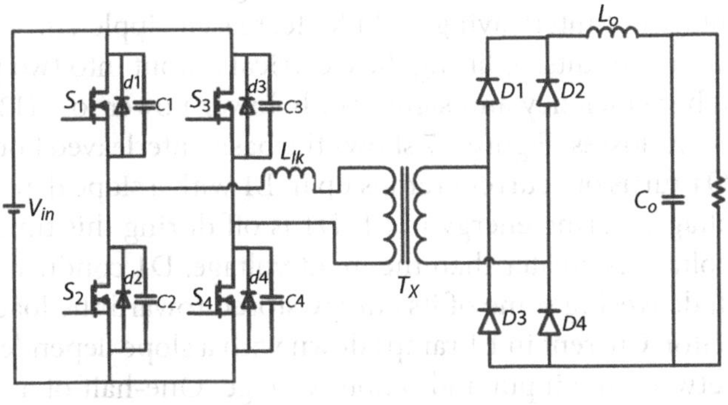

Figure 4.6 Full-bridge isolated DC-DC converter topology.

to power system disturbances, such as voltage and frequency changes. National or international Electro Magnetic Compatibility (EMC) laws are the best way to prevent interference and enhance power quality. The most significant EMC laws are determined and published by International Special Committee on Radio Interference (CISPR), International Electrotechnical Commission (IEC), Federal Communications Commission (FCC), and Verband der Elektrotechnik (VDE), some of the most renowned organizations from Germany, other European countries, and the US. By comparing the measured or computed interference level in the specified frequency range with the regulations' stated requirements, compliance with regulations is assessed. Regulation compliance is required in the European Union, and products must have approved labels to demonstrate that they meet the standards. The maximum interference level for EVs must adhere to DIN VDE 0879. The requirements and conducted emission cap in this standard are almost identical to those in class B of VDE 0871. A line impedance stabilization network is a tool used to produce known impedance on the power lines of electrical equipment during electromagnetic interference testing. It is normally designed to allow for measurements of the electromagnetic interference present on the power line [6].

ii. Interleaved Multichannel DC-DC Converter

A basic interleaved multichannel DC-DC converter topology permits to reduce the input and output current and voltage ripples, to reduce the volume and weight of the inductors, and to increase the efficiency. These structures, however, cannot work efficiently when a high-voltage step-up ratio is required because the duty cycle is limited by circuit impedance, leading to a maximum step-up ratio of 4 approximately. As a result, two series-connected step-up converters would be necessary to meet the application-specific voltage gain.

A simple boost converter raises the DC voltage from a lower DC voltage. Other advantages of interleaving include decreased ripple currents in the input and output circuits. Splitting the electrical output into two pathways results in higher efficiency and significantly lower power loss (I2R) losses and inductor AC losses. Figure 4.7 shows the basic interleaved boost topology. When Q1 turns on, current ramps up in L1 with a slope depending on the input voltage, storing energy in L1. D1 is off during this time because the output voltage is greater than the input voltage. D1 conducts once Q1 is turned off, delivering some of its energy stored toward the load and the output capacitor. Current in L1 ramps down with a slope dependent on the difference between the input and output voltage. One-half of a switching period later, Q2 also turns on completing the same cycle of events [5, 6].

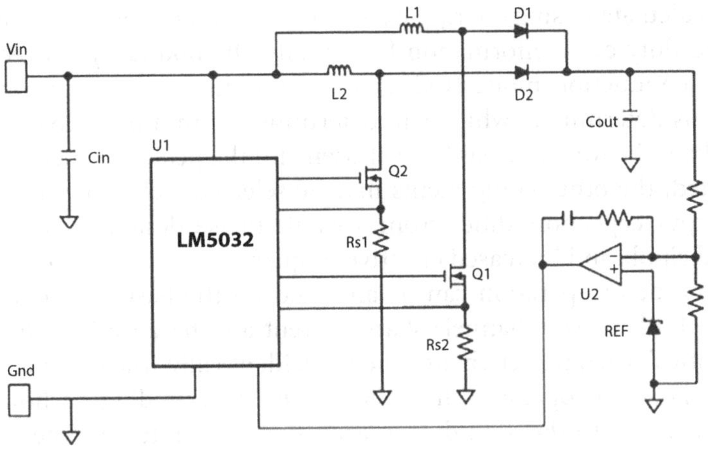

Figure 4.7 Interleaved basic circuit configuration with LM5302 IC.

Because both power channels are combined at the output capacitor, the effective ripple frequency is twice that of a conventional single-channel boost regulator.

The interleaved boost converter design involves the selection of the inductors, the input and output capacitors, the power switches, and the output diodes. Both the inductors and diodes should be identical in both channels of an interleaved design. Knowing the following factors is important in order to choose these components: the duty cycle range and peak currents. Because the output power is channeled through two power paths, a good starting point is to design the power path components using half the output power. Basically, the design starts with a single boost converter operating at half the power. However, a trade-off exists that will depend on the goals of the design. Because currents are smaller at each phase, the designer may employ smaller components. Or, larger components may be selected to minimize losses. In particular, this design operates at ambient temperature across the whole range of input voltages without requiring airflow. Obviously, there are many trade-offs possible, such as requiring external airflow that would allow the use of smaller components and more power per watt.

Knowing the maximum and minimum input voltages, the output voltage, and the voltage drops across the output diode and switch, the maximum and minimum duty cycles are calculated. Next, the average inductor current can be estimated from the load current and duty cycle. It is possible to calculate the peak inductor current by assuming that the peak-to-peak inductor current ripples the average inductor current. The inductor value is then calculated using the ripple current, switching frequency, input voltage, and duty cycle information [7]. Finally, the boundary between continuous conduction mode (CCM) and discontinuous conduction mode (DCM) is determined, which will determine the minimum load current. Once the inductor value has been chosen and the peak currents have been calculated, the other components may be selected. Selection of the input and output capacitors differ from a single-phase design because of the reduced ripple and increased effective frequency.

The mode of operation can be analyzed on the basis of one channel. Because both power channels share current and because both inductors are identical, each power channel behaves identically. The boost converter can be classed as operating in CCM or DCM depending on how much energy is given to the load during each switching interval. The mode of operation is categorized as DCM if all of the energy stored in the inductor is transferred to the load during each switching cycle (DCM). The inductor current ramps down to zero during switch-off time in this manner. When only a portion of the energy is given to the load, the converter is said to be working in CCM. The mode of operation is a fundamental factor in determining the electrical characteristics of the converter. The characteristics vary significantly from one mode to the other.

4.4 Battery in Electric Vehicles

4.4.1 Types of Battery in Electric Vehicles

An essential part of automobiles, which advance the field of sustainable mobility, is the battery. Batteries are being widely applied in high-power applications such as EVs where they play a vital role. In electric vehicle applications, many types of battery can be adopted as the power supply. In recent days, some popular types of battery are mostly used in EVs. The battery is generally categorized into two groups based on the ability of recharging. They are the primary battery and the secondary battery. The secondary battery can be recharged after being entirely exhausted; however, the primary battery can only be used once after being totally drained.

For the applications of EVs and HEVs, the secondary battery with a long cycle life, small energy loss, high power density, and enough safety level is indispensably required. Some popular types of batteries used in EVs include lithium-ion (Li-ion), lead acid, nickel-cadmium (NiCd), and nickel metal hydride (NiMH). Table 4.2 illustrates some key characteristics for these popular battery types. It is clearly shown that Li-ion battery is

Table 4.2 Key characteristics of types of battery.

Key characteristics	Battery type			
	Li-ion battery	Lead acid battery	NiCd battery	NiMH battery
Service life in cycle	600–3000	200–300	1,000	300–600
Nominal voltage in volts	3.2–3.7	2.0	1.2	1.2
Energy density in Wh kg^{-1}	100–270	30–50	50–80	60–120
Power density in W kg^{-1}	250–680	180	150	250–1,000
Charging efficiency in %	80–90	50–95	70–90	65
Self-discharge rate in % $month^{-1}$	3–10	5	20	30
Charging temperature in °C	0 to 45	–20 to 50	0 to 45	0 to 45
Discharging temperature in °C	–20 to 60	–20 to 50	–20 to 65	–20 to 65

significantly better than other types of battery, especially in terms of large cycle life, which is key to long service of EVs (e.g., 6–10 years of service life). Also, Li-ion batteries have a high level of safety and are made of eco-friendly components that do not emit hazardous gases [8]. Furthermore, it is now widely accepted that lithium technology is the preferred technique for power storage in EVs. Therefore, the Li-ion battery becomes the most popular power supply for EVs.

4.4.2 Traditional Battery Charging Approach

When a battery energy source is exhausted or its terminal voltage drops below the cutoff voltage or SoC, standing for the remaining battery capacity, declines to 20% or lower, the discharging process should be stopped, and the battery needs to be recharged. Incorrect operations such as over-discharging, over-charging, or improper charging will speed up the degradation process of the battery dramatically. Compared with other types of battery, the Li-ion battery has fairly stable performance but less cycle life at high-temperature conditions, while no permission is allowed for being charged below freezing [9]. According to the accurate enough estimation of battery SoC, State of Health (SoH), and temperature, proper battery charging approaches can be effectively designed, further to charge battery from the initial state to the final SoC target value. Meanwhile, the charging approaches can also protect batteries from overheating, prolong the service life, and improve the capacity utilization.

There are some traditional but popular charging approaches to solving the battery charging problem with numerous objectives and termination conditions. There are four conventional charging methods that have been commonly used to power EV batteries:

- Constant current (CC) charging
- Constant voltage (CV) charging
- Constant current–constant voltage (CC–CV) charging
- Multi-stage constant current (MCC) charging

4.4.2.1 Constant Current (CC) Charging Approach

As a basic unrefined method, CC charging uses a modest CC rate to charge the battery throughout the charging procedure. The CC charging is terminated when the time to charge reaches a predefined threshold. This charging approach is first introduced to charge NiCd or NiMH batteries and is also widely used for Li-ion batteries. However, the behaviors of batteries are highly dependent on the current rate in CC charging; hence, the main challenge for CC charging approach is to search a suitable charging current rate which is capable of equilibrating battery charging speed and capacity utilization. The charging speed is increased for high current rates during CC charging; however, the battery aging process will be accelerated. For small current rate in CC charging, high-capacity utilization is achieved, but a too low current rate will slow down the battery charging speed and further have a negative effect on the convenience of EV usage [8, 9].

4.4.2.2 Constant Voltage (CV) Charging Approach

Batteries are charged using CV charging, which completely adopts a constant voltage. The main advantage of CV charging is that it prevents over-voltage and other potentially harmful side effects during the charging process and significantly extends battery cycle life. When the CV charging is applied, the charging current will gradually reduce due to the low acceptance with progressing recharge. This approach, however, needs a high current rate in order to keep constant terminal voltage at the early stage of the charging process, which is easy to cause the battery lattice collapse and battery poles broken. The common problem of a CV charging approach is also to select a proper value for CV to obtain a good trade-off among the charging speed, electrolyte decomposition, and capacity utilization. In summary, CV charging approach is capable of effectively improving the charging speed but brings great damage to the battery capacity.

4.4.2.3 Constant Current–Constant Voltage (CC-CV) Charging Approach

This approach is obtained by integrating CC charging and CV charging. In this approach, a battery is firstly charged by a predefined CC in the CC phase, and the battery voltage will increase to the maximum safe threshold. Afterward, the battery enters into the CV phase with a predefined constant voltage, entailing the continuous step-down of the charging current. This CV phase will end until a terminal value of the decreasing current or a goal capacity is reached [14]. The standard CC-CV approach is first utilized to charge a lead acid battery with the preset values of constant current as well as CV, which are recommended by battery manufacturers, and is also extended to charge a Li-ion battery with some modifications. Because of higher terminal voltage and charging acceptance for a Li-ion battery, CC in the applications of Li-ion battery CC-CV charging should be much larger than that of a lead acid battery, which is usually chosen from 0.5 to 3.0 C. In CC-CV charging process, the CC stage and the CV stage can be complementary in some way; the capacity loss caused by the large electrochemical polarization in the CC stage will be effectively compensated by CV stage.

Because of this, the CC-CV charging technique has been chosen as a standard to compare and is better than the sole CC as well as the sole CV charging in applications of EVs with other newly developed battery charging approaches. Although standard CC-CV charging approach is easy to apply, the challenging issue is to set the appropriate CC rate at the CC stage and CV value at the CV stage [10]. Battery charging speed of CC-CV approach is primarily determined by the CC rate, while the capacity utilization of battery charging is mainly affected by the values of CV and termination. For CC rate in CC-CV, on the one hand, a high value of current rate may cause lithium plating, further to cause low efficiency of energy conversion, and battery temperature may exceed in high power applications, in particular. On the other hand, low charging current may result in slower battery charge and reduced usability of EVs. Consequently, it is essential to develop an appropriate CC-CV strategy in order to enhance overall charging performance and ensure the battery's operational safety [11].

4.4.2.4 Multi-Stage Constant Current (MCC) Approach

This popular MCC approach has been successfully developed to charge numerous types of battery such as lead acid battery, NiMH battery, and Li-ion battery. The main difference between MCC charging and CC-CV charging is that, in MCC charging, the multi-stage series of monotonic charging currents are injected into the battery during the total charging process [11]. This series of charging currents should be gradually reduced as the form of various CC stages (ICC1>ICC2>……>ICC_N). When terminal voltage goes up to a default voltage threshold by the CC in one stage, charging procedure will turn into another CC stage, and then a new less CC rate will be utilized accordingly. This decrease process of charging current will continue until battery terminal voltage reaches the last default voltage threshold under the condition of minimum current [12, 13]. The charging speed for standard MCC approach will be usually a bit slower than the standard CC-CV approach with the same initial current.

Table 4.3 illustrates the comparison of all kinds of traditional charging approach for the battery used in EVs.

All in all, for rough charging approaches including sole CC charging and CV charging, the implementation costs are relatively low with just a few parameters needing to be considered. However, these simple charging approaches would cause many charging problems such as battery lattice collapse and battery poles broken [15]. It is significantly difficult to equilibrate battery capacity utilization and charging speed by using just sole CC or CV charging approach. In order to further improve charging performance such as avoiding over-voltage, enhancing capacity utilization, and achieving fast charging, some hybrid charging approaches including CC-CV and MC Care developed [16, 18, 19]. The open problem for using these hybrid approaches is to search the proper current and voltage values to efficiently equilibrate conflicting objectives such as charging speed, energy loss, temperature variation, and battery lifetime [20, 21]. Besides, the analysis of electrochemical reaction such as lithium plating during these charging processes is still at its primitive stage and will be a thriving area of research in the field of EV applications [17, 22].

Table 4.3 Comparison of conventional methods for battery charging.

Approach	Merits	Demerits	Key elements
CC	Easy to implement	Capacity utilization is low	Charging constant current rate; terminal condition
CV	Easy to implement; stable terminal voltage	Easy to cause the lattice collapse of battery	Charging constant voltage; terminal condition
CC-CV	Capacity utilization is high; stable terminal voltage	Difficult to balance objectives such as charging speed, energy loss, and temperature variation	Constant current rate in CC phase; constant voltage in CV phase; terminal condition
MCC	Easy to implement; easy to achieve fast charging	Difficult to balance objectives such as charging speed, capacity utilization, and battery lifetime	The number of CC stages; constant current rates for each stage

4.5 Conclusion

This chapter presents the major components of EV power systems. Firstly, it dealt with the various types of electric motors that can be adopted by an EV, especially DC and AC motors with its various features. Secondly, it dealt with the major types of power electronics converters named isolated and non-isolated converters and also discussed which kind of DC-DC converter is most suitable for high–power transmission purposes. Moreover, it covered the types of battery and the traditional battery charging approaches. The advantages and limitations in the battery and its charging approach are also discussed. This chapter will provide researchers with the basic information on electric vehicles in selecting the motor, the controller, the battery type, and the charging approach whenever they are developing a model in real time.

References

1. https://electricvehicles.in/what-is-the-future-of-electric-vehicles-in-india-by-2030/
2. https://circuitdigest.com/article/different-types-of-motors-used-in-electric-vehicles-ev
3. https://www.epsilonelectronics.in/portfolio/three-phase-bldc-motor-driver/
4. https://electricvehicle.wiki/terms/electric-motor/
5. M. Al, J. Van, and H. Gualous (2011). DC/DC Converters for Electric Vehicles. Electric Vehicles - Modelling and Simulations, DOI: 10.5772/17048.
6. Kumar, Rahul, Ramani Kannan, Nursyarizal Bin Mohd Nor, and Apel Mahmud (2021). A High Step-Up Switched Z-Source Converter (HS-SZC) with Minimal Components Count for Enhancing Voltage Gain. *Electronics*, 10(8), 924.
7. T. Meenakshi, N. Suthanthira Vanitha, K. Radhika, M. Maheshwari (2020). Parametric impact analysis on transformer based dual output Quasi Z DC-DC converter. *Mater. Today Proc.*, 33(7), 4030-4039.
8. P Singh, K Arora, UC Rathore, Control Strategies for Improvement of Power Quality in Grid Connected Variable Speed WECS with DFIG–An Overview, *J. Phys. Conf. Ser.*, 2022.
9. Rahul Kumar *et al.* (2023). Design and Simulation of Interleaved Boost Converter, *IJRASET*, 11(1),1761-1772, https://doi.org/10.22214/ijraset.2023.48924
10. Kailong Liu, Kang LI, Qiao Peng, Cheng Zhang. (2019). A brief review on key technologies in the battery management system of electric vehicles. *Front. Mech. Eng.*, 14(1): 47–64, https://doi.org/10.1007/s11465-018-0516-8.
11. M. Brandl *et al.* (2012). Batteries and battery management systems for electric vehicles, *Design, Automation & Test in Europe Conference & Exhibition (DATE)*, Dresden, Germany, pp. 971-976, doi: 10.1109/DATE.2012.6176637.
12. Gupta, Ragini & Rathore, Bhuvnesh. (2020). A Review of Improvement in the Capacity of Battery used in Electric Vehicle. Vol. 2, 27-30.
13. Arora, K.; Kumar, A.; Kamboj, V.K.; Prashar, D.; Jha, S.; Shrestha, B.; Joshi, G.P. Optimization Methodologies and Testing on Standard Benchmark Functions of Load Frequency Control for Interconnected Multi Area Power System in Smart Grids. *Mathematics*, 2020, *8*, 980.
14. Pawaskar Gayatri Nilesh *et al.* (2012). Estimation of Automotive Li-Ion Battery Parameters Using MATLAB Simulink. *International Journal for Research in Applied Science & Engineering Technology (IJRASET)*, Vol. 10 Issue XII.
15. Buccolini, Luca & Ricci, Adrianna & Scavongelli, Cristiano & De Maso-Gentile, Giuseppe & Orcioni, Simone & Conti, Massimo. (2016). Battery Management System (BMS) simulation environment for electric vehicles. 1-6. 10.1109/EEEIC.2016.7555475.

16. Li, Lian-xing & Tang, Xin-cun & Qu, Yi & Liu, Hong-tao. (2011). CC-CV charge protocol based on spherical diffusion model. *J. Cent. South Univ. Technol.*, 18. 319-322. 10.1007/s11771-011-0698-2.
17. Liu, Kailong & Li, Kang & Peng, Qiao & Zhang, Cheng. (2018). A brief review on key technologies in the battery management system of electric vehicles. *Front. Mech. Eng.* 14. 1-18. 10.1007/s11465-018-0516-8.
18. T. Meenakshi and N. S. Vanitha. (2017). Modified multilevel Quasi Z inverter for solar irrigation systems with improved control technique. *International Conference on Advances in Electrical Technology for Green Energy (ICAETGT)*, 139-148, doi: 10.1109/ICAETGT.2017.8341467.
19. N Suthanthira Vanitha, R. Sivakumar. (2010). A Novel Approach Grid-Connected PV System Employing PID Controller. *I-Manager's J. Electr. Eng.*, 3(4),15-21.
20. S. L. Tripathi, S. Padmanabhan, D. Kumar, P. Raja (Eds.) (2021) *Design and Development of Energy Efficient Systems*, Scrivener Publishing & Wiley, USA.
21. R. Sahani, M. Daga, I. Jamatia, P. Kumar and S. L. Tripathi (2023) Design of a self-charging electric vehicle with embedded multiple natural energy sources. *2023 IEEE Devices for Integrated Circuit (DevIC)*, Kalyani, India, pp. 296-301, doi: 10.1109/DevIC57758.2023.10134774.
22. N. Suthanthira Vanitha, J. Karthikeyan, G. Kavitha, K. Radhika (2020). Modelling of Intelligent Transportation System for Human Safety using IoT. *Mater. Today Proc., 33(7)*, 4026-4029.

5

Electric Vehicle and Design Using MATLAB

Vinay Anand[1] and Himanshu Sharma[2]*

[1]Department of Examinations, BBD University, Lucknow, U.P., India
[2]School of Electronics and Electrical Engineering Lovely Professional University, Jalandhar, Punjab, India

Abstract

Electric vehicles, also known as EVs, use an electric motor powered by rechargeable batteries rather than an internal combustion engine fueled by gasoline or diesel. Electric vehicles are gaining popularity in automotive industries due to their environmental benefits, lower operating costs, and improved technology. Electric vehicles have a battery pack that stores energy to power the electric motor. One of the main benefits of EVs is their lower operating costs. EVs are cheaper to maintain and have lower fuel costs than traditional gasoline-powered cars. Additionally, they produce zero emissions, making them a more environmentally friendly option. There are different types of EVs, including battery electric vehicles (BEVs) that are powered solely by electricity and plug-in hybrid electric vehicles that use a combination of electricity and gasoline. While EVs still make up a relatively small percentage of vehicles on the road, they are becoming more accessible and affordable to consumers. As technology improves and more charging infrastructure is developed, EVs will likely become an increasingly popular option for transportation in the future. The BEVs have a battery pack that requires continuous charging, and charging infrastructure and solutions are a significant focus of the researchers. Also, instead of brushless DC motors, the possibility of another type of motor like an induction motor is an area of research. This article enlightens us about the limitations and solutions and also spreads general awareness of EVs.

***Keywords*:** Electric vehicle, sustainability, infrastructure, automotive, solar, photovoltaic

**Corresponding author*: himanshu.23441@lpu.co.in

Krishan Arora, Suman Lata Tripathi and Himanshu Sharma (eds.) Electric Vehicle Design: Design, Simulation and Applications, (101–128) © 2024 Scrivener Publishing LLC

List of Abbreviations

EV	Electric vehicle
BEVs	Battery Electric Vehicles
PHEVs	Plug-in hybrid electric vehicles
BLDC	Brushless DC motor
RSE	Renewable source of energy
ICV	Internal combustion engine vehicle
NEVs	Neighborhood electric vehicles
FCEVs	Fuel cell electric vehicles
HVAC	Heating, ventilation, and air conditioning
BMS	Battery management system
SOC	State of charge
SOH	State of health
V2G	Vehicle-to-grid

5.1 Introduction

Most vehicles are used by working professionals for their daily office-going purpose. They drive approximately 30–40 miles per day as they commute daily in India. Now, the automakers are moving toward the fully electrified vehicle [1]. An electric vehicle (EV) is the most attractive choice because of its zero emission, light weight, and comfort; it is also supposed to reduce most of the emissions [2, 3]. But there is a big challenge in charging the battery pack of any BEV [4]. There is a better alternative charging to the battery pack of any battery electric vehicle (BEV).

A BEV is a renewable source of energy (RSE) like solar and wind energy, which is the future of power generation systems, and, using these, we can take help to charge BEV batteries [5]. Nowadays, RSE is highly in-demand and is neither going to bear a cost nor going to be exhausted over the next decades. Different sources of renewable energy are freely available in the environment like wind and solar energy. In addition, the BEV is taking the place of an internal combustion engine vehicle (ICV) that is solely responsible for carbon emissions; many countries have taken initiatives to decarbonize the environment by 2030. [6] In addition to costing less than the fuel consumed by ICV, it has been observed that the alternative source of energy can reduce the load on the grid and help to improve the life of the battery [4]. This may lead to a significant reduction of the battery problem in the near future [7]. The BEV requires a charging station to charge the battery pack, and researchers are moving toward various options of charging systems even though robustness and durability of motor used in BEV are also a limitation. So there is an investigation of the suitability of an induction motor as a prime mover of an EV. This chapter is divided into sections. Section 5.2 describes the motivation of the theme and the history of EVs; Section 5.3 identifies the fundamentals of EVs, and Section 5.4 explains why we should go for EVs; Section 5.5 compares the features of ICV and of EV, and Section 5.6 classifies the different EVs. Section 5.7 elaborates the design and structure of EV, Section 5.8 describes the mathematical model, and the control strategy is shown in Section 5.9. Design methodology for EVs is explained in Section 5.10, and the latest emerging technology in EVs is discussed in Section 5.11. Section 5.12 offers a performance evaluation of a BLDC motor and induction motor for EV propulsion application, and the last section provides a conclusion to the article.

5.2 Motivation

From various discussions of the research article, the major observations are as follows.

- The electric motor for the EVs propulsion application should be robust, reliable, and compatible with the long-range run, and so another type of motor performance can be evaluated for EV propulsion applications.
- Shortcomings of charging stations limit the reach of EV drivers.
- Overloading the power grid in daytime especially can say "Sunny Days" so renewable energy may be an alternative solution.

5.2.1 History of EVs

The history of EVs dates back to the early 19th century, with the invention in 1834 of the first electric motor, which was more economical and efficient [8], by Thomas Davenport, an American blacksmith. In 1837, Robert Anderson, a Scottish inventor, built the first EV prototype, which was powered by non-rechargeable primary cells. However, it was not until the late 1800s that EVs gained popularity and started being produced on a larger scale. In 1891, William Morrison, an American inventor, built the first successful EV in the United States. The vehicle had a top speed of 14 miles per hour and was powered by a rechargeable battery. In the same year, a French engineer, Gustave Trouvé, built an electric tricycle that he exhibited at the International Exposition of Electricity in Paris. In the early 1900s, EVs became more popular and were widely used for transportation in cities. Electric taxis, buses, and delivery vehicles were common in cities like New York and London. However, the invention of the gasoline-powered internal combustion engine (ICE) and the mass production of cheap gasoline-powered cars by companies like Ford and General Motors led to the decline of EVs in the 1920s and 1930s. It was not until the 1960s and 1970s that EVs started to regain popularity due to concerns over air pollution and the oil crisis. In 1966, the first battery-powered EV was introduced by General Motors, the "Electrovair." In 1971, the first modern electric car was built by Victor Wouk, an electrical engineer, using a converted Buick Skylark. In the 1990s, with the development of more efficient and powerful batteries, companies like General Motors, Toyota, and Honda started producing electric cars for the mass market. In 1997, the Toyota Prius became the first mass-produced hybrid electric vehicle (HEV), which combined an electric motor with a gasoline engine for an extended range. In the 2000s and 2010s, with the development of lithium-ion batteries and advances in electric motor technology, EVs became more popular and started to compete with gasoline-powered cars in terms of range and performance. Companies like Tesla, Nissan, and Chevrolet introduced mass-market electric cars with ranges of over 200 miles per charge. From Perth to Sydney, Australia, a solar-powered EV was driven in 1982 by the two partners, Hans and Larry Perkins [9]. Today, EVs are becoming increasingly popular as consumers are more concerned about the environment and the need to reduce greenhouse gas emissions. Governments around the world are also promoting the use of EVs through incentives, subsidies, and regulations to reduce air pollution and combat climate change.

5.3 Basic Fundamentals of EVs

Electric vehicles (EVs) are vehicles that are powered by electric motors instead of ICEs. They rely on batteries to store and supply energy to the electric motor. Here are some fundamental aspects of EVs:

1. **Battery:** The battery is the heart of an EV, and it is responsible for storing energy that powers the electric motor. Lithium-ion batteries are commonly used in EVs, but other types of batteries are also being researched and developed. The mass of the lithium-ion battery is 75 times more compared to petrol which is a limitation.
2. **Electric Motor:** The electric motor is the main source of propulsion for an EV. It converts electrical energy from the battery into mechanical energy that drives the wheels.
3. **Charging Infrastructure:** Unlike gasoline-powered vehicles that can be refueled at any gas station, EVs require charging infrastructure to recharge their batteries. There are various types of charging stations, including Level 1, Level 2, and DC fast-charging stations.
4. **Regenerative Braking:** EVs use regenerative braking, which converts the kinetic energy of the vehicle into electrical energy that is stored in the battery. This process helps extend the vehicle's range.
5. **Range:** The range of an EV refers to the distance that it can travel on a single charge. The range depends on several factors, including the battery size, driving style, and environmental conditions.
6. **Efficiency:** EVs are more efficient than gasoline-powered vehicles because they convert more of the energy stored in the battery into motion. EVs also do not waste energy through idling, as the electric motor only consumes power when the vehicle is moving.
7. **Cost:** The cost of an EV is typically higher than that of a gasoline-powered vehicle, but the operating costs are generally lower. This is because electricity is cheaper than gasoline, and EVs require less maintenance.

Overall, EVs offer several advantages over gasoline-powered vehicles, including lower emissions, higher efficiency, and lower operating costs. As the technology continues to advance and battery costs come down, EVs are likely to become even more widespread in the future.

5.4 Why Electric Vehicles?

There are several reasons why people are increasingly turning to EVs, including the following:

- **Environmental Benefits:** Electric vehicles emit significantly fewer greenhouse gases and pollutants compared to traditional gasoline-powered vehicles, helping to reduce the negative impact of transportation on the environment.
- **Cost Savings:** While the initial cost of purchasing an EV may be higher, the cost of fuel and maintenance over the life of the vehicle is typically lower than that of a gasoline-powered car.
- **Performance:** Electric vehicles can offer instant torque, resulting in quick acceleration and smoother driving experiences.
- **Noise Reduction:** Electric vehicles are much quieter than gasoline-powered vehicles, which can lead to a reduction in noise pollution.
- **Government Incentives:** In many countries, governments offer tax incentives, rebates, and other perks to encourage the adoption of EVs.
- **Technological Advancements:** As technology continues to advance, EVs are becoming increasingly efficient and offering longer ranges on a single charge, making them more practical for everyday use.

5.5 Comparison Between ICV and EV

Here are some comparisons between EVs and internal combustion engine (ICE)–based vehicles, where the HEVs and their types are clearly explained in section 5.7 [11]

- **Environmental Impact:** EVs produce fewer greenhouse gas emissions and pollutants than ICE vehicles, making them a more environmentally friendly option.
- **Energy Efficiency:** EVs are typically more energy efficient than ICE vehicles, meaning they can travel farther on the same amount of energy.
- **Operating Costs:** EVs generally have lower operating costs than ICE vehicles because they require less maintenance, have lower fuel costs, and may be eligible for tax incentives.

- **Performance:** In terms of acceleration and overall performance, EVs can often match or exceed ICE vehicles.
- **Range:** ICE vehicles typically have a longer range than EVs, although the range of EVs is improving rapidly as battery technology improves.
- **Refueling/Recharging Time:** Refueling an ICE vehicle takes only a few minutes, whereas recharging an EV battery can take several hours. However, with the growth of fast-charging infrastructure, this gap is shrinking.
- **Availability:** ICE vehicles are more widely available and have a more established infrastructure for refueling. However, the availability of EVs and charging infrastructure is rapidly increasing.

Overall, EVs offer several advantages over ICE vehicles in terms of environmental impact, energy efficiency, and operating costs. While the range and charging times are still a consideration, advances in battery technology and the growth of charging infrastructure are helping to make EVs an increasingly viable option for many drivers.

5.6 Classification of EVs

There are several types of EVs, each with its unique characteristics and applications. Here are some of the most common types of EVs:

- **Battery Electric Vehicles (BEVs):** These vehicles are powered entirely by an electric motor and a battery pack. They do not have an ICE and do not require any gasoline or other fuels. They are charged by plugging into an electric power source.
- **Plug-in Hybrid Electric Vehicles (PHEVs):** These vehicles have both an electric motor and an ICE. They can be charged from an external power source, and they can also run on gasoline. PHEVs can operate in all-electric mode for a limited distance before switching to gasoline.
- **Hybrid Electric Vehicles (HEVs):** These vehicles have an electric motor and an ICE, but they cannot be charged from an external power source. Instead, they rely on regenerative braking and the gasoline engine to charge the battery.

HEVs typically have better fuel economy than traditional gasoline-powered vehicles.
- **Fuel Cell Electric Vehicles (FCEVs):** These vehicles use a fuel cell to convert hydrogen into electricity, which powers an electric motor. They do not produce any greenhouse gas emissions, but they do require a source of hydrogen to operate.
- **Neighborhood Electric Vehicles (NEVs):** These are small EVs designed for short trips and low speeds, typically used in neighborhoods or on golf courses.
- **Electric Buses:** These are larger EVs designed to transport passengers. They are becoming increasingly popular in urban areas because they produce no emissions and are quieter than traditional buses.

Overall, there are a variety of EV types to suit different needs and applications, from small personal vehicles to large commercial buses [10].

5.7 Design and Structure of EV

A BEV is an EV that is powered solely by an electric motor and a rechargeable battery pack. BEVs do not have an ICE and are designed to run entirely on electricity. Here are some key features and characteristics of BEVs:

- **Battery:** The battery is the primary power source for a BEV. It is typically made up of many individual battery cells connected to achieve the desired voltage and capacity. BEVs can have different types of batteries, such as lithium-ion, nickel-metal hydride, or solid-state batteries.
- **Electric Motor:** The electric motor is the main propulsion system for a BEV. It converts electrical energy from the battery into mechanical energy to propel the vehicle. The motor is usually located close to the wheels and can be either AC or DC, with a permanent magnet or an induction design.
- **Range:** The range of a BEV refers to how far it can travel on a single charge of the battery. This can vary depending on factors such as battery size, driving style, weather conditions, and terrain. Most BEVs have a range of between 100 and 300 miles on a single charge.
- **Charging:** BEVs can be charged using a variety of methods, including Level 1 (120 V) or Level 2 (240 V) charging

stations, or DC fast-charging stations. The time it takes to charge a BEV can vary depending on the charging method and the size of the battery. DC fast charging can typically provide an 80% charge in around 30–45 min.

- **Maintenance:** BEVs typically require less maintenance than ICE vehicles because they have fewer moving parts and do not require oil changes or other traditional maintenance tasks. However, they still require regular maintenance of components such as brakes, tires, and suspension.
- **Environmental Impact:** BEVs have a lower environmental impact compared to ICE vehicles because they produce zero tailpipe emissions. However, the environmental impact of a BEV depends on how the electricity used to charge the battery is generated. If the electricity comes from renewable sources such as solar or wind power, then the environmental impact is much lower than of the electricity that comes from coal or natural gas [9].

Overall, BEVs, the basic structure of which is shown in Figure 5.1, offer several advantages over ICE vehicles, including lower operating costs, reduced emissions, and a quieter and smoother driving experience. However, they also face challenges such as range anxiety, limited charging infrastructure, and higher upfront costs. As technology continues to improve and costs come down, BEVs are expected to become increasingly popular in the coming years.

In Figure 5.1, the electric motor is connected to the wheels of the vehicle and is powered by the battery. The battery is connected to a charging port, which allows the battery to be charged from an external power source. When the vehicle is being driven, the electric motor converts electrical energy from the battery into mechanical energy to propel the vehicle forward.

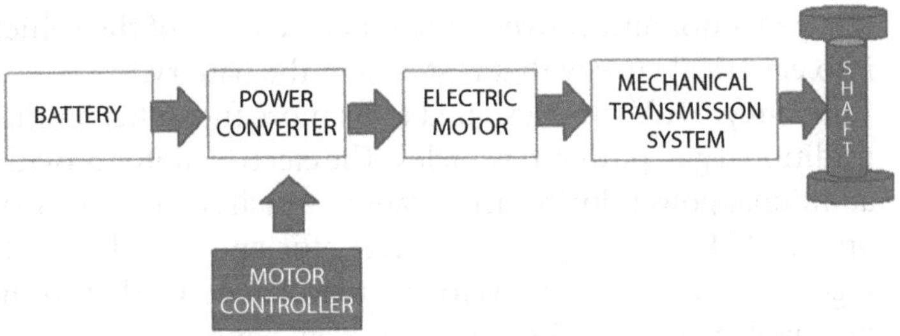

Figure 5.1 BEV basic structure [10].

The battery is made up of multiple individual battery cells, which are connected to form a battery pack. The battery pack is typically located on the floor of the vehicle, which helps to provide a low center of gravity and improve handling.

The charging port allows the battery to be charged using a charging station or other external power source. Charging times can vary depending on the charging method and the size of the battery.

Overall, this diagram provides a basic overview of the components and layout of a BEV.

5.7.1 HEV

An HEV is a vehicle that combines an ICE with an electric motor and a battery pack. HEVs can be designed in a variety of different configurations, but here are some key features and characteristics:

- **Electric Motor and Battery:** HEVs have an electric motor and a battery pack that works together with the ICE to provide propulsion. The electric motor is usually smaller than the motor in a BEV and is used primarily to assist the ICE during acceleration and to regenerate energy during braking. The battery is typically smaller than in a BEV and is used to store energy for use by the electric motor.
- **ICE:** HEVs also have an ICE that can be fueled with gasoline or diesel. The ICE is used to power the vehicle when the battery is depleted or when more power is required than the electric motor can provide. The ICE is typically smaller than in a traditional gas-powered vehicle because it works in conjunction with the electric motor to provide propulsion.
- **Regenerative Braking:** HEVs use regenerative braking to capture the energy that is normally lost during braking. When the driver applies the brakes, the electric motor acts as a generator and converts the kinetic energy of the vehicle into electrical energy that is stored in the battery.
- **Efficiency:** HEVs are designed to be more fuel-efficient than traditional gas-powered vehicles. The electric motor provides additional power during acceleration, which reduces the load on the ICE and improves overall efficiency. Additionally, regenerative braking helps to capture the energy that would normally be lost, further improving efficiency.

- **Charging:** Unlike BEVs, HEVs do not need to be charged from an external power source because the battery is charged by the ICE and through regenerative braking. However, some HEVs may have the option to plug in and charge the battery for short periods to improve efficiency.
- **Maintenance:** HEVs require maintenance similar to traditional gas-powered vehicles, including oil changes, air filter replacements, and other routine maintenance tasks.

Overall, HEVs offer a balance between the efficiency and environmental benefits of an EV and the range and convenience of a gas-powered vehicle. They are ideal for drivers who want to reduce their fuel consumption and emissions without the range limitations of a BEV.

HEV diagram

In Figure 5.2, the electric motor is connected to the transmission, which is responsible for transferring power from the engine and the electric motor to the wheels. The battery provides power to the electric motor, which can assist the ICE during acceleration or operate the vehicle on electric power alone at low speeds. The ICE is connected to the transmission as well and can be used to power the vehicle when the battery is depleted or when more power is required than the electric motor can provide. The engine can also be used to charge the battery if needed. The battery is usually smaller than in a BEV because it is used to store energy for short periods rather than to power the vehicle exclusively. The battery can be charged through regenerative braking when the electric motor acts as a generator to convert kinetic energy into electrical energy that is stored in the battery.

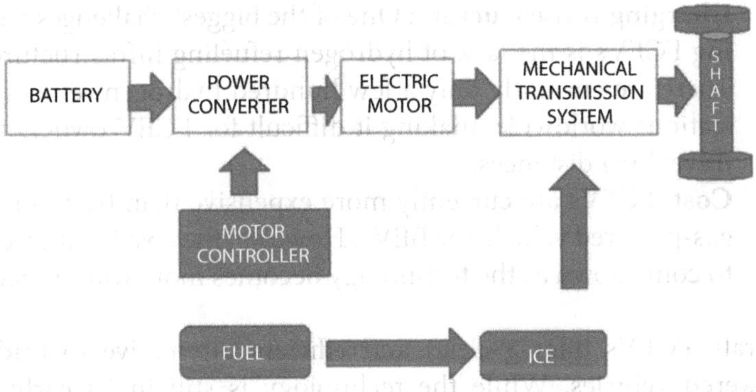

Figure 5.2 Basic structure of HEV [10].

Overall, this diagram provides a basic overview of the components and layout of an HEV.

5.7.2 FCEV

An FCEV is an EV that uses a fuel cell to generate electricity to power an electric motor. Here are some key features and characteristics of FCEVs:

- **Fuel Cell:** The fuel cell is the heart of the FCEV. It uses hydrogen and oxygen from the air to generate electricity through an electrochemical reaction. The only byproduct of this reaction is water, making FCEVs a very clean and environmentally friendly option.
- **Electric Motor:** Like other EVs, FCEVs use an electric motor for propulsion. The motor is powered by the electricity generated by the fuel cell.
- **Hydrogen Storage:** FCEVs require hydrogen to generate electricity. The hydrogen is stored in high-pressure tanks located in the vehicle. The tanks are designed to be safe and durable, but they do add weight and cost to the vehicle.
- **Range:** FCEVs have a similar range to traditional gas-powered vehicles, with some models capable of traveling up to 300–400 miles on a single tank of hydrogen. Refueling times are comparable to traditional gas-powered vehicles, taking only a few minutes to fill up.
- **Efficiency:** FCEVs are very efficient, with a fuel economy of around 60–70 miles per kilogram of hydrogen. This makes them more efficient than traditional gas-powered vehicles and on par with BEVs.
- **Charging Infrastructure:** One of the biggest challenges facing FCEVs is the lack of hydrogen refueling infrastructure. There are currently only a few hundred hydrogen refueling stations worldwide, making it difficult for FCEV owners to travel long distances.
- **Cost:** FCEVs are currently more expensive than traditional gas-powered vehicles or BEVs. However, the cost is expected to come down as the technology becomes more widespread.

Overall, FCEVs offer a clean and efficient alternative to traditional gas-powered vehicles. While the technology is still in its early stages,

FCEVs have the potential to revolutionize the transportation industry and play a major role in reducing greenhouse gas emissions. The fuel cell is located between the electric motor and the hydrogen tank. The fuel cell generates electricity through an electrochemical reaction between hydrogen and oxygen. The electric motor is powered by the electricity generated by the fuel cell. The hydrogen tank is located at the rear of the vehicle and stores the hydrogen that is used to power the fuel cell. The tank is made of lightweight and high-strength materials and is designed to be safe and durable.

5.7.3 PHEV

A PHEV is a type of HEV that can be charged from an external power source as depicted in Figure 5.3, in addition to using an ICE and regenerative braking to charge its battery. PHEVs offer several benefits over traditional gas-powered vehicles and conventional HEVs. Here is a detailed explanation of how PHEVs work and their key features:

- **Electric Motor:** Like other EVs, PHEVs use an electric motor for propulsion. The motor is powered by a rechargeable battery that can be charged from an external power source.
- **Internal Combustion Engine:** PHEVs also have an ICE that can be used to power the vehicle when the battery is depleted. The engine can also be used to charge the battery while driving.
- **Battery:** The battery in a PHEV is larger than the battery in a conventional HEV but smaller than the battery in a BEV. The

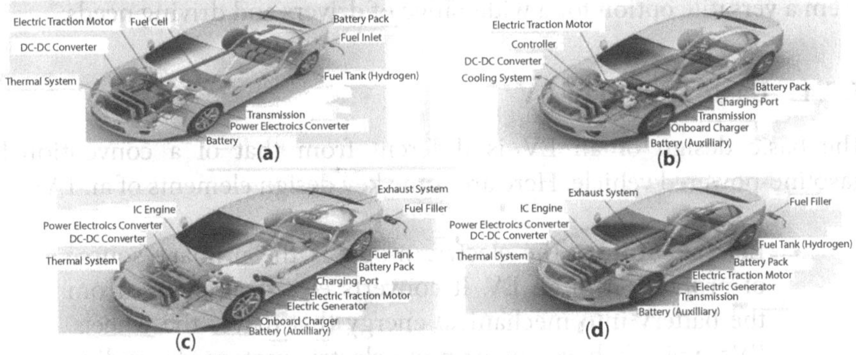

Figure 5.3 Hybrid electric vehicle and battery electric vehicles technology [11].

larger battery allows the PHEV to operate in electric-only mode for longer distances.

- **Charging:** PHEVs can be charged from an external power source, such as a standard electrical outlet or a dedicated EV charging station. Charging times vary depending on the type of charger used and the size of the battery but can range from a few hours to overnight.
- **Range:** PHEVs have a limited electric-only range, typically between 20 and 50 miles, depending on the size of the battery. However, the ICE can be used to extend the range of the vehicle beyond the electric-only range.
- **Fuel Efficiency:** PHEVs are more fuel efficient than traditional gas-powered vehicles and conventional HEVs. The electric-only range allows the PHEV to operate without using any gasoline, whereas the ICE provides additional range and power when needed.
- **Regenerative Braking:** PHEVs use regenerative braking to recharge the battery while driving. When the driver applies the brakes, the electric motor acts as a generator and converts the kinetic energy of the vehicle into electrical energy, which is then stored in the battery.
- **Cost:** PHEVs are typically more expensive than conventional HEVs, but less expensive than BEVs. However, PHEVs can offer significant cost savings over time, as they require less fuel and maintenance than traditional gas-powered vehicles.

Overall, PHEVs offer a good balance between the range and convenience of traditional gas-powered vehicles and the efficiency and environmental benefits of EVs. The ability to charge from an external power source makes them a versatile option for a wide range of drivers and driving needs.

5.7.4 Basic Design of EV

The basic design of an EV is different from that of a conventional gasoline-powered vehicle. Here are some key design elements of an EV:

- **Electric Motor:** The electric motor is the primary source of propulsion in an EV. It converts electrical energy from the battery into mechanical energy that drives the wheels. EVs typically have one or more electric motors, depending on their size and design.

- **Battery:** The battery is the heart of an EV, and it is responsible for storing energy that powers the electric motor. The battery pack is typically located under the floor of the vehicle, which lowers the center of gravity and improves handling. Lithium-ion batteries are commonly used in EVs, but other types of batteries are also being researched and developed.
- **Charging Port:** EVs have a charging port that allows them to be connected to a chargingstation or a home charging unit. The port is typically located on the side or rear of the vehicle.
- **Regenerative Braking:** EVs use regenerative braking, which converts the kinetic energy of the vehicle into electrical energy that is stored in the battery. This process helps extend the vehicle's range.
- **Power Electronics:** EVs have power electronics which is nothing but an inverter which converts DC into AC that manages the flow of electrical energy between the battery, the electric motor, and other electrical components in the vehicle. The power electronics also control the speed and torque of the electric motor.
- **Inverter:** The inverter is an important component of an EV's power electronics. It convertsthe DC power from the battery into AC power that is used by the electric motor.
- **HVAC System:** The HVAC (heating, ventilation, and air conditioning) system in an EV is electrically powered, and it is designed to be energy efficient to maximize the vehicle's range.
- **Onboard Charger:** EVs have an onboard charger that converts AC power from a charging station or a home charging unit into DC power that can be stored in the battery.

Overall, the design of an EV is focused on maximizing efficiency, minimizing weight, and optimizing the use of electrical energy. As battery technology continues to improve, the design of EVs will likely become even more efficient and streamlined in the future.

5.8 Mathematical Model of an Electric Vehicle

A mathematical model for an EV typically includes equations to describe the dynamics of the vehicle, the battery, and the electric motor. Here is a general overview of what might be included in such a model:

- **Vehicle Dynamics:** The motion of the vehicle can be described using Newton's laws of motion. This includes the acceleration of the vehicle, the forces acting on it (such as air resistance and friction), and the vehicle's weight.
- **Battery Model:** The battery can be modeled as an electrical circuit with a voltage source and internal resistance. The voltage and current of the battery depending on its state of charge (SOC) and its temperature.
- **Motor Model:** The electric motor can be modeled using equations that describe its torque, speed, and power output. The motor's performance depends on its efficiency, the voltage and current supplied to it, and the load that it is driving.
- **Controller Model:** The control system for the EV can be modeled using control theory. The controller regulates the power output of the motor to maintain a desired speed and acceleration while also protecting the battery from overcharging or discharging.
- **Energy Consumption:** The energy consumption of the vehicle can be calculated on the basis of the power output of the motor, the efficiency of the motor and drivetrain, and the driving conditions (such as speed, terrain, and temperature).

Overall, the mathematical model for an EV is complex and requires knowledge of physics, electrical engineering, and control theory. However, it is a valuable tool for designing and optimizing EVs for efficiency, performance, and range.

5.9 Control Strategy of EVs

Control strategy refers to the methods used to manage and optimize the performance of an EV. The main objective of the control strategy is to improve the efficiency, range, and performance of the EV. Here are some common control strategies used in EVs:

- **Motor Control:** The motor controller is responsible for regulating the speed, torque, and power output of the electric motor. The control strategy for the motor involves adjusting the voltage and current supplied to the motor to optimize its efficiency and performance. This can be achieved through techniques such as field-oriented control, direct torque control, and pulse width modulation.

- **Battery Management:** The battery management system (BMS) is responsible for monitoring the SOC and state of health (SOH) of the battery, as well as protecting it from overcharging and overheating. The control strategy for the BMS involves regulating the charging and discharging of the battery to optimize its performance and prolong its life span.
- **Regenerative Braking:** Regenerative braking is a technique that allows the EV to recover some of the energy that is typically lost during braking. The control strategy for regenerative braking involves adjusting the amount of regenerative braking based on the speed and braking force applied by the driver.
- **Thermal Management:** Thermal management involves controlling the temperature of the battery, motor, and other components to ensure optimal performance and safety. The control strategy for thermal management involves monitoring the temperature of these components and adjusting cooling or heating as needed.
- **Energy Management:** Energy management involves optimizing the use of energy from the battery and other sources to maximize the range and efficiency of the EV. The control strategy for energy management involves managing the power flow between the battery, motor, and other components based on driving conditions and driver behavior.

Overall, control strategy plays a critical role in the performance, efficiency, and safety of an EV. By optimizing the control strategy, EV manufacturers can improve the range, performance, and overall driving experience for consumers.

5.10 Design Methodology for Electric Vehicles (EVs)

The design methodology for EVs typically involves several stages, from concept development to production. Here are some of the key steps in the design methodology for EVs:

- **Concept Development:** This stage involves defining the basic requirements for the EV, such as range, performance, and cost. The design team will typically develop multiple concept designs and evaluate them on the basis of factors such as feasibility, performance, and cost.

- **System Architecture:** Once a concept design is selected, the design team will develop a system architecture that defines the main components of the EV, such as the battery, motor, and power electronics. The system architecture will also define the power flow and control strategy for the EV.
- **Component Design:** The design team will then focus on the detailed design of each component of the EV, such as the battery pack, motor, and power electronics. This will involve selecting appropriate materials, dimensions, and manufacturing processes.
- **Integration:** Once the individual components are designed, the design team will integrate them into a complete EV system. This will involve testing and optimizing the performance of the individual components and the overall system.
- **Validation and Testing:** The final stage of the design methodology involves validating the performance and safety of the EV through a series of tests and simulations. This may include testing the EV under various driving conditions, such as high speed, low temperature, and heavy load.
- **Production:** Once the design is validated, the EV can move into production. This will involve setting up a production line and developing quality control processes to ensure consistency and reliability.

Overall, the design methodology for EVs is complex and requires expertise in a variety of disciplines, such as electrical engineering, mechanical engineering, and control systems. However, by following a systematic approach, EV manufacturers can develop high-performance, reliable, and cost-effective EVs.

5.11 Latest Emerging Technology in EV

There are several emerging technologies in EVs that are expected to play a significant role in the future of the industry. Here are a few examples:

- **Solid-State Batteries:** Solid-state batteries are a type of battery that use solid electrolytes instead of liquid ones.

They have the potential to offer higher energy density, faster charging, and improved safety compared to traditional lithium-ion batteries.

- **Wireless Charging:** Wireless charging technology allows EVs to charge their batteries without needing to physically connect to a charging station. This could make charging more convenient and reduce the need for public charging infrastructure.
- **Vehicle-to-Grid (V2G) Technology:** V2G technology enables EVs to transfer energy back to the grid when they are not in use. This could help balance the grid and provide additional revenue streams for EV owners.
- **Advanced Materials:** Advances in materials science are leading to the development of lighter and stronger materials for EVs, such as carbon fiber and aluminum. This could help improve range and performance while reducing weight and cost.
- **Autonomous Driving:** While not specific to EVs, autonomous driving technology could have a major impact on the EV industry. Self-driving EVs could potentially reduce accidents and traffic congestion, as well as offer new mobility options for people who cannot drive.

These are just a few examples of emerging technologies in the EV industry. As technology continues to evolve, we will likely see even more advances that improve the performance, range, and convenience of EVs. Several electric motor types are widely used in EVs for propulsion applications [12]. The motor for an EV is used on the basis of its overall performance.

5.12 Performance Valuation of BLDC Motor and Induction Motor for Electric Vehicle Propulsion Application

Performance evaluation is an important aspect of selecting a motor for EV propulsion application. In this article, we will discuss the performance evaluation of two types of motors: brushless DC (BLDC) motor and induction motor.

BLDC Motor: BLDC motors are becoming popular for EV propulsion applications due to their high efficiency, high power density, and low maintenance requirements. The following are the key parameters used to evaluate the performance of BLDC motors:

- Power Output: The power output of a BLDC motor is measured in watts. The power output of the motor determines the maximum speed and acceleration of the EV.
- Torque: Torque is the rotational force produced by the motor. It is measured in Newton meters (Nm). The torque of a BLDC motor determines the ability of the motor to accelerate the vehicle from a standstill and climb uphill [13, 14].
- Efficiency: Efficiency is the ratio of the output power of the motor to the input power. The efficiency of a BLDC motor is typically between 85% and 95%, which means that only a small amount of energy is lost as heat during the operation of the motor [13, 14].
- Temperature Rise: The temperature rise of a BLDC motor is an important parameter as it affects the performance and life span of the motor. The temperature rise is determined by measuring the temperature of the motor before and after the operation.
- Regenerative Braking: Regenerative braking is the process of converting the kinetic energy of the vehicle into electrical energy and storing it in the battery. BLDC motors are capable of providing regenerative braking, which increases the range of the vehicle [14].

5.12.1 A Mathematical Model for a Brushless DC (BLDC) Motor–Driven Electric Vehicle

This can be expressed using the following equations:

Electrical equations: The voltage equation for a BLDC motor can be written as follows:

$$V = L(di/dt) + e \tag{5.1}$$

where V is the applied voltage, L is the inductance of the motor winding, di/dt is the rate of change of current in the winding, and e is the back-emf of the motor.

The back-emf of the motor can be expressed as follows:

$$e = k\omega \tag{5.2}$$

where k is a constant representing the motor's electrical and magnetic properties and ω is the motor speed.

Mechanical equations: The torque developed by the motor can be expressed as follows:

$$T = kI \tag{5.3}$$

where T is the torque developed by the motor, k is a constant representing the motor's electrical and magnetic properties, and I is the current flowing through the motor winding.

The mechanical power developed by the motor can be expressed as follows:

$$P_m = T\omega \tag{5.4}$$

where Pm is the mechanical power developed by the motor.

Vehicle dynamics equations: The motion of the EV can be described using thefollowing equations:

$$F = ma \tag{5.5}$$

where F is the net force acting on the vehicle, m is the mass of the vehicle, and a is the acceleration of the vehicle.

The force acting on the vehicle can be expressed as follows:

$$F = F_w + F_r + F_a \tag{5.6}$$

where Fw is the force due to the rolling resistance of the tires, Fr is the force due to air resistance, and Fa is the force due to the motor.

The force due to the motor can be expressed as follows:

$$F_a = T/r \tag{5.7}$$

where r is the radius of the tire.

By combining the above equations, we can derive a mathematical model for a BLDC motor–driven EV. This model can be used to analyze the performance of the vehicle and to design control systems for the motor and the vehicle.

5.12.2 Induction Motor

Induction motors are commonly used in EV propulsion applications due to their robustness and reliability. The following are the key parameters used to evaluate the performance of induction motors:

- **Power Output:** The power output of an induction motor is measured in watts. The power output of the motor determines the maximum speed and acceleration of the EV [15].
- **Torque:** Torque is the rotational force produced by the motor. It is measured in Newton meters (Nm). The torque of an induction motor determines the ability of the motor to accelerate the vehicle from a standstill and climb uphill and achieve maximum output torque [16].
- **Efficiency:** Efficiency is the ratio of the output power of the motor to the input power. The efficiency of an induction motor is typically between 80% and 95%, which means that only a small amount of energy is lost as heat during the operation of the motor [16, 17].
- **Temperature Rise:** The temperature rise of an induction motor is an important parameter as it affects the performance and life span of the motor. The temperature rise is determined by measuring the temperature of the motor before and after the operation.
- **Regenerative Braking:** Induction motors are also capable of providing regenerative braking, which increases the range of the vehicle.
- **Volume:** Induction motors have smaller volumes comparatively [16].

In conclusion, both BLDC and induction motors have their advantages due to no magnetic eddy current loss and low windage and iron loss [16] and disadvantages when it comes to performance evaluation for EV propulsion applications. While BLDC motors have a higher efficiency and power density, induction motors are known for their robustness and reliability. Ultimately, the choice between the two depends on the specific requirements and needs of the application [16].

5.12.3 Mathematical Model for an Induction Motor–Driven Electric Vehicle

This can be represented by a set of equations that describe the behavior of the motor, battery, and vehicle dynamics. Here is a simplified mathematical model for an induction motor–driven electric vehicle, the modeling of induction motor for EV propulsion applications. For the investigation analysis purpose, a parameter of the induction motor was identified on the basis of the rating of 5 hp and 60 Hz. In this model, speed varies at a constant speed and torque base.

5.12.4 Induction Motor Design with Their Specifications

- **Motor Equations:**

The equations for an induction motor are based on the fundamental principles of electromagnetism and describe the relationship between the stator and rotor currents, the electromagnetic torque, and the rotor speed. For a three-phase induction motor, the equations are as follows:

- **Stator Equation:**

$$V_{sa} = R_s i_{sa} + L_s di_{sa}/dt\ e_{as} \tag{a}$$

$$V_{sb} = R_s i_{sb} + L_s di_{sb}/dt - e_{bs} \tag{b}$$

$$V_{sc} = R_s i_{sc} + L_s di_{sc}/dt - e_{cs} \tag{c}$$

- **Rotor Equation:**

$$e_{ar} = R_r i_{ra} + L_r di_{ra}/dt - (w_r * L_m * i_{sb}) \tag{d}$$

$$e_{br} = R_r i_{rb} + L_r di_{rb}/dt - (w_r * L_m * i_{sc}) \tag{e}$$

$$e_{cr} = R_r i_{rc} + L_r di_{rc}/dt - (w_r * L_m * i_{sa}) \tag{f}$$

- **Electromagnetic Torque:**

$$T_e = (3/2) * P * L_m * (i_{sa} * i_{rb} - i_{sb} * i_{ra}) \tag{i}$$

- **Rotor Speed:**

$$w_r = (P/2) * [(i_{sa} * i_{rb} - i_{sb} * i_{ra})/(L_m * i_m)] \quad \text{(j)}$$

- **Battery Equations:**

The equations for the battery model are based on the battery's voltage and current characteristics. For a typical lead–acid battery, the equations are as follows:

- **Terminal Voltage:**

$$V_{batt} = E - I_{batt} * R_{int} - K * SoC \quad \text{(k)}$$

- **Battery Current:**

$$I_{batt} = I_{load} + I_{loss} \quad \text{(l)}$$

- **State of Charge:**

$$SoC = (Q - Q_r) / Q \quad \text{(m)}$$

where E is the open-circuit voltage, Rint is the internal resistance, K is a constant that depends on the battery, Q is the battery capacity, Q_r is the remaining capacity, I_{Load} is the current drawn by the motor, and I_{Loss} is the current lost to internal resistance.

- **Acceleration:**

$$a = (Te - Fd) / m \quad \text{(n)}$$

where T_e is the electromagnetic torque from the motor, F_d is the drag force, and m is the mass of the vehicle.

- **Drag Force:**

$$Fd = (1/2) * \rho * A * C_d * v^2 \quad \text{(o)}$$

where ρ is the air density, A is the frontal area of the vehicle, C_d is the drag coefficient, and v is the vehicle speed.

- **Rolling Resistance:**

$$F_{rr} = m * g * C_{rr} \quad (p)$$

where g is the gravitational constant and Crr is the coefficient of rolling resistance.

These equations provide a basic model for an induction motor–driven EV, but many factors can affect the performance and efficiency of the vehicle, such as temperature, motor control strategy, and battery management. More complex models can be developed to take these factors into account. Induction motors are widely attractive to the EV world due to their robustness, reliability, and control characteristics [18–20].

5.13 Conclusion

This article concludes that EVs are the only alternative to zero-emission transportation, with comparatively lower operating and maintenance costs. Additionally, they are more environmentally friendly. The absence of a charging infrastructure and a solar-based stand-alone charging system can be an added advantage, and the body of any passenger car can utilize photovoltaic-based thin film that converts fallen photons from the sun into electrical energy. As technology improves and more charging infrastructure is developed, EVs will likely become an increasingly popular option for transportation in the future. Also, instead of BLDC, the possibility of an induction motor as a propulsion motor can be used to make an EV a robust, reliable, low-cost passenger vehicle.

References

[1] V. Anand, "Photovoltaic actuated induction motor for driving electric vehicle," *Int. J. Eng. Adv. Technol.*, vol. 8, no. 6 Special Issue 3, pp. 1612–1614, Sep. 2019, doi: 10.35940/ijeat.F1298.0986S319.

[2] Y. Wu and E. Kontou, "Designing electric vehicle incentives to meet emission reductiontargets," *Transp. Res. Part D*, vol. 107, no. May, p. 103320, 2022, doi: 10.1016/j.trd.2022.103320.

[3] V. Anand and B. S. Ram, "A comprehensive investigation of the design of solar-powered induction motor-driven electric vehicle (SIM-EV)," *Mater. Today Proc.*, vol. 56, no. 6, pp. 3682–3686, 2022, doi: 10.1016/j.matpr.2021.12.438.

[4] T. I. Of, I. M.- Driven, H. Sharma, and B. S. Ram, *J. Optoelectron. Laser*, vol.41, no. 10, pp. 615–629, 2022.
[5] V. Anand, "Power Generation on Highway by Harnessing Wind Power," *Int. J. Innov. Technol. Explor. Eng.*, vol. 8, no. 12S, pp. 330–331, 2019, doi: 10.35940/ijitee.l1085.10812s19.
[6] L. Maybury, P. Corcoran, and L. Cipcigan, "Mathematical modelling of electric vehicle adoption: A systematic literature review," *Transp. Res. Part D Transp. Environ.*, vol. 107, no. May, p. 103278, 2022, doi: 10.1016/j.trd.2022.103278.
[7] Arora, K.; Kumar, A.; Kamboj, V.K.; Prashar, D.; Jha, S.; Shrestha, B.; Joshi, G.P. Optimization Methodologies and Testing on Standard Benchmark Functions of Load Frequency Control for Interconnected Multi Area Power System in Smart Grids. *Mathematics,* 2020, *8*, 980.
[8] D. F. de Souza, F. A. M. Salotti, I. L. Sauer, H. Tatizawa, A. T. de Almeida, and A. G. Kanashiro, "A Performance Evaluation of Three-Phase Induction Electric Motors between1945 and 2020," *Energies*, vol. 15, no. 6, pp. 1–31, 2022, doi: 10.3390/en15062002.
[9] P. O. Babalola and O. E. Atiba, "Solar powered cars - a review," *IOP Conf. Ser. Mater. Sci. Eng.*, vol. 1107, no. 1, p. 012058, 2021, doi: 10.1088/1757-899x/1107/1/012058.
[10] Arora, K.; Kumar, A.; Kamboj, V.K.; Prashar, D.; Shrestha, B.; Joshi, G.P. "Impact of Renewable Energy Sources into Multi Area Multi-Source Load Frequency Control of Interrelated Power System." *Mathematics,* 2021, *9*, 186.
[11] T. Selmi, A. Khadhraoui, and A. Cherif, "Fuel cell–based electric vehicles technologiesand challenges," *Environ. Sci. Pollut. Res.*, vol. 29, no. 52, pp. 78121–78131, 2022, doi:10.1007/s11356-022-23171-w.
[12] A. I. S. Juhaniya, A. A. Ibrahim, M. A. A. Mohd Zainuri, M. A. Zulkifley, and M. A. Remli, "Optimal Stator and Rotor Slots Design of Induction Motors for Electric VehiclesUsing Opposition-Based Jellyfish Search Optimization," *Machines*, vol. 10, no. 12, 2022, doi: 10.3390/machines10121217.
[13] J. Jimenez-Gonzalez, F. Gonzalez-Montañez, V. M. Jimenez-Mondragon, J. U. Liceaga- Castro, R. Escarela-Perez, and J. C. Olivares-Galvan, "Parameter identification of BLDC motor using electromechanical tests and recursive least-squares algorithm: Experimental validation," *Actuators*, vol. 10, no. 7, 2021, doi: 10.3390/act10070143.
[14] C. Huang, F. Lei, X. Han, and Z. Zhang, "Determination of modeling parameters for abrushless DC motor that satisfies the power performance of an electric vehicle," *Meas. Control (United Kingdom)*, vol. 52, no. 7–8, pp. 765–774, 2019, doi: 10.1177/0020294019842607.
[15] Dipl.-Ing. (FH) Werner Ness and Dr. Klaus Raggl, "Neß-Raggl2022_Article_E-motor Types for Secondary Electr," *Mtz*, pp. 40–45, 2022.
[16] P Singh, K Arora, UC Rathore, "Control Strategies for Improvement of Power Quality in Grid Connected Variable Speed WECS with DFIG – An Overview", *J. Phys. Conf. Ser.*, 2022.

[17] P. Tích *et al.*, "Analysis and Comparison of Electric Motor Types for Electric Vehicles," vol. 2, pp. 20–24, 2022.
[18] M. Aishwarya and R. M. Brisilla, "Design of Energy-Efficient Induction motor usingANSYS software," *Results Eng.*, vol. 16, no. June, p. 100616, 2022, doi: 10.1016/j.rineng.2022.100616.
[19] B. S. Ram, S. L. Tripathi, "Operations of Doubly Fed Induction Generators Applied in Green Energy Systems," Chapter 18 in *Green Energy: Solar Energy, Photovoltaics, and Smart Cities*, Scrivener Publishing & Wiley, USA, 2021.
[20] S. L. Tripathi, S. Padmanabhan, D. Kumar, P. Raja (Eds.) *Design and Development of Energy Efficient Systems*, Scrivener Publishing & Wiley, USA, 2021.

6

Model Order Reduction of Battery for Smart Battery Management System

Dheeraj Kumar Dhaked[1]*, Aswant Kumar Sharma[2], Dhanesh Kumar Sambariya[1] and Dinesh Birla[1]

[1]Electrical Engineering, Rajasthan Technical University, Kota, India
[2]Electrical Engineering, Finolex Academy of Management and Technology, Ratnagiri, India

Abstract

A battery management system needs to have an indefinite accurate mathematical model. A full-order battery model is not suitable for real-time application as it usually takes time to discretise the system with a finite difference method and it is also difficult to solve the system numerically. Therefore, a fast, reliable, and computationally approximation model is required. In this chapter, the higher-order model of a battery management system is reduced into a lower-order model. The lower order of battery management system maintains the approximation response as of the original battery. The frequency-domain method is taken into consideration for order reduction. The result is compared in terms of step response, phase, and magnitude response. The reduced order is an approximation of the original model. The error is calculated as the absolute error and the relative error. The relative error between the original and reduced order is of the lowest value of –246 dB after the frequency of 10^{-6} rad/s.

Keywords: Model order reduction, battery model, electrical vehicle, step response, bode plot

6.1 Introduction

The future belongs to renewable energy and hybrid electric vehicle (HEV) systems because of the pollution issues created by the conventional

**Corresponding author*: ddhakar9@gmail.com

Krishan Arora, Suman Lata Tripathi and Himanshu Sharma (eds.) Electric Vehicle Design: Design, Simulation and Applications, (129–146) © 2024 Scrivener Publishing LLC

energy sources. In renewable energy circuitry, the battery plays an essential role in power delivery and storage. The understanding of the performance of the battery system under a wide range is required before designing an efficient battery management system (BMS). Battery modeling is one of the most important tasks for electric and hybrid vehicle control. This requires a model that can simulate in real time to make them compatible with estimation algorithms embedded in on-board electronic control units. For example, a BMS in an HEV has to estimate the battery state of charge (SOC) in real time to control the electrical power coming in and out of the battery as well as to prevent the battery pack from excessive heating.

Various algorithms are generated for the efficient and reliable working of BMS. These algorithms control the operation and maintain the performance of battery packs. The ultimate aim is to prolong battery life and ensure reliable safety. Battery modeling is done in many ways depending on the types of battery.

The electrical modeling of the battery enables the visualization of the process occurring inside the rechargeable batteries. The highly detailed model results in considerable computational complexity, and it takes hours to simulate a charge-discharge cycle of a detailed battery model. The state-space model for transfer function calculation of BMS of higher-order is used here. Therefore, the need of a lower-order model of a battery without affecting the response of battery is required to have fast and simple simulation for designing the BMS. The concept of model order reduction (MOR) of battery is described in [1–4].

The literature on modeling of batteries is extensively available, including the full-order and simplified model. Simplification of the higher order usually follows a systematic procedure from the frequency-domain techniques such as time moment matching [5, 6], pole-zero simplification [7], and the time-domain methods.

For the above reasons, considerable research efforts have been devoted to apply MOR techniques to first-principle models of lithium-ion batteries, including polynomial approximation [8], Padé approximation [9], residue grouping [10], balanced truncation (BT) [4], and proper orthogonal decomposition (POD) [11]. The objective of applying MOR techniques is to approximate the governing partial differential equation (PDE) into the low-order systems of ordinary differential equation (ODE) that can be solved with standard simulation tools. In recent times, the Galerkin projection method has also been applied to analytically approximate the diffusion PDEs in electrochemical models [12–14]. The main advantage of this method is that it can provide accurate predictions of the concentration profile in the whole-cell domain while retaining the physical meaning of all model parameters. This is especially important for SOC/SOH estimation and cycle life prediction, because the analytical form of the reduced-order models (ROMs) allows one to perform state and parameter estimation. In addition, the accuracy of the Galerkin projection method is comparable to numerical MOR methods such as POD and BT, at a fraction of the computational cost [4].

In [15], a nonlinear diffusion model is developed to predict the behavior of a lithium-ion battery cell particularly for low-temperature and high-current conditions. However, it is not suitable for real-time control and estimation applications due to significant computational complexity. In this chapter, authors extended the aforementioned research by developing a ROM for smart BMS through a systemic application of the frequency-domain method. The proposed method is the reduction of battery using the BT method. The proposed result in the third order and the second order quite matches with the original and represented in the step response for result analysis.

The chapter is divided into six sections. The second section presents the problem formulation. The third section is related to the modeling of battery followed by the methodology of reduction in the fourth section. The fifth section contains the results and discussion, and the last section presents the conclusion of the chapter.

6.2 Problem Formulation

The battery is considered a higher-order single-input–single-output system transfer function with unknown order of n, and it can be represented by the following equation:

$$G_n(s) = \frac{N_{n-1}(s)}{D_n(s)} = \frac{\sum_{a=0}^{n-1} N_i s^i}{\sum_{a=0}^{n} D_i s^i} \tag{6.1}$$

where N_i and D_i are the numerator and denominator constants of the original system transfer function and G is the gain of battery. In some cases, $N_0 = D_0$ for the steady-state output response to a unit step input that will be unity. To find the unknown scalar constant of the reduced [mth (m<n)] order system from the original higher-order system the obtained reduced order has following transfer function:

$$R_n(s) = \frac{Nr_{n-1}(s)}{Dr_n(s)} = \frac{\sum_{a=0}^{m-1} Nr_i s^i}{\sum_{a=0}^{m} Dr_i s^i} \tag{6.2}$$

6.3 Modeling of Battery

The mathematical modeling of batteries is discussed in [18–22], which existed over the past year. Here, a state-space model with state variables V_{Cb}, V_{Cs}, and V_0 is taken. The voltage and current are illustrated in Figure 6.1. In the model, the bulk capacitor C_{bk} acts as an energy storage component in the form of charge. The surface capacitance and diffusion effects within the cell $C_{surface}$, a terminal surface R_t, surface resistance R_s, and end resistance R_e. The voltage across both capacitors are denoted as V_{Cb} and V_{Cs}, + correspondingly.

The result of both time and charging current leading to the time and current reliant on the model for C_{bk} that makes the state space equations to be nonlinear. The formula is given as follows [16, 17]:

$$C_{bk} = \begin{cases} C_0 & t \le t_0 \\ -K_1 I(t - t_0) - K_2(t - t_0) + C_0 & t \ge t_0 \end{cases} \tag{6.3}$$

They C_{bk} can be included in the existing state variable. The value of C_{bk} shows the ability of battery store changing. The C_{bk} is a factor that got consideration in determining the state of health of a battery. The rate of charge

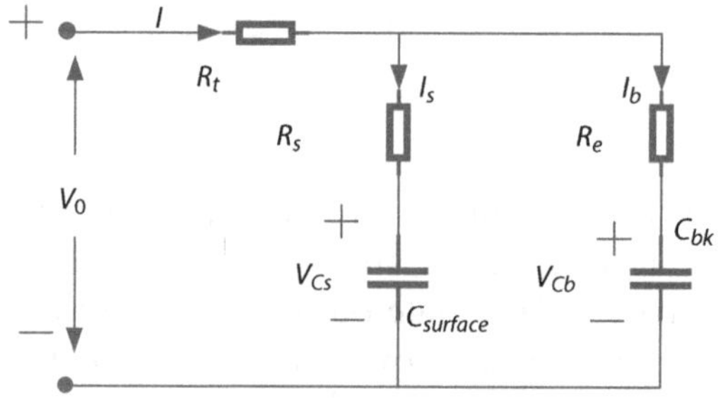

Figure 6.1 Basic battery model.

oversampling is negligible; in other words, $\dot{C}_{bk} = \frac{\partial C_{bk}}{\partial t} = 0$. Therefore, the battery model can be written as follows:

$$\begin{bmatrix} \dot{V}_{Cb} \\ \dot{V}_{Cs} \\ \dot{V}_{0} \\ \dot{C}_{bk} \end{bmatrix} = \begin{bmatrix} -A & A & 0 & 0 \\ B & -B & 0 & 0 \\ (-A+B) & 0 & (A-B) & 0 \\ 0 & 0 & 0 & 0 \end{bmatrix} \begin{bmatrix} V_{Cb} \\ V_{Cs} \\ V_{0} \\ C_{bk} \end{bmatrix} + \begin{bmatrix} A.R_s \\ B.R_s \\ A(0.5R_s - R_t - D) + B(0.5R_e + R_t + D) \\ 0 \end{bmatrix} I, \tag{6.4}$$

The output $y(t)$ is given as follows:

$$y(t) = \begin{bmatrix} V_{Cb} \\ V_{Cs} \\ V_{0} \\ C_{bk} \end{bmatrix} \tag{6.5}$$

Based on control theories, a lumped network can be written as follows:

$$\begin{aligned} \dot{x}(t) &= Ax(t) + Bu(t) \\ y(t) &= Cx(t) + Du(t) \end{aligned} \tag{6.6}$$

The state variable and substituting all resistor and capacitor values. The values of $A, B, C,$ and D are as follows:

$$A = \begin{bmatrix} -0.001508 & 0.001508 & 0 & 0 \\ 1.6238379 & -1.6238379 & 0 & 0 \\ 1.6223291 & 0 & -1.6223291 & 0 \\ 0 & 0 & 0 & 0 \end{bmatrix} \tag{6.7}$$

$$B = \begin{bmatrix} 0.000005657847553 \\ 0.006089392278651 \\ 0.010542685882214 \\ 0 \end{bmatrix} \tag{6.8}$$

$$C = \begin{bmatrix} 0 & 0 & 1 & 0 \end{bmatrix} \tag{6.9}$$

$$D = \begin{bmatrix} 0 \end{bmatrix} \tag{6.10}$$

The state-space variable is transformed to transfer function, $G(s)$ yielding

$$G(s) = \frac{0.01054s^3 + 0.0171s^2 + 2.981 \times 10^{-5}s}{s^4 + 3.248s^3 + 2.637s^2 - 1.144 \times 10^{-18}s} \tag{6.11}$$

The transfer function of the battery is of higher-order; therefore, the equivalent lower-order models (third, second, and first) is proposed by using the following methodology. The error in the original system and in the lower order is also calculated and presented as minimum as possible.

6.4 Methodology for Model Order Reduction

The BT method is from [18]. The state coordinate basis is selected such that the controllability and observability Gramian is equal to a diagonal matrix, Σ, with the diagonal entries of Σ in decreasing order. The state vector entries contributing the least are associated with the smallest σ_i. Two Lyapunov equations give the relation to system matrices $A, B,$ and C for balanced realization.

$$\begin{aligned} A\Sigma + \Sigma A^T + BB^T &= 0 \\ \Sigma A + A^T \Sigma + C^T C &= 0 \end{aligned} \tag{6.12}$$

Hence, $\Sigma = diad\left[\sigma_i\right]$ and $\sigma_1 \succ \sigma_2 \succ \cdots \succ \sigma_n$. σ_i represents the Hankel singular values of the original system.

Let P be the controllability and Q be the observability Gramian associated with an arbitrary minimal realization, $\{A, B, C\}$, of a stable transfer function. The orthogonal transformations U_c and U_o are given as follows:

$$\begin{aligned} P &= U_c S_c U_T \\ Q &= U_o S_o U_T \end{aligned} \tag{6.13}$$

The construction of the diagonal matrix S_c and S_o is given as follows:

$$H = S_O^{1/2} U^T U_C S_C^{1/2} \tag{6.14}$$

Singular value decomposition is given as follows:

$$H = U_H S_H V_H^T \tag{6.15}$$

The balancing transformation is given by

$$T = U_O^{1/2} S_O^{-1/2} U_H S_H^{1/2} \tag{6.16}$$

The balanced realization is

$$\begin{bmatrix} A & B \\ C & D \end{bmatrix} = \begin{bmatrix} T^{-1}AT & T^{-1}B \\ TC & D \end{bmatrix}$$

$$T^T Q T = T^{-1} P \left(T^T\right)^{-1} S_H \tag{6.17}$$

From a stable system $G \in RH\infty$ and

$$G = \begin{bmatrix} \bar{A} & \bar{B} \\ C & D \end{bmatrix} \tag{6.18}$$

that is a balanced realization denoting the balanced Gramians by Σ, we have

$$\begin{aligned} \bar{A}\Sigma + \Sigma\bar{A}^* + \bar{B}\bar{B}^* &= 0 \\ \Sigma\bar{A} + \bar{A}^*\Sigma + \bar{C}^*\bar{C} &= 0 \end{aligned} \tag{6.19}$$

Now, partition the Gramian system as follows:

$$\Sigma = \begin{pmatrix} \Sigma_1 & 0 \\ 0 & \Sigma_2 \end{pmatrix} \tag{6.20}$$

Also, partition system accordingly:

$$G = \left[\begin{array}{cc|c} A_{11} & A_{12} & B_1 \\ A_{21} & A_{22} & B_2 \\ \hline C_1 & C_2 & D \end{array}\right] \tag{6.21}$$

Then, the equation can be written in terms of their partitioned matrices as follows:

$$\begin{aligned} &\bar{A}_{11}\Sigma_1 + \Sigma_1 \bar{A}_{11}^* + \bar{B}_1\bar{B}_1^* = 0 \\ &\Sigma_1 \bar{A}_{11} + \bar{A}_{11}^* \Sigma_1 + \bar{C}_1^*\bar{C}_1 = 0 \end{aligned} \tag{6.22}$$

$$\begin{aligned} &\bar{A}_{21}\Sigma_1 + \Sigma_2 \bar{A}_{12}^* + \bar{B}_2\bar{B}_1^* = 0 \\ &\Sigma_2 \bar{A}_{21} + \bar{A}_{12}^* \Sigma_1 + \bar{C}_2^*\bar{C}_1 = 0 \end{aligned} \tag{6.23}$$

$$\begin{aligned} &\bar{A}_{22}\Sigma_2 + \Sigma_2 \bar{A}_{22}^* + \bar{B}_2\bar{B}_2^* = 0 \\ &\Sigma_2 \bar{A}_{22} + \bar{A}_{22}^* \Sigma_2 + \bar{C}_2^*\bar{C}_2 = 0 \end{aligned} \tag{6.24}$$

The method is adopted to construct Σ, the most energetic mode of the system is in Σ_1, and the less energetic one is in Σ_2. The system Σ_1 is chosen because its balanced Gramian would be a good approximation of the original system. The procedure to obtain a ROM would be as follows:

1. to obtain the balanced realization of the system and
2. to select an appropriate order r of the ROM and partition these system matrices accordingly.

The method BT from model reducer application is used in MATLAB.

6.5 Result and Discussion

The obtained reduced second- and third-order transfer functions from the BT are as follows:

$$R(s)=\frac{0.01054s^2+0.01712s+2.98e^{-05}}{s^3+3.251s^2+2.645s-1.201e^{-18}} \tag{6.25}$$

$$R(s)=\frac{2.246e^{-07}s^2+0.01054s+1.84e^{-05}}{s^2+1.627s+2.12e^{-07}} \tag{6.26}$$

The bode plot in Figure 6.2 represents the original battery model as well as the reduced third order and reduced second order of the battery.

The characteristic response of the original battery order with reduced order is given in Table 6.1. Table 6.1 gives bode response characteristics of the original battery and the reduced battery in the third order and second order. Interestingly, it is noted that, in the closed-loop system, the proposed reduced order is stable. However, stability margins, i.e., minimum and all, are approximately the same as that of the original battery order. This table justified the definition of MOR, and the proposed reduced order

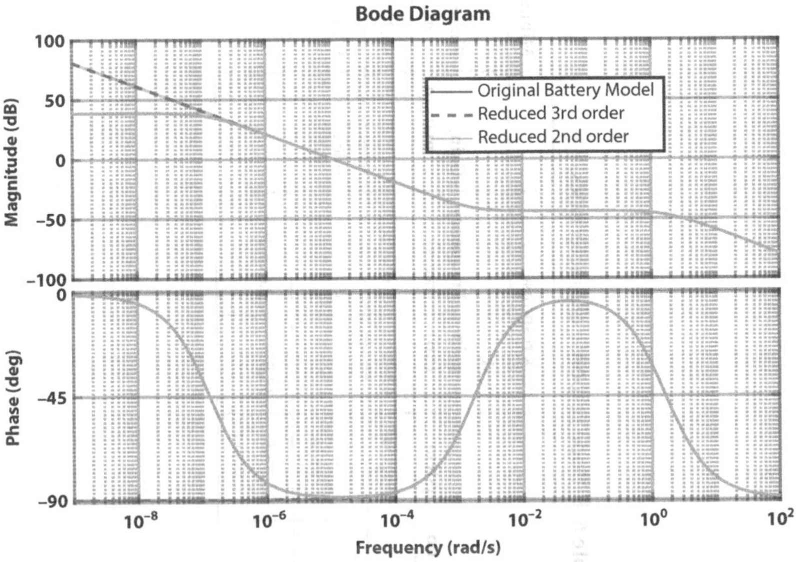

Figure 6.2 Bode plot of original battery model with the third and second reduced orders.

Table 6.1 Bode response characteristics.

		Peak response			Minimum stability margin				All stability margin		
Sr. no	**System**	**Peak gain (dB)**	**Frequency (rad/s)**	**Gain margin**	**Phase margin**	**Delay margin**	**Frequency**	**Closed-loop stability**	**Phase margin**	**Delay margin**	**Frequency**
1	Original	49.4	7.63e-20	-	90.2	1.39e+05	1.13e-05	No	90.2	1.39e+05	1.13e-05
2	Third order		0	−262	90.6	2.76e+05	5.72e-06	yes	90.6	2.76e+05	5.72e-06
3	Second order	38.8	2.61e-19	-	91	1.41e+05	1.13e-05	Yes	91	1.41e+05	1.13e-05

preserves the important characteristics. The step response of the battery with the reduced order is given in Figure 6.3.

Figures 6.4 to 6.7 give the representation of the absolute error and the relative error of the reduced second order and third order. In MOR, there is always a possibility of approximation error. These errors in Figures 6.4 to 6.7 are absolute errors that are the approximate difference between the original and the proposed order. The less is the error, the better is the reduced order. The relative error measures the uncertainty ratio of absolute error and is represented in the figures. The value of absolute error is 0.9903 for the third-order and 0.8715 for the second-order reduced model.

By analyzing the bode plot (Figure 6.2) and the step response (Figure 6.3) of the battery and its reduced order, it is noted that the approximation has an error. The error is absolute and relative. The absolute error is the difference between the original battery value and the true value. The absolute error plot with Hankel singular values is represented in Figures 6.3 and 6.4.

Figure 6.5 represents the absolute error between the original battery models and the obtained reduced order. The error sharply declines from singular values in dB from −252 up to some −303 dB and frequency of 10^{-5} (rad/s) and further goes down to singular values of −380 dB and frequency of 10^{2} (read/s). The Hankel singular value contribution of state energy with respect to state is 3.10×10^{-3}. The reduced order in the second form is stable. The observation of the third order from Figure 6.5 gives the nearly

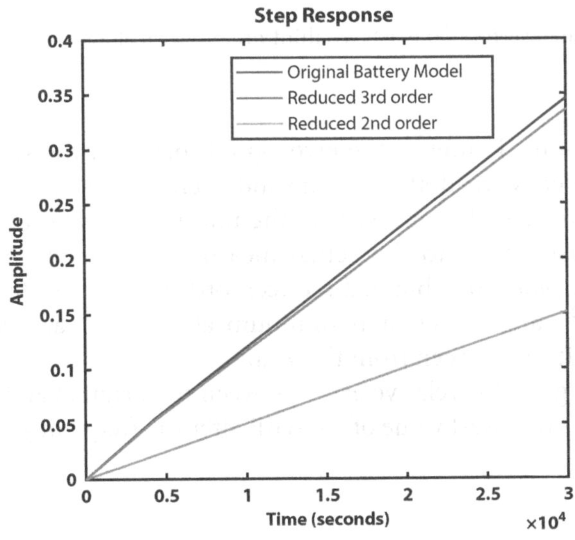

Figure 6.3 Step response of original battery and reduced second and third order.

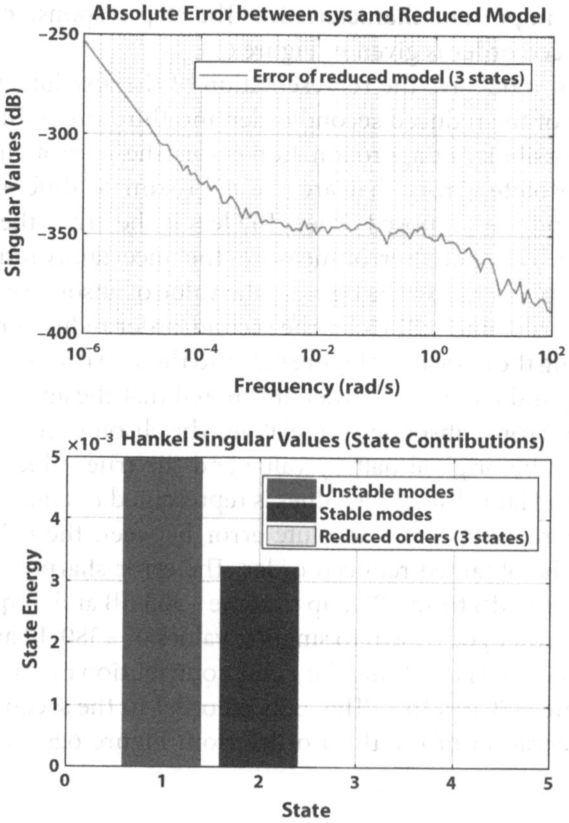

Figure 6.4 Absolute error of the reduced third-order system along with Hankel singular values.

same indications. However, the error graph between the singular values (dB) vs. frequency (read/s) is floating and decreasing.

The relative error is a measure of the uncertainty ratio of the absolute error of measurement to the actual measurement. Figure 6.6 gives the relative error plot. The obtained reduced order has the state energy level of 3.22×10^{-2}, and the error is mininum at −311 dB at a frequency of 40 (rad/s) as it can be seen from the graph.

Figure 6.7 gives the relative error between the original and the reduced order which is of lowest value of −246 dB after the frequency of 10^{-6} rad/s.

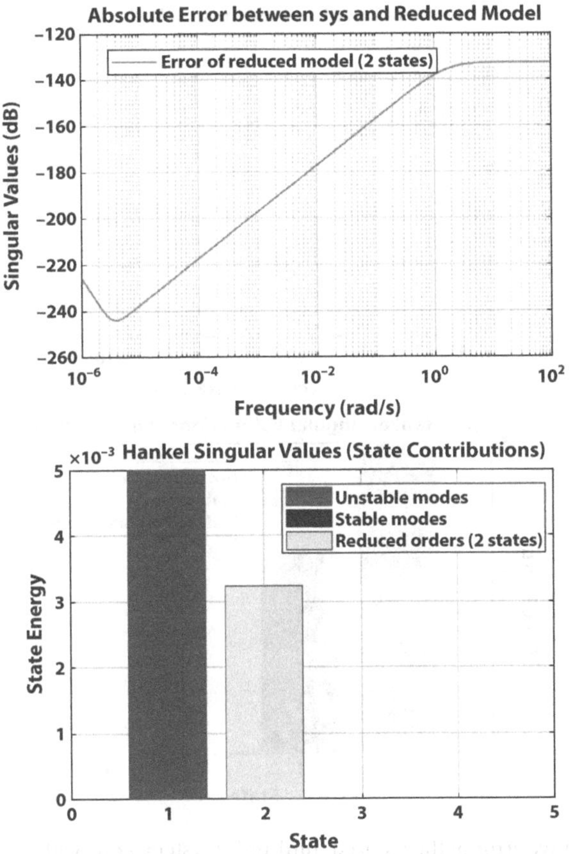

Figure 6.5 Absolute error of the reduced second-order system along with Hankel singular values.

6.6 Conclusion

The BT method is used to reduce the large-scale system and obtained a reduced order. On evaluating the impact of the proposed reduced order of battery, it is clear from Figure 6.2 that it follows the approximate same path as that of the original battery system. The reduced-order battery in the closed loop is stable. The BT uses diagonalization of controllability and observability Gramians for the large-scale model obtained with the frequency-division method. The all-stability margin for the original system including the phase margin is 90.2, 90.6 for the third order, and 91 for the

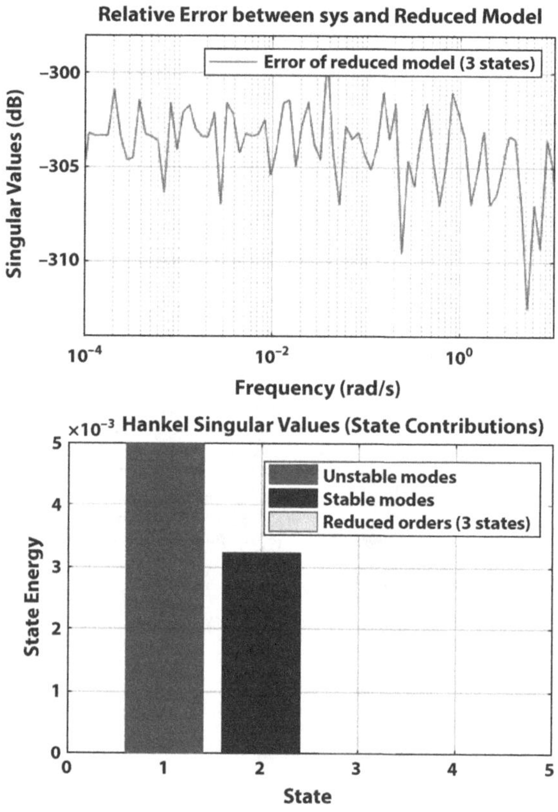

Figure 6.6 Relative error of the reduced third-order system along with Hankel singular values.

second order. Delay margin of the original is 1.39e+05, 2.76e+05 for the third order, and 1.41 e+05 for the second order. The overall study concludes that ROM follows the original order and hence defines accurately the definition of reduced-order system mathematically.

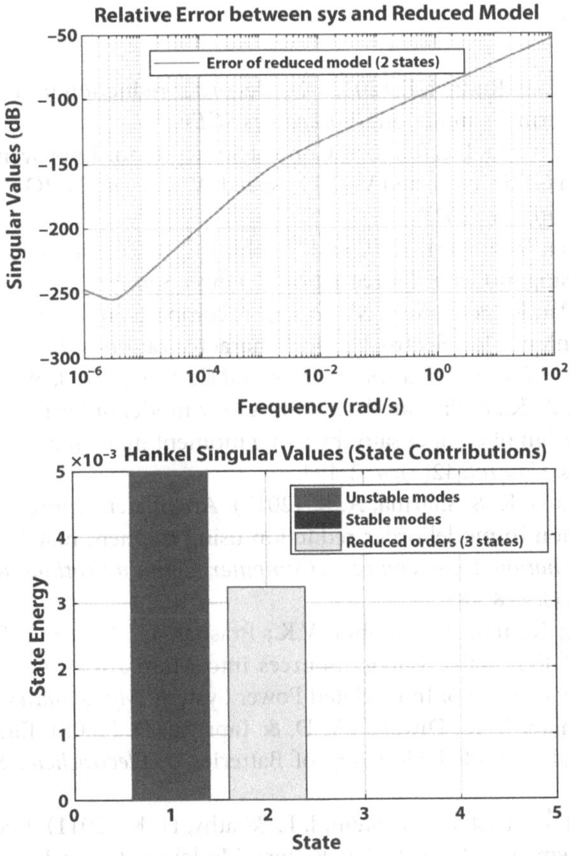

Figure 6.7 Relative error of the reduced second-order system along with Hankel singular values.

Appendix

C_{bk}	$C_{surface}$	R_e	R_s	R_t
888372.83F	82.11F	0.00375Ω	0.00375Ω	0.002745Ω

References

[1] Kostetzer L. & Rudnyi, E. (2012). Model order reduction for efficient battery electro-thermal simulation. *Proceedings SCEE.*

[2] P Singh, K Arora, UC Rathore, Control Strategies for Improvement of Power Quality in Grid Connected Variable Speed WECS with DFIG – An Overview, *J. Phys. Conf. Ser.*, 2022.

[3] Esfahanian, V., Ansari, A. B. and Torabi, F. (2015). Simulation of lead-acid battery using model order reduction. *J. Power Sources*, 279(1), 294-305.

[4] Fan, G., Pan K., & Canova, M. (2015). A comparison of model order reduction techniques for electrochemical characterization of Lithium-ion batteries. *54th IEEE Conference on Decision and Control (CDC)*, 3922-3931.

[5] Prajapati, A. K., & Prasad, R. (2020). A new model order reduction method for the design of compensator by using moment matching algorithm. *Trans. Inst. Meas. Control*, 42(3), 472-484.

[6] Sambariya D. K. & Sharma, A. K. (2017). An efficient approach for stability preservation in model order reduction using moment matching technique. 2017 *International Conference on Computer, Communications and Electronics (Comptelix)*, 478-483.

[7] Arora, K.; Kumar, A.; Kamboj, V.K.; Prashar, D.; Shrestha, B.; Joshi, G.P. Impact of Renewable Energy Sources into Multi Area Multi-Source Load Frequency Control of Interrelated Power System. *Mathematics*, 2021, *9*, 186.

[8] Subramanian, V. R., Diwakar, V. D. & Tapriyal, D. (2005). Efficient Macro-Micro Scale Coupled Modeling of Batteries. *J. Electrochem. Soc.*, 152(10), A2002.

[9] Forman, J. C., Bashash, S., Stein, J. L. &Fathy, H. K. (2011). Reduction of an Electrochemistry-Based Li-Ion Battery Model via Quasi-Linearization and Padé Approximation. *J. Electrochem. Soc.*, 158(2), A93.

[10] Smith, K. A. Rahn, C. D. & Wang, C.Y. (2008). Model Order Reduction of 1D Diffusion Systems Via Residue Grouping. *J. Dyn. Syst. Meas. Contr.*, 130(1), 1-8.

[11] Arora, K.; Kumar, A.; Kamboj, V.K.; Prashar, D.; Jha, S.; Shrestha, B.; Joshi, G.P. Optimization Methodologies and Testing on Standard Benchmark Functions of Load Frequency Control for Interconnected Multi Area Power System in Smart Grids. *Mathematics*, 2020, *8*, 980

[12] Dhaked, D. K., Gopal, Y., & Birla, D. (2019). Battery charging optimization of solar energy based telecom sites in India. *Engineering, Technology & Applied Science Research (ETASR)*, 9(6), 5041-5046.

[13] Kehs, M. A., Beeney, M. D. & Fathy, H. K. (2014). Computational efficiency of solving the DFN battery model using descriptor form with Legendre polynomials and Galerkin projections. *American Control Conference (ACC)*, 260-267.

[14] Fan G. & Canova, M. (2015). Model order reduction of electrochemical batteries using Galerkin method. *Dynamic Systems and Control Conference, Am. Soc. Mech. Eng.*, 57243, V001T13A006.

[15] Fan, G. Pan, K. Canova, M., Marcicki, J. & Yang, X. G. (2016). Modeling of li-ion cells for fast simulation of high C-rate and low-temperature operations. *J. Electrochem. Soc.*, 163(5), A666.

[16] Ting, T., Man, K. L., Zhang, N., Lei, C.-U., &Lu, C. (2014). State-space battery modeling for smart battery management system. *International MultiConference of Engineers and Computer Scientists (IMECS)*, 2(1), 12-15.

[17] Saberi H. & Salmasi, F. R. (2007). Genetic Optimization of Charging Current for Lead-Acid Batteries in Hybrid Electric Vehicles. *International Conference on Electrical Machines and Systems (ICEMS)*, 2028-2032.

[18] Kotsalis G. & Rantzer, A. (2010). Balanced Truncation for Discrete Time Markov Jump Linear Systems. *IEEE Trans. Autom. Control*, 55(11), 2606-2611.

[19] Dhaked, D. K., Gopal, Y., & Birla, D. (2019). Battery charging optimization of solar energy based telecom sites in India. *Engineering, Technology & Applied Science Research (ETASR)*, 9(6), 5041-5046.

[20] Li, Y., Karunathilake, D., Vilathgamuwa, D. M., Mishra, Y., Farrell, T. W., & Zou, C. (2021). Model order reduction techniques for physics-based lithium-ion battery management: A survey. *IEEE Ind. Electron. Mag.*, 16(3), 36-51.

[21] Karunathilake, D., Vilathgamuwa, M., Mishra, Y., & Farrell, T. W. (2020). Capacity loss reduction using smart-battery management system for li-ion battery energy storage systems. *IEEE 29th International Symposium on Industrial Electronics (ISIE)*, 997-1002.

[22] R. Sahani, M. Daga, I. Jamatia, P. Kumar and S. L. Tripathi, Design of a self-charging electric vehicle with embedded multiple natural energy sources, *2023 IEEE Devices for Integrated Circuit (DevIC)*, Kalyani, India, 2023, pp. 296-301, doi: 10.1109/DevIC57758.2023.10134774.

7

Power Electronic Converters for Electric Vehicle Application

P. Swati Patro[1]*, Sarat Kumar Sahoo[1†] and Fernando Yanine[2‡]

[1]*Department of Electrical Engineering, Parala Maharaja Engineering College, Berhampur, Odisha, India*

[2]*Faculty of Engineering, Universidad Finis Terrae, Santiago, Chile*

Abstract

Due to the extensive usage of fossil fuels in the conventional vehicles system the supply is gradually running out. Additionally, improvements in the transportation sector are gradually increasing the need for fossil fuels. Hence, Electric motors are gradually taking the place of Internal Combustion Engines (ICE) in modern vehicles because they are more efficient and emit fewer greenhouse pollutants. The main features of an Electric vehicles are simplicity of operation, reduced number of moving parts that reflect increased efficiency, capability of frequent starting and stopping operation, and the high starting torque.

Power Electronics plays a vital role in this revolution of EVs. Modernized Power Electronic Converters (PECs) and motor drives are crucial to vehicular technology breakthroughs in recent years. The batteries of electric vehicles that are parked or inactive can be charged using the promising technology known as vehicle-to-grid (V2G) to function as distributed resources, allowing them to emit or store energy as needed, creating a bidirectional power is transferred from the DC EV batteries to the AC grid. Power is sent back and forth in this way is made possible with the use of bidirectional power electronic converters that link the grid with the EV battery. The battery pack that drives the propulsion of electric vehicles (EVs) is dependable, powerful, and high-energy.

Automobile system has become very much advanced due to the applications of luxurious loads demanding high power with individual voltage ratings, leads to involvement of PECs in EV systems which also act as power modulator according

**Corresponding author*: patropswati03@gmail.com
†*Corresponding author*: sksahoo.ee@pmec.ac.in
‡*Corresponding author*: fyanine@uft.cl

Krishan Arora, Suman Lata Tripathi and Himanshu Sharma (eds.) Electric Vehicle Design: Design, Simulation and Applications, (147–166) © 2024 Scrivener Publishing LLC

to the supply and load requirements. Conventional EVs are only viewed as a load on the power grid. The development of power electronic converters that are appropriate for electric vehicles. With a focus on power electronics for EV propulsion, battery charging, and power accessories, this article has assessed the current state of interdisciplinary technologies in EVs. It suggests that power electronics technology is crucial to the advancement of electric vehicles. EV market will grow significantly in the upcoming years.

The main objective of this chapter is to review the importance of Power Electronics converters in Electric Vehicles. This chapter consists of six sections as follows:

1. Introduction
2. Types of electrical vehicle and role of power electronics converter
3. Recent development in power electronics for EV applications
4. Power electronics converters in electric, hybrid and fuel cell vehicle
5. Challenges in power electronics for EV applications
6. Conclusion

***Keywords*:** Power electronics converter, electric vehicles, DC to DC converters, internal combustion engine, solar cell, grid to vehicle, electric motor

7.1 Introduction

Energy consumption is increasing day by day as there is an increase in human population. Vehicles using petrol and diesel can also be considered as one of the main reasons for the increase in energy consumption. Due to the extensive usage of fossil fuels in conventional vehicles, their efficiency is around 20% because the rest of the energy is dissipated to the surrounding atmosphere in the form of heat; moreover, the supply of fossil fuels is gradually running out. Nevertheless, improvements in the transportation sector are gradually increasing the need for fossil fuels. Large-scale global automobile usage gives rise to serious problems in human life as well as in the environment like air pollution and global warming. The petroleum resources on the earth are depleting gradually, leading to a critical issue. This can be considered as a serious drawback that requires looking for an environmentally friendly, greener alternative. Hence, electric motors are gradually taking the place of internal combustion engines (ICEs) in modern vehicles because they are more efficient and emit comparatively fewer greenhouse pollutants [2]. The main features of electric vehicles (EVs) are simplicity of operation and decreased quantity of moving components that translate to greater economy capacity of frequent starting and stop operations and high starting torque. Electric vehicles can be considered as a replacement of vehicles using ICEs because of their advantages like higher efficiency (85%–90%), less noise, and effective reduction in the production of greenhouse gases, as well as regenerative braking capabilities.

Coming to the historical background of EVs, the first battery electric vehicle (BEV) was developed in 1828, but there was no rechargeable EV until 1881 before the secondary battery invention. By 1900, EVs were made by companies in countries like France, the US, and England, and, as their popularity begun to grow, production increased slowly [3]. In 1912, the number of manufactured EVs registered in the US was 33,842. ICE vehicles took off shortly after this peak, due to, first, the development of the Ford Model T and, second, to the development of the automobile starter motor. The revitalization of EVs around 1950 was due to the need to reduce smog, air pollution, etc.

Electric vehicles are of various types such as BEV, plug-in hybrid electric vehicle (plug-in HEV), hybrid electric vehicle (HEV), fuel cell electric vehicle (FCEV), and solar cell electric vehicle (SCEV). With the development of the majority of renewable energy sources, the EV industry will eventually see a large increase in demand. In recent years, EVs like cars, trucks, vans have drawn a lot of attention and have become increasingly popular.

In recent years also, there has been tremendous progress in the development of EVs. According to the International Energy Agency (IEA) reports, recent years have seen the rapid growth in EVs. Figure 7.1 shows the worldwide sales of EVs from 2017 to 2021. The sale of EVs in 2017 was 1.18 million globally, which increased to 6.6 million in 2021. In just a year, 2020–2021, EV sales rose to 120%. Additionally, according to the IEA EV data explorer report, 12,000 electric cars (BEVs) were sold in 2021 in India particularly, which is approximately four times the 3,100 that were sold in 2020. This shows how fast the EV market is growing in the world as well as in India. This will reduce the price of the battery in the upcoming years.

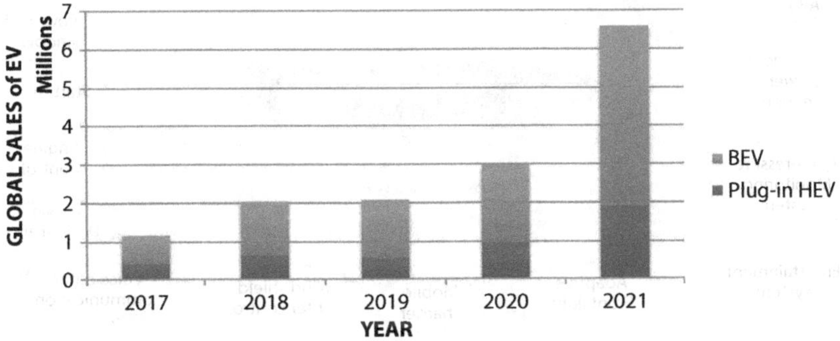

Figure 7.1 Sales of EVs (plug-in HEV and BEV) from 2017 to 2021 in the world.

Power electronics (PE) and electric motor drives are used in EVs to regulate the flow of energy from the source to the load and vice versa, i.e., the flow of electrical energy from grids that have charging stations externally to the vehicle or from vehicles to the charging stations. Also, it controls the energy flow within the vehicle. Power electronics has a vital role in the revolution of EVs. Modernized power electronic converters (PECs) and motor drives have been crucial to vehicular technology breakthroughs in recent years. The batteries of EVs that are parked or inactive can be charged using the promising technology known as vehicle-to-grid (V2G). This will enable them to function as distributed resources that can emit or store energy as needed, creating a bidirectional power that is transferred from the DC EV batteries to the AC grid. Power sent back and forth in this way is made possible with the use of bidirectional DC-to-DC converters that link the grid to the battery of EV. The battery pack that drives the propulsion of EVs is dependable, is powerful, and is of high energy.

An automobile system has become very advanced due to the applications of luxurious loads demanding high power with individual voltage ratings. This leads to the involvement of PECs in EV systems that also act as power modulator according to the supply and load requirements. Traditional EVs are considered to be a load on the electrical grid. The development of PE converters is appropriate for EVs. This article evaluates the present state of multidisciplinary technologies in EVs, with special importance on converters for driving EV, powering various accessories and battery charging,

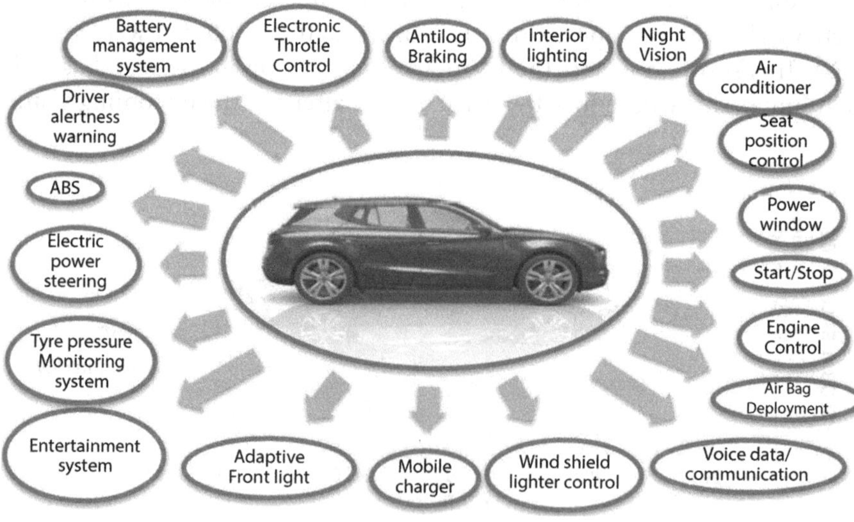

Figure 7.2 The various electrical/electronic appliances (luxurious loads) in EV.

and it assesses the current state of interdisciplinary technologies in EVs. Electrically driven systems replace hydraulic systems (e.g., brake) and mechanically driven systems (e.g., steering), which makes for a more effective and safer vehicle. Figure 7.2 shows the schematic diagram for various appliances used in EVs.

The advancement in automobile vehicular systems like high-power head lamp, power windows, and auto start-ups can be considered as luxurious loads that demand high powers that are possible through PECs. It suggests that PE technology is crucial to the advancement of EVs. Research on converter technology for EVs must be very alluring because the EV market will grow significantly in the upcoming years.

The main objective of this chapter is to review the importance of PECs in EVs. This chapter consists of seven sections as follows:

Section 7.1 Introduction

Section 7.2 Types of electrical vehicle and role of power electronic converter

Section 7.3 Recent development in power electronic converter

Section 7.4 Power electronic converters in electric hybrid and fuel cell vehicle

Section 7.5 Challenges in power electronic vehicular system

Section 7.6 Conclusion

Section 7.7 References

7.2 Types of Electrical Vehicle and Role of Power Electronic Converter

More electric car types are constantly becoming available, and the market for them is increasing. Nowadays, different types of EVs like scooter, sporty bikes, SUVs, pickup trucks, and buses are available in the market [3]. But for the different kinds of EVs, handling options are also different, according to the configuration. Some run solely on battery power, whereas some HEVs include an ICE and an electric motor through a variety of ways. Solar-powered and fuel cell electric automobiles are also available. Considering different factors, EVs are mainly classified as follows:

- Battery electric vehicles (BEVs)
- Plug-in hybrid electric vehicles (plug-in HEVs)
- Hybrid electric vehicles (HEVs)
- Fuel cell electric vehicles (FCEVs)
- Solar cell electric vehicles (SCEVs)

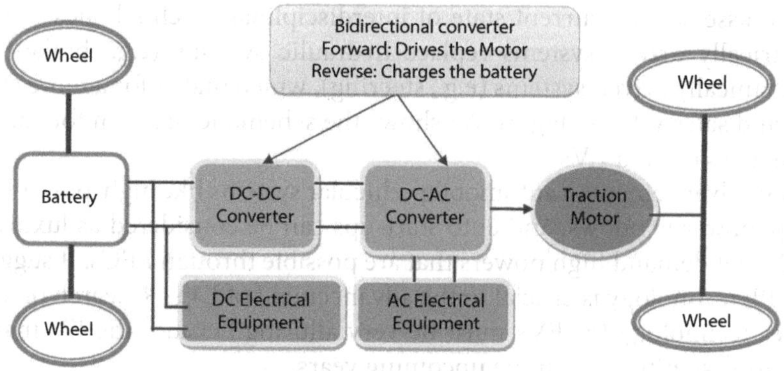

Figure 7.3 Block diagram of battery electric vehicle (BEV).

7.2.1 Battery Electric Vehicles (BEVs)

Battery electric vehicles (BEVs), often known as all EV/pure EV, have no gasoline engine and run solely on electricity. Figure 7.3 shows a block diagram for a BEV. The rechargeable batteries inside BEVs power one or more electric motors. Lithium batteries are used in almost all modern vehicles. The automobile sector prefers lithium because it has a high power–to-weight ratio, a wider range of uses, and a lower total footprint. An electric car charger provides power to charge the battery packs that is known as an EV charger. These can be classified into many types based on the charging speeds, with level 1 being the slowest and level 3 being the fastest.

In contrast to vehicles powered by gasoline, BEVs are emission-free vehicles that do not produce any hazardous gases or induce any pollution in the air. Despite the introduction of new models with longer ranges, the BEVs used nowadays have a driving range of 50 to 350 miles when charged once.

7.2.1.1 Power Electronics in BEVs

A unidirectional DC-to-DC boost converter is used in a BEV to raise voltage in response to the requirements of the propulsion system and the electrical load. The battery provides electricity to low-voltage or power-rated devices like mobile chargers. The step-up converter used in a BEV transmits the battery's DC voltage to an HVDC bus. Projector lamps and other equipment with higher voltage ratings are powered by the HVDC bus. To be able to operate an AC motor, the DC-to-AC inverter must convert DC voltage with a fixed value into a three-phase AC with variable voltage and frequency. The battery's DC power source is used as fuel to run the vehicle and other electronic devices. To maintain the battery's charging level between the highest and lowest values, a voltage controller regulates DC-to-DC converter.

7.2.2 Plug-In Hybrid Electric Vehicles (Plug-In HEVs)

Plug-in hybrid electric vehicles (plug-in HEVs) have the components that combine an ICE that refuels the batteries with an electric motor to extend their driving ranges [12]. Depending on the vehicle model and battery capacity, the plug-in HEV normally operates on power that is stored in the battery until it is completely depleted, usually for 10 to 50 miles. After the power is fully depleted, the vehicle switches to the ICE automatically and may cover a distance of several miles on a single tank of petrol. This can significantly cut down on fuel use and emissions, especially for short distances. A plug-in hybrid vehicle can function only on fuel when power is not available [13]. The majority of plug-in HEVs batteries are charged by using either a level 1 type EV charger or level 2 type EV chargers. Level 3 DC rapid chargers are not always possible.

7.2.2.1 Power Electronics in Plug-In Hybrid Electric Vehicles (Plug-In HEV)

When an EV is of the plug-in HEV type, power is required for battery charging as well as to propel it. As a plug-in HEV, energy injection can also be done to the grid from the battery by regenerative braking operation. 3-Ø/1-Ø rectifiers can be used to convert fixed AC to variable DC so that the battery can be charged through a DC-to-DC buck converter. The flow of energy in either direction can be controlled by bidirectional DC-to-DC converters, whose mode operation is controlled by voltage controllers as V2G mode requires step-up while G2V mode requires step-down converter.

7.2.3 Hybrid Electric Vehicles (HEVs)

Hybrid electric vehicles, also known as self-charging HEVs, draw power from a battery pack that combines an electrical motor and an ICE. An HEV cannot be charged by connecting it to an EV charger, which is the primary difference between a plug-in HEV and an HEV. As an alternative, the engine powered the battery pack, and regenerative braking was generated. The electric motor's greater power may enable the car to use a smaller engine. When compared to a car with a traditional engine, the battery pack's ability to function as an auxiliary battery and vehicle helps to reduce its running, when it is stopped, which improves the fuel efficiency and lowers emissions [9]. A HEV can combine the power of an ICE and an electric motor in a variety of ways. According to this, HEVs are again classified into three types: series hybrid EV, parallel hybrid EV, and series–parallel hybrid EV.

7.2.3.1 Series Hybrid Electric Vehicles (SHEVs)

The electric motor is used only to turn the wheels in series hybrids, commonly called range extended EV/extended range EV. In a SHEV, the ICE and battery are both simulated, so it can produce the required power to move the vehicle and all other electric or electronics equipment around it. In a SHEV, a generator is used to produce electrical energy from the mechanical energy of ICE. Using an AC-to-DC converter, we transferred the generated AC power into DC electricity to charge the battery [4]. The block diagram for SHEV is shown in Figure 7.4.

7.2.3.2 Parallel Hybrid Electric Vehicles (PHEVs)

When operating in parallel hybrid mode, the vehicle is propelled by both the engine and the electric motor. In a PHEV, to produce traction power, an ICE and an electric motor are connected in parallel. Here, it uses an ICE as its main source of power, with a motor acting as a backup. This strategy reduces the drive train's fuel usage [4]. Using this drive train has the major advantage that a little battery backup is required for lowering the cost of operating the car. The efficiency of a PHEV is higher than a SHEV because ICE drive train has a single-step intermediate conversion stage. A PHEV is a practical option for long-distance range because there is no intermediate

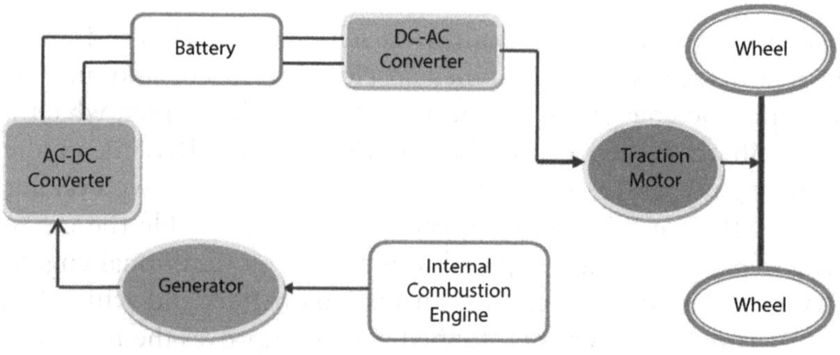

Figure 7.4 Block diagram of series hybrid electric vehicle (SHEV).

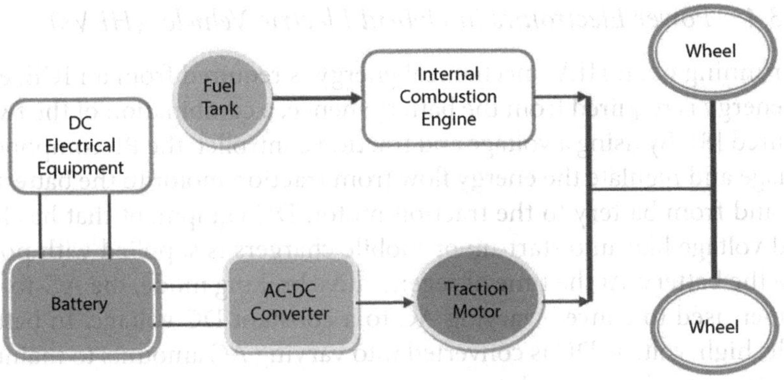

Figure 7.5 Block diagram of parallel hybrid electric vehicle (PHEV).

conversion state like in a SHEV. The block diagram of a PHEV is shown in Figure 7.5.

7.2.3.3 Series–Parallel Hybrid Electric Vehicles (SPHEVs)

The SHEV and PHEV configurations are combined in the series–parallel hybrid electric vehicle (SPHEV). As opposed to a PHEV, in a SPHEV, the generator is placed between the battery and the ICE for battery charging, and, in contrast to a SHEV, mechanical shaft and ICE are directly connected for the purpose of driving the vehicle [4]. The block diagram for SPHEV is shown in Figure 7.6.

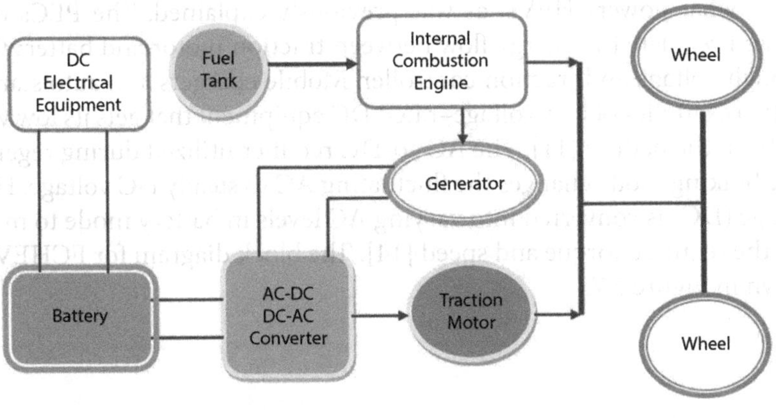

Figure 7.6 Block diagram of SPHEV.

7.2.3.4 Power Electronics in Hybrid Electric Vehicles (HEVs)

For running of an HEV, mechanical energy is required from an ICE, electric energy is required from the battery; hence, a combination of the two is required [8]. By using a voltage and traction controller, the PE components manage and regulate the energy flow from traction motor to the battery or ICE and from battery to the traction motor. DC equipment that has low-rated voltage like auto start-up or mobile chargers is supplied with power from the battery. At the time of regenerative braking mode, the AC-to-DC rectifier used to convert varying AC to a constant DC voltage. In battery mode, high-voltage DC is converted into varying AC amounts to maintain necessary speed along with torque.

7.2.4 Fuel Cell Electric Vehicles (FCEVs)

Fuel cell electric vehicles, which run on hydrogen, have no exhaust emissions. They run on a similar principle to that of BEVs. An FCEV stores energy as hydrogen, in which the fuel cell then converts to electricity to power the vehicle. FCEVs include a gas/fuel tank that is used to store pure hydrogen [10]. It requires a few minutes to fill the tank, exactly like a modern ICE vehicle. FCEVs also utilize regenerative braking, to charge the battery. Overall, fewer FCEVs are available in the market, and also the refueling infrastructure is not available broadly; hence, more support is needed for a mass launch of FCEVs.

7.2.4.1 Power Electronics in Fuel Cell Electric Vehicles (FCEVs)

The ICE's mechanical energy, the battery's electric energy, or a mix of the two is what powers HEVs, as was previously explained. The PECs control and regulate the energy flow between traction motor and battery/ICE through voltage and traction controller. Mobile chargers and auto starters are two examples of low-voltage–rated DC equipment that gets its low voltage from the battery [11]. The AC-to-DC rectifier utilized during regenerative braking mode changes the fluctuating AC to steady DC voltage. High voltage (DC) is converted into varying AC levels in battery mode to maintain the required torque and speed [14]. The block diagram for FCHEVs is shown in Figure 7.7.

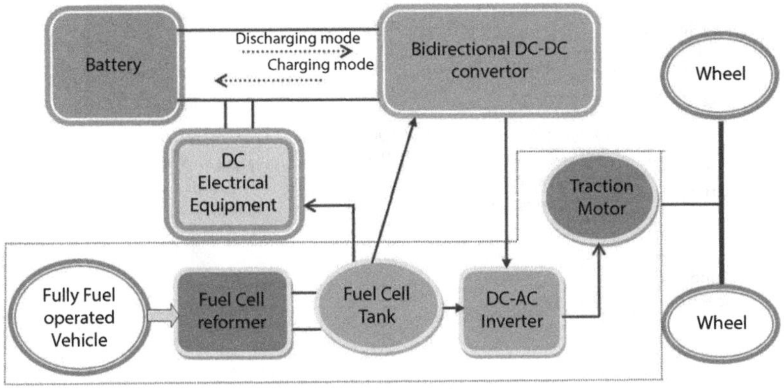

Figure 7.7 Block diagram of fuel cell hybrid electric vehicle (FCHEV).

7.2.5 Solar Cell Electric Vehicles (SCEVs)

Electric vehicles, known as SCEVs, employ photovoltaic (PV) cells, which are similar to those found in solar panels, to convert solar energy into electricity. When compared to a regular BEV, the battery pack is recharged using solar energy from the sun, extending the vehicle's range. Solar energy produced is free of cost and is completely free, so the ownership cost of SCEVs is potentially reduced. Nevertheless, for SCEVs to function as an all-electric car, the sun must be inside and not out. Level 3- to level 1-type EV chargers can be used to charge SCEVs [16].

7.2.5.1 Power Electronics in Solar Cell Electric Vehicles (SCEVs)

Electricity is transformed to different parts through different forms using PE. A commonly used PEC is an inverter. It transforms electricity produced in the form of DC by solar PV panels into the electrical energy in the form of AC, used in electrical grid. An additional illustration is a bidirectional converter, which transforms energy in the form of DC from battery storage into AC electrical energy during discharge for use on grid and from AC to DC during battery charge.

The electric power system contains a wide variety of additional PE equipment. Converters are needed for the generated electrical energy from PV to pass through. More inverters are being connected to the grid in comparison to other types of EVs. Due to additional inverters, as additional solar energy systems, the device becomes vital to ensure that the robust grid is stable. One type of converter that is widely used in renewable energy systems, such as grid-connected DC microgrids and grid-connected solar

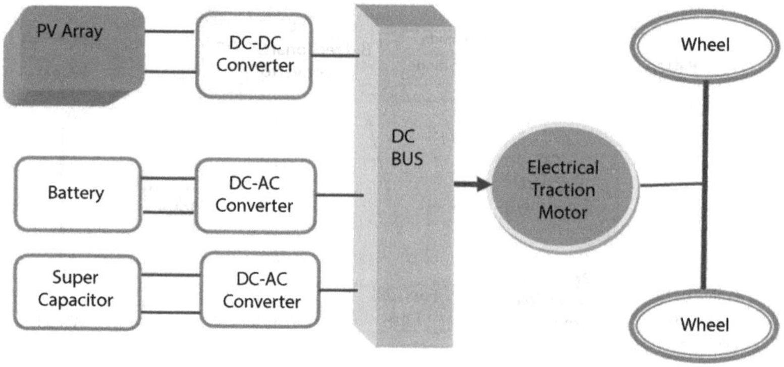

Figure 7.8 Block diagram of solar cell hybrid electric vehicle (SCEV).

PV energy conversion systems, is the AC-to-DC converter, also referred to as a rectifier. The block diagram for a solar electric vehicle is shown in Figure 7.8.

7.3 Recent Development in Power Electronic Converter

Power electronics plays a vital role for an electric drive system to make it more affordable and effective. Due to this, more research and developments are going on. The main objective is to reduce the cost and enhance the performance of a converter in an EV system. In plug-in HEV and HEV, the flow of electrical energy is processed and controlled primarily by vehicle PE. Moreover, they manage the motor's speed and the required torque as output [2]. Last but not least, the lighting system, entertainment system, heating and ventilation system, and chargers, in which PE distribute and convert the required electrical power, are all automotive systems; hence, DC-to-DC converters and inverters are essential parts of PE for plug-in EVs.

An inverter is required to convert DC energy from a battery to AC power to drive the motor. An inverter functions as a motor controller, a filter, and a protective barrier against stray currents that can damage the battery. For this reason, upgrading of inverters in PE is going on day by day in various research and development projects. Continuously upgrading the inverters results in reduction of the number of inverters being used; hence, the size and cost are being decreased significantly. These days, mainly silicon-based semiconductors are used in EVs as a component of PE, but it is found that wide-bandgap (WBG) semiconductors are more efficient as they are able to withstand higher temperature than a silicon component [1]. The ability to operate at higher temperature can reduce the system cost. For implementation of these features, a less complex thermal management system is needed. Due to the above advantages, WBG has a significant potential to fulfill the goals partially.

Silicon carbide and gallium nitride are the two most widely used WBG semiconductors. Recent research development has been carried out on these semiconductors. The motors and vehicle system that are running at a higher voltage range, typically from 220 V to 450 V, require DC-to-DC converters to boost the battery voltage. It means that, when the internal permanent magnet of motor requires a greater voltage, a boost DC-to-DC converter is needed [15]. The motors and vehicle system that are running at lower voltage range, typically from 15 V to 50 V, require DC-to-DC converters to buck the battery voltage. It means that, when the internal permanent magnet of the motor requires a lower voltage, a buck DC-to-DC converter is needed, such as the majority of automotive systems (lighting system, entertainment system, etc.).

When battery recharge is needed, onboard vehicle chargers that transform the AC power into the DC power from the power grid are used. Currently, for plug-in electric vehicles, battery chargers are using a high-frequency charger circuit, which can be either mounted on the vehicle or may be used as a part of DC fast charger for the vehicle. Researchers are also investing in onboard solutions, so that recharge uses the inductance of electric motor and includes the charging function in the present PE. These strategies reduce parts that are required in numbers as well as the weight, price, and volume of current chargers. Chargers, like other converters, must be lightweight, have a small physical footprint and offer high durability and efficiency with a reasonable price. Power semiconductor technologies, powerful computer aided design (CAD), manufacturing technologies, and complicated converter topologies all contribute to the advancement of PE in electrical vehicles. Power semiconductor device technology has evolved dramatically in recent decades. The power rating and performance of these power devices have risen as a result of an evolutionary process. The insulated-gate bipolar transistor (IGBT), static-induction transistor, static-induction thyristor (SIT), bipolar-junction transistor (BJT), power metal oxide semiconductor (MOS) field-effect transistor (MOSFET), gate turn-off thyristor (GTO), and MOS-controlled thyristor (MCT) are some of the recently introduced PE devices. In comparison to the BJT and MOSFET equivalents, the IGBT, which combines properties of bipolar and field-effect transistors, has clear advantages. Also, the recently released MCT is so distinct that it combines the possibility for high voltage and current ratings with the high switching speed.

To attain high power density, high efficiency, and reliable power converters, the progress of power converter topologies typically follows that of power semiconductor devices as shown in Figure 7.9. Typically, DC commutator motors and AC motors are powered by DC-to-DC converters (also known as DC choppers) and DC-to-AC converters (also known as inverters).

The newest topologies in inverters, known as the resonant DC link inverter, provide another option to traditional pulse width–modulated inverters. To decrease switching losses, resonant DC link inverters comprise either a parallel or a series resonant circuit that produce either a zero-voltage–switching or zero-current–switching state [6]. Moreover, the recently proposed inverters are quasi-resonant DC link inverters, which are another possible converter topology also known as high-frequency resonant AC link inverter. By using this topology, we are able to minimize the switching losses by only allowing switching instants that occur at link-voltage zero-crossing points [7].

An EV uses numerous sources of energy, such as supercapacitors, batteries, and fuel cells, to power it, as shown in Figure 7.9. In an electric car, one or more energy storage devices assist the primary energy source. The EV's cost, volume, and mass may all be decreased, whereas its performance is considerably increased.

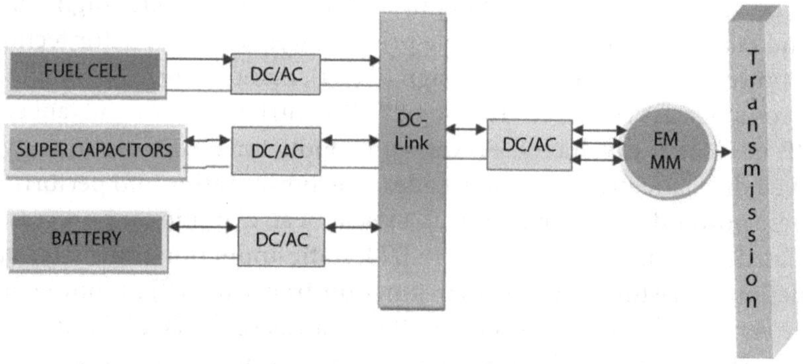

Figure 7.9 Block diagram of power converter topology.

7.4 Power Electronic Converters in Electric, Hybrid, and Fuel Cell Vehicles

7.4.1 Power Electronic Converters in EVs

To perform the functions of electrical and electronics load and to drive the vehicle, EVs use fuel cell, ultra-capacitors, batteries, or ICE for the power supply. Nowadays, solar cell EVs are also used. In EVs, the unidirectional boost converter can be used to boost the voltage for driving the vehicle as voltage is not sufficient to drive in fuel mode. Different electrical loads require different voltage ratings. Among the various luxurious loads, some require 42 V, whereas other loads require 12 V for its operation. As many different-rated loads are implemented in a vehicle, they require DC-to-DC converters proportionally. It may be considered as a drawback to supply the voltage to the various loads according to their requirement by single battery; the efficiency also drops. A hybrid automobile system uses two forms:

1. motor vehicle that runs on ICE or a fuel cell with one battery (36 V) and
2. motor vehicle that runs on ICE or a fuel cell with a dual battery (14 V and 42 V).

A low-voltage application uses a 12-V battery, a medium-voltage application uses a 36-V battery, and a high-voltage application uses a 42-V battery for its operations. Subject to the type of EV, dual voltage can be obtained from either from generator or from the grid for a double-battery system.

Driving operation of FCEV and plug-in HEV and charging of the batteries of SHEV and SPHEV can be done by the generated voltage of generator either rectifying through a rectifier or by a unidirectional DC-to-DC converter. A bidirectional DC-to-DC converter boosts the voltage and uses 3-θ inverter for variable voltage and frequency. During regenerative braking (3-θ inverter → rectifier) and deceleration, wasted energy can charge the battery. Battery and fuel, in combination and in split mode, have PECs, as inverter may be a single-phase or three-phase; however, regenerative mode has PECs as rectifier. Traction control unit selects the control mode. Mode of operation as inverter or rectifier is determined by control pulse.

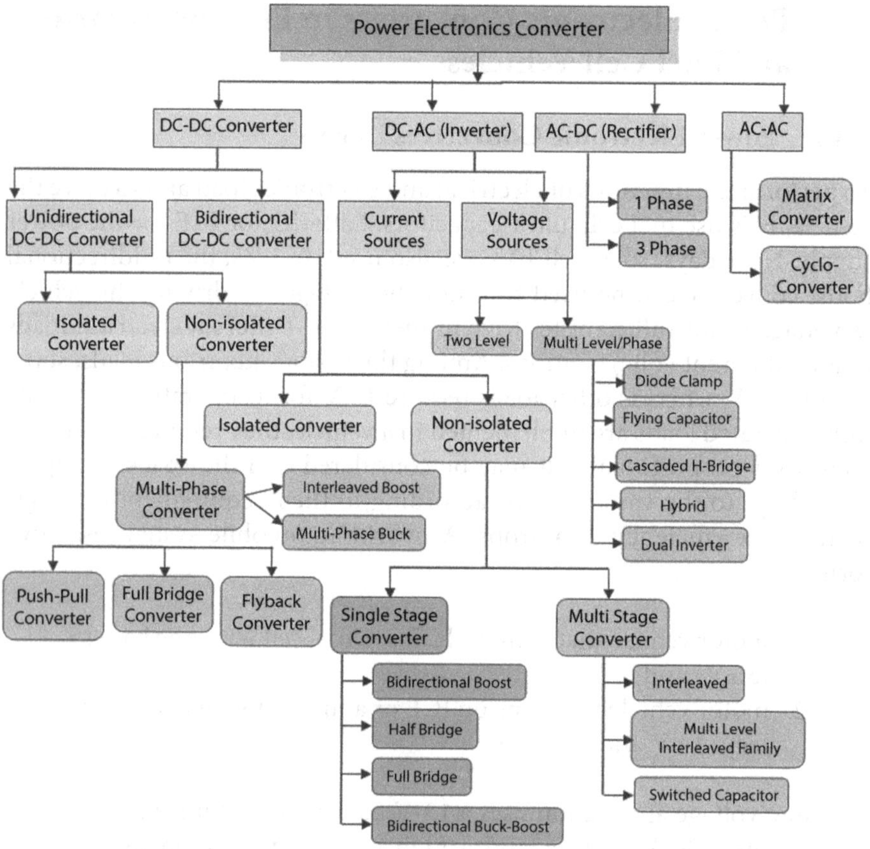

Figure 7.10 Categorization of power electronic converters.

7.4.2 Categorization of Power Electronic Converters

Figure 7.10 shows the categorization of PECs. EVs use a bidirectional DC-to-DC converter commonly for their various applications. Figure 7.10 shows various types of DC-to-DC, DC-to-AC, AC-to-DC, and AC-to-AC converter classification. This converter can step down or step up the voltage with non-inverting output as per switching mode. Power flow in both directions can also be achieved by multi-stage, multi-input, single-input, and multi-phase non-isolated converter.

7.5 Challenges in Power Electronic Vehicular System

Power electronic converters play an important role in EVs, so it is obvious that PECs have to deal with the various challenges in vehicular systems [5].

7.5.1 Efficiency

The electromagnetic interference effects, the number of luxurious loads, and the control strategies decide the efficiencies of PECs. Electric motors were replaced by hydraulic and mechanical shafts for propelling vehicles with proper PEC selection for perfect combination of fuel cell, battery, and ICE, which can improve efficiency. Switching methods of PECs, individual unit pacing and integrations, are key factors to improve efficiency.

7.5.2 Longevity

Switches of PECs decide its durability. By steadily managing the charging levels and discharging level with the aid of the voltage controller, the battery's longevity rises. The converter should be able to withstand high vibration and extreme heat conditions. The difficulties come from choosing a converter with the right combination of efficiency, rigidity, affordability, and size. Power wastage can be avoided, and, hence, longevity can be increased by maintaining battery state of charge provided continuously to the voltage controller and comparing with the reference voltage.

7.5.3 Performance of EV

PE approach is used with digital signal processing (DSP) to obtain performance improvement and also for industrial motion control, which are a trend in high-power applications.

7.5.4 Luxurious Features

The modernized EVs in the automobile sector have increased demand of vital luxurious mobile charger, air conditioner, air bag deployment, seat position control, driver alertness warning, interior lighting, etc., which cannot be operated with a single battery as different loads require different voltage. This problem can be overcome by a multi-output or multi-stage DC-to-DC converter.

7.5.5 Safety

The automobile sector requires advancement in all possible ways as the matter of safety comes into the picture. Even if propelling control and power conversion are essential factors like detection of failure of rotor, bearing and stator are needful. DSP in integration with PEC can raise the level of safety by including air bags and ABS as they need high-power actuators.

7.5.6 Overall Cost

The rising number of luxurious loads in EVs will raise the number of PECs, which, in turn, may increase the cost. Choosing fewer power conversion units for a larger proportion of expensive loads can lower the cost of the vehicle. Use of EVs in the transportation sector [17] greatly reduces the emission of GHGs, which, in turn, reduces the cost of transportation that is a major element in the reduction of the cost of EVs.

7.5.7 Noise

All the EVs produce less noise due to the replacing of mechanical systems by the electrical system. So, the EVs on the roads may not be heard by pedestrians, and this may cause problems. However, the problems can be overcome by an acoustic vehicle alerting system that emits sounds so that pedestrians can be warned.

7.6 Conclusion

This chapter encapsulates the various types of PECs; classification of EVs; and the use of PECs in BEVs, plug-in HEVs, HEVs, FCEVs, and SCEVs. The role of PEC in every type of EV has been discussed briefly. This also summarizes the recent developments of PE in the EV sector. Due to the advancement of EVs, luxurious loads increase, resulting in an increase of PECs proportionally as different electrical loads require different voltage ratings that cannot be supplied by a single battery. A hybrid automobile system uses two vehicular systems (single-battery and double-battery) with a converter for different voltage. Most of the EVs use a bidirectional DC-DC converter for driving and braking function. The challenges that arise in the automobile vehicular system caused by the application in PECs in EV can be resolved, and this can be considered an advantage. Factors like safety, performance, and cost are the challenges that can be overcome by the use of PECs.

References

[1] W. Choi, D. Han, C.T. Morris, B. Sarlioglu, "Achieving high efficiency using SiC MOSFETs and reduced output filter for grid-connected V2G Inverter", *IECON 2015 - 41st Annu. Conf. IEEE Ind. Electron. Soc.*, 2015, pp. 3052–3057.

[2] C. C. Chan and K. T. Chau, "Electric Vehicle Technology - an Overview of Present status and Future Trends", *Proceedings of the 11th International Electric Vehicle Symposium*, vol. 1, September, 1992, no. 1.02.

[3] Arora, K.; Kumar, A.; Kamboj, V.K.; Prashar, D.; Jha, S.; Shrestha, B.; Joshi, G.P. Optimization Methodologies and Testing on Standard Benchmark Functions of Load Frequency Control for Interconnected Multi Area Power System in Smart Grids. *Mathematics*, 2020, *8*, 980.

[4] Akbarian, H., Pillay, P., & Lopes, L. "Design of a power electronic emulator for series hybrid electric vehicle powertrains", In: *Proceedings of the IEEE Vehicle Power and Propulsion Conference (VPPC)*, 2015, pp. 1–5.

[5] A.K. Jaura, A.G. Grabowski, J.M. Miller; "Developmental Challenges of a Hybrid Electric Technology Demonstrator Vehicle", Global Powertrain Congress, GPC2002 Alternative Fuel/Lightweight Vehicle Public Show, Sheraton Inn, Ann Arbor, MI., 2002, pp. 24-26.

[6] C. P. Henze, H. C. Martin, and D. W. Parsley, "Zero-voltage switching in high frequency power converters using pulse width modulation," in *Applied Power Electronics Conference and Exposition APEC '88*, New Orleans, LA, 1988, pp. 33–40.

[7] D. M. Divan, "The Resonant DC Link Converter – A New Concept in Static Power Conversion", *IEEE Trans. Ind. Appl.*, vol. 25, no. 2, March/April, 1989, pp. 317-325.

[8] Arora, K.; Kumar, A.; Kamboj, V.K.; Prashar, D.; Shrestha, B.; Joshi, G.P. Impact of Renewable Energy Sources into Multi Area Multi-Source Load Frequency Control of Interrelated Power System. *Mathematics*, 2021, *9*, 186.

[9] J. Drobnik and P. Jain, "Electric and hybrid vehicle power electronics efficiency, testing and reliability," in: *Proc. IEEE EVS27*, Nov. 2013, pp. 1–12.

[10] O. Hegazy, J. Van Mierlo, and P. Lataire, "Analysis, modeling, and implementation of a multidevice interleaved DC/DC converter for fuel cell hybrid electric vehicles," *IEEE Trans. Power Electron.*, vol. 27, no. 11, Nov-2012, pp. 4445–4458.

[11] Van Mierlo, G. Magatto, and P. Van den Bossche, "Models of energy sources for EV and HEV: Fuel cells, batteries, ultra-capacitors, flywheels and engine-generators," *J. Power Sources*, vol. 28, no. 28, 2004, pp. 76–89,

[12] O. Hegazy, J. Van Mierlo, and P. Lataire, "Control and analysis of an integrated bidirectional DC/AC and DC/DC converters for plug-in hybrid electric vehicle applications," *J. Power Electron.*, vol. 11, no. 4, 2011, pp. 408–417.

[13] P Singh, K Arora, UC Rathore, Control Strategies for Improvement of Power Quality in Grid Connected Variable Speed WECS with DFIG – An Overview, *J. Phys. Conf. Ser.*, 2022.

[14] C. C. Chan, A. Bouscayrol, and K. Chen, "Electric, hybrid, and fuel-cell vehicles: Architectures and modeling," *IEEE Trans. Veh. Technol.*, vol. 59, no. 2, Feb. 2010, pp. 589–598.
[15] C. C. Chan, W. S. Leung, and K. T. Chau, "A new permanent magnet motor drive for mini electric vehicles," in: *Proc. Int. Electric Vehicle Symposium*, 1990, pp. 165–174.
[16] Sarat Kumar Sahoo, Sukruedee Sukchai, Franco Fernando Yanine, "Review and comparative study of single-stage inverters for a PV system", *Renew. Sustain. Energy Rev.*, 91(2018), pp. 962-986.
[17] R. Sahani, M. Daga, I. Jamatia, P. Kumar and S. L. Tripathi, "Design of a self-charging electric vehicle with embedded multiple natural energy sources," *2023 IEEE Devices for Integrated Circuit (DevIC)*, Kalyani, India, 2023, pp. 296-301, doi: 10.1109/DevIC57758.2023.10134774.

8

Integrating Electric Vehicles Into Smart Grids Through Data Analytics: Challenges and Opportunities

Vikram Kulkarni[1]*, Sarat Kumar Sahoo[2], Ketan Shah[1] and Prapita Thanarak[3]

[1]Department of Information Technology, SVKM's NMIMS Mukesh Patel School of Technology Management and Engineering, Mumbai, India
[2]Department of Electrical Engineering, Parala Maharaja Engineering College, BPUT, Berhampur, Odisha, India
[3]School of Renewable Energy and Smart Grid Technology, Naresuan University, Phitsanulok, Thailand

Abstract

The emergence of electric vehicles (EVs) is transforming the traditional concept of the power grid. As more and more EVs are added to the grid, they provide valuable data that can be analyzed to optimize grid performance and promote sustainability. This chapter explores the impact of EVs on smart grids based on data analytics. Specifically, it highlights how EVs can help to manage load, integrate renewable energy, support demand response programs, optimize charging infrastructure, and provide insights into user behavior. By leveraging the data generated by EVs and their charging infrastructure, utilities and stakeholders can work to balance load, integrate renewable energy sources, support demand response programs, optimize charging infrastructure, and develop more targeted outreach and education programs. This chapter provides an overview of the opportunities and challenges associated with EV-based data analytics for smart grids and highlights some of the key strategies that can be employed to promote sustainable energy usage. Overall, the chapter concludes that EV-based data analytics can help to optimize grid performance, promote sustainability, and encourage more sustainable energy usage.

***Keywords*:** Electric vehicle, smart grid, data analysis, time of use, electric tariff systems

**Corresponding author*: vikram.kulkarni@ieee.org

Krishan Arora, Suman Lata Tripathi and Himanshu Sharma (eds.) Electric Vehicle Design: Design, Simulation and Applications, (167–196) © 2024 Scrivener Publishing LLC

8.1 Introduction

The conventional grid is typically designed to deliver power from a few large power plants to many consumers, with little or no two-way communication or interaction between the grid and the end-users. The smart grid is a highly advanced electrical infrastructure that incorporates digital technologies, smart sensors, and communication systems to improve the effectiveness, dependability, and eco-friendliness of power transmission. This system is designed to allow bidirectional communication and interaction between the power grid and consumers, as well as to integrate and regulate various distributed energy resources, energy storage units, and EVs.

The smart grid is characterized by several notable features that include the following:

1. Advanced metering infrastructure: Smart grids use advanced metering systems that enable real-time observation of energy usage and allow customers to participate in demand response programs [1].
2. Distribution automation: Smart grids incorporate intelligent sensors and automation technologies to oversee and manage the flow of electricity in the distribution system, thereby minimizing the duration of power outages and enhancing overall system reliability [2].
3. Renewable energy integration: The smart grid system is tailored to incorporate numerous renewable energy sources (RESs), including solar, wind, and hydroelectric power, while effectively regulating the fluctuating and intermittent nature of these sources.
4. Energy storage: Smart grid technology employs energy storage systems to stockpile excess energy when electric usage is low and release it when usage is high. This approach reduces dependence on peaker plants and elevates the stability of the grid [3].
5. Electric vehicles (EVs): Smart grids are capable of efficiently regulating the charging of EVs, allowing them to charge during periods of low demand, which minimizes the influence of EVs on the grid.

The smart grid is intended to be an improved version of the traditional grid, with increased efficiency, dependability, and sustainability. One of its primary objectives is to foster two-way communication and interaction

between the grid and end-users while also integrating and managing distributed energy resources.

8.2 Smart Grid and Electric Vehicle

The smart grid technologies are very helpful for making a success of EV. Several factors influence the effect of EV charging on power tariffs, such as the charging location and time of day, the accessibility of RES, as well as the power tariff's pricing structure [4].

In general, EV charging can increase the demand for electricity, particularly during peak hours when the demand for electricity is already high. This increased demand can result in higher electricity prices during these peak periods, as well as potentially higher demand charges, which are fees based on the peak power usage during a billing cycle. However, EV charging can also have a positive impact on power tariffs by providing a source of demand flexibility. This means that EV owners can adjust their charging behavior to match periods of low electricity demand, such as overnight when rates are typically lower. In some cases, utilities may offer special EV tariffs or time-of-use (TOU) tariffs that incentivize off-peak charging through lower rates during these periods [5].

RES such as wind and solar power can have an influence on reducing the impact of EV charging on power tariffs. By integrating RES into the power grid, utilities can reduce the reliance on fossil fuel–based power plants, which can be more expensive to operate during peak periods. Moreover, electricity providers have the option to provide rewards or inducements to EV owners who charge their vehicles when renewable energy is abundantly available [6]. This approach can further decrease the ecological effect of EV charging and potentially decrease electricity costs during these timeframes.

Overall, the effect of EV charging on power tariffs is complex and depends on a variety of factors. By incentivizing off-peak charging, integrating RES, and implementing smart charging technologies, utilities can help to manage the influence of EV charging on power tariffs and promote a more sustainable and affordable transportation system.

8.3 Impact of Electric Vehicle–Based Data Analytics for Smart Grids

The impact of EV-based data analytics for smart grids can be significant. By leveraging the data generated by EVs and their charging infrastructure, utilities and grid operators can gain valuable insights into usage patterns and load management strategies, which can help to optimize grid performance and promote a more sustainable and reliable energy system [7].

Some of the key impacts of EV-based data analytics for smart grids include the following:

1. Improved grid stability: By analyzing data from EVs, utilities and grid operators can gain insights into load management strategies that can help to maintain grid stability. This can include optimizing charging schedules to avoid peak demand periods and leveraging battery storage to balance load.
2. Enhanced demand response: Electric vehicle–based data analytics can also help to support demand response programs by providing insights into user behavior and load patterns. This data can be utilized to create improved pricing models and motivation plans that promote the use of sustainable energy.
3. Optimized charging infrastructure: By analyzing data from EVs, utilities and charging infrastructure providers can gain insights into usage patterns and adjust their infrastructure accordingly [8]. This can help to optimize charging infrastructure and ensure that it is meeting the needs of EV users.
4. More targeted outreach and education: Electric vehicle–based data analytics can also provide insights into user behavior that can be used to develop more targeted outreach and education programs. By understanding when and where users are charging their vehicles, utilities and stakeholders can develop more effective education campaigns and outreach programs to promote sustainable energy usage [9].

Overall, EV-based data analytics can have a substantial influence on the performance and sustainability of smart grids. By leveraging the data generated by EVs and their charging infrastructure, stakeholders can work to optimize grid stability, enhance demand response, optimize charging infrastructure, and develop more targeted outreach and education programs.

Example of how time-of-use (TOU) electric vehicle (EV) energy charging can work [10]:

- Let us say that a utility company offers a TOU tariff for EV owners. This tariff includes a higher rate for electricity usage during peak hours (e.g., 2 p.m. to 8 p.m.), and the rate for usage is reduced during non-peak hours (for example, 8 p.m. to 6 a.m.).

- An EV owner who signs up for this TOU tariff can adjust their charging behavior to take advantage of the lower off-peak rates. For example, they can schedule their EV charging to occur during off-peak hours, such as overnight or early in the morning. This can help the EV owner save money on their electricity bill, as they will be paying the lower off-peak rates for their EV charging [11].
- In addition to saving money, TOU EV charging can also help to manage the demand for electricity during peak periods. By incentivizing off-peak charging, utilities can reduce the strain on the power grid during peak hours, which can help to improve grid stability and potentially reduce the need for new power plant construction [12].
- Overall, TOU of EV energy charging is an example of how utilities can use pricing structures to incentivize sustainable and cost-effective energy usage. Utilities can manage the effect of EV charging on the power grid and encourage a more eco-friendly transportation system by offering EV owners the option to charge their vehicles during non-peak hours.

8.4 Importance of Resource Availability, Price, and Load for EV

Resource availability, price, and load are important considerations for EVs because they can have a substantial influence on the overall sustainability, affordability, and reliability of EVs as a transportation option [13].

- Resource availability: Resource availability refers to the availability of resources such as RES that can be used to power EVs. The availability of these resources can affect the overall sustainability of EVs, as using RES can help to lessen greenhouse gas emissions and promote a more sustainable transportation system.
- Price: Price is a critical consideration for EVs, as it can impact the affordability and accessibility of EVs for consumers.

The cost of EVs, charging infrastructure, and electricity rates can all have an effect on the adoption of EVs and the expansion of the EV market.
- Load: Load refers to the amount of electricity demand that is placed on the grid, which can fluctuate depending on factors such as time of day and season. Managing this load is important for maintaining grid stability and ensuring that EV charging does not place undue strain on the grid [14].

Stakeholders can facilitate the adoption of EVs in an ecological, economical, and dependable manner by taking into account resource availability, pricing, and load. This can include initiatives such as promoting the use of RES, implementing pricing structures that incentivize sustainable energy usage, and managing the load from EV charging to ensure grid stability [15]. Overall, addressing these key considerations can help to support the development of the EV market and promote a more sustainable transportation system. Figure 8.1 displays the grid load and indicates the optimal time for charging an EV battery at a lower tariff [16]. The arrows in Figure 8.1 show that the opportunity for EV users to charge their vehicles as peak load is less; it means that the per-hour tariff is lesser or optimal in this case.

There are several opportunities for EV charging data analysis and its impact on grid performance, which can help to provide insights into the usage patterns and trends of EV charging. Some of these opportunities include the following [17, 18]:

1. Optimize charging infrastructure: By analyzing EV charging graphs, utilities and charging infrastructure providers can gain insights into usage patterns and adjust their infrastructure accordingly. This can help to optimize charging

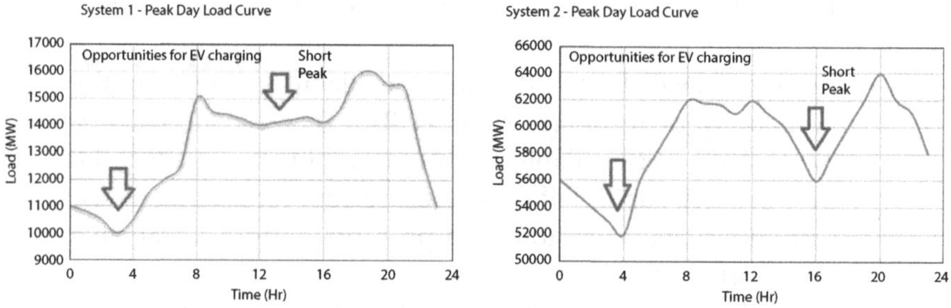

Figure 8.1 Times with existing grid capacity for new loads display optimal time for charging an EV battery at a lower tariff.

infrastructure and ensure that it is meeting the needs of EV users.

2. Manage grid load: EV charging can place a significant strain on the grid, particularly during peak periods. By analyzing EV charging graphs, utilities can gain insights into when and where EVs are charging and adjust their grid management strategies accordingly to ensure that grid stability is maintained.
3. Understand user behavior: EV charging graphs can provide insights into user behavior, such as the times of day when users are most likely to charge their vehicles, how long they typically charge for, and how far they typically travel between charges. This information can be used to design more effective pricing structures and to develop more targeted outreach and education programs.
4. Support fleet management: EV charging graphs can be particularly useful for fleet management, as they can provide insights into the usage patterns of fleet vehicles and help to optimize charging schedules. By doing so, it is possible to lower the operating expenses of the fleet and guarantee that the vehicles are fully charged and available when required.
5. Promote transparency: EV charging graphs can help to promote transparency and accountability by providing users with a clear and detailed breakdown of their charging usage. This can help to build trust with users and promote more sustainable energy usage behaviors.

Overall, EV charging graphs provide a valuable tool for understanding and managing the usage patterns of EVs. By leveraging the insights provided by EV charging graphs, stakeholders can work to optimize charging infrastructure, manage grid load, understand user behavior, support fleet management, and promote transparency and accountability.

8.5 Electric-Tariff Design Based on Impact of Electric Vehicle Usage

The principles of tariff design for EV usage depend on the specific goals of the utility or regulatory body [17]. However, here are some general principles that can be applied when designing tariffs for EV usage:

- Encourage off-peak charging: Off-peak charging can help to lessen the stress on the grid, which can help to improve grid stability and decrease the need for new power plant construction. According to reference [18], TOU tariffs or other similar tariffs that provide incentives for charging EVs during off-peak hours can be a useful strategy to encourage EV owners to charge their vehicles when electricity demand is low.
- Reflect the true cost of electricity: Tariffs should redirect the true cost of providing electricity to customers. This can include costs associated with generation, transmission, distribution, and other factors. By accurately reflecting these costs, tariffs can help to ensure that all customers are paying their fair share of the costs of maintaining the grid.
- Considering the impact of EVs on the grid: EVs can have an important impact on the grid, particularly during peak periods. Tariffs should account for this impact by charging EV owners higher rates during peak periods or by introducing demand charges that are based on the peak power usage during a billing cycle [19].
- Provide transparency and predictability: Tariffs should be transparent and predictable so that customers can understand how their usage will impact their bills. This can help to promote trust in the utility and encourage customers to adopt more sustainable energy usage behaviors.
- Promote energy equity: Tariffs should be designed to promote energy equity by ensuring that all customers have access to affordable and reliable electricity. Ways to encourage energy efficiency and sustainability may involve implementing low-income rate initiatives and other incentivizing programs [20].

8.6 Data Analytics for Electric Vehicles

The paper [21] discusses the challenges and opportunities associated with implementing these technologies, including regulatory and policy considerations, technical aspects, and potential benefits to grid operators, utilities, EV owners, and the environment. The authors argue that smart EV charging and V2G technologies have the ability to transform the energy consumption and management and would play a significant role in enabling a more ecological and resilient energy system. However, they also emphasize the need for coordinated action from policymakers, regulators, utilities, and other stakeholders in order to realize the full potential of these technologies. Overall, the paper delivers a useful overview on the present state of smart EV charging and V2G technologies and highlights the importance of collaboration and innovation in order to drive progress in this area. The authors propose an algorithm in [22] that enables plug-in EVs to respond to dynamic pricing signals in real time using data analytics, optimizing the charging process to minimize costs for both the vehicle owner and the grid operator. The proposed algorithm takes into account various factors such as the vehicle's battery state of charge, charging requirements, and grid conditions. The paper also presents real-time algorithm in reducing the overall cost of electricity consumption. The authors conclude that smart grid agents have the potential to enable more efficient and cost-effective charging of EVs while also contributing to the stability and resilience of the grid. The paper provides useful insights into the technical aspects of integrating EVs into the smart grid and highlights the importance of developing innovative solutions to address the challenges of scaling up EV adoption.

The article [23] describes a big data analytics platform based on IoT technology, consisting of five interconnected layers and multiple components for acquisition, integration, storage, analysis, and visualization of data. The platform integrates advanced technologies in a systematic and interoperable way, ensuring clear information flows. The proof-of-concept solution showcases how several tools and methods can be harmonized to facilitate decision-making and enhance product and process quality in smart manufacturing environments. In the article [24], a big data analytics approach is used to conduct an empirical case study that investigates the sentiment of Indian consumers toward EVs. The study analyzes a vast amount of data from different sources, such as online forums and social media platforms, to examine consumers' attitudes and sentiment toward EVs in India. The study's outcomes provide valuable insights into the factors that impact Indian consumers' acceptance and adoption of EVs, which can be beneficial for policymakers, automakers, and marketers to develop effective strategies for promoting the use of EVs in the country.

The article in [25] presents a unified analytical hierarchy process-quality function deployment approach to elucidate the requirements of Indian customers for the adoption of EVs. The study aims to identify the key factors that influence Indian consumers' decision-making regarding the adoption of EVs and to prioritize these factors based on their importance. The research findings provide useful insights into the preferences and needs of Indian consumers and can be beneficial for policymakers, automakers, and marketers in developing appropriate strategies for encouraging the use of EVs in India. The article [26] presents a state-level analysis of the role of EVs in decarbonizing and maintaining carbon neutrality. The study assesses the potential benefits of EVs in reducing air pollution and energy consumption at the state level in India. The results of the study provide insight into the potential of EVs in achieving carbon neutrality and improving air quality in India and may be useful to policymakers and stakeholders alike in developing effective strategies to promote the use of EV and establish a sustainable transport system in the country.

The authors in [27] discuss the concept of Vehicle to Anything (V2X), which refers to the use of EV batteries as a source of energy for other applications, such as providing grid support during peak demand periods. The paper also discusses the Cyber-Physical Power System (CPPS), which refers to the integration of ICT with power systems. The authors describe how bidirectional EV charging systems can be integrated with CPPS to provide various benefits, such as improving the reliability and stability of the power grid.

8.7 Machine Learning for EV Analytics

Machine learning (ML) is a facet of artificial intelligence that empowers computer systems to learn and enhance from experience without requiring commendable programming. It has been widely used in data analytics for EVs to improve their performance, reliability, and efficiency. ML techniques can analyze large amounts of data collected from EV sensors, battery management infrastructure, and user behavior, to provide understanding and predictions that can optimize EV operations. ML has shown promising results in data analytics for EV, and its use is expected to increase in the future. However, there are still challenges associated with data quality, privacy, and scalability that increase the potential of ML in EV analytics. Further research is required to develop more robust and efficient ML models for EV analytics.

Below is a literature survey on the use of ML for EV analytics.

The research in [28] presents different ML techniques used for EV charging and discharging. It covers different data sources, data processing techniques and ML models, and presents various applications of ML in EV charging and discharging, such as charging optimization and demand prediction. The paper [29] provides different ML techniques used for EV battery management. It discusses different data sources, data processing techniques, and ML models and presents various applications of ML in battery management, such as state estimation and fault diagnosis.

The study in [30] proposes different ML techniques that can be adapted for EV routing optimization. It covers different data sources, data processing techniques, and ML models and presents various applications of ML in routing optimization, such as energy consumption prediction and route planning. In paper [31], ML techniques are used for EV user behavior analysis. It discusses different data sources, data processing techniques, and ML models and presents various applications of ML in user behavior analysis, such as driving pattern analysis and user segmentation.

The paper [32] is based on EV integration in smart grids. It covers different data sources, data processing techniques, and ML models and presents various applications of ML in smart grid integration, such as load forecasting and energy trading. The paper [33] reviews the use of different ML techniques for EV battery management, comprising battery state-of-charge, battery-health prediction, and its life optimization. The article [34] presents different ML techniques used for EV charging station optimization, including charging demand prediction, charging station recommendation, and charging station deployment. It also discusses the challenges and opportunities associated with the integration of ML in charging station optimization. The paper [35] reviews the use of different ML techniques for EV range estimation, including range prediction and range anxiety reduction. It discusses the challenges associated with range estimation and the potential of ML to improve accuracy and reduce uncertainty.

The paper [36] reviews the use of different ML techniques for EV energy management, including energy consumption prediction, charging scheduling, and vehicle-to-grid interaction. It also discusses the difficulties and opportunities associated with the integration of ML in EV energy management. The paper [37] is a review of ML techniques used for EV fleet management, including fleet scheduling, routing optimization, and battery swapping. It also deliberates the difficulties associated with fleet management and the potential of ML to improve efficiency and reduce costs.

8.8 What are the Different ML Algorithms Used by Authors for EV Analytics?

Some of the commonly used ML algorithms for EV-based data analysis include the following:

1. Artificial Neural Networks (ANNs): ANN has been used for various EV applications such as predicting EV range, charging time, and state-of-charge estimation.
2. Support Vector Machines (SVMs): SVM has been used for predicting EV range and charging demand.
3. Random Forest (RF): RF has been used for predicting EV energy consumption and range.
4. Decision Tree (DT): DT has been used for predicting EV range and energy consumption.
5. K-Nearest Neighbor (KNN): KNN has been used for predicting EV charging demand and energy consumption.
6. Linear Regression (LR): LR has been used for predicting EV range and energy consumption.
7. Gradient Boosting Machine (GBM): GBM has been used for predicting EV charging demand.
8. Long Short-Term Memory (LSTM): LSTM has been used for predicting EV charging demand and battery state-of-charge estimation.
9. Convolutional Neural Network (CNN): CNN has been used for EV image recognition and traffic flow prediction for EV charging demand.
10. Reinforcement Learning (RL): RL has been used for EV routing optimization.

Selection of ML algorithm is based on the specific application and the data available for analysis. Different algorithms have different strengths and weaknesses, and selecting the appropriate algorithm is crucial for accurate and effective EV analytics.

8.9 Importance of Data Analysis in the EV Industry Using an Open Source Data

The popularity of EVs is on the rise because of their decreased carbon footprint, boosted energy efficiency, and lowered fuel expenses. However, one of the important apprehensions of EV buyers is the higher upfront cost when compared to traditional vehicles. Therefore, predicting the prices of EVs and identifying the factors that influence the prices are crucial.

In this section, we will explore the data analysis and price prediction of EVs. We will begin by discussing the importance of data analysis in the

EV industry, followed by a description of the dataset used for the analysis. Then, we will explore the various features and factors that influence the prices of EVs. Data analysis plays a significant role in the EV industry as it helps manufacturers and buyers make informed decisions. By analyzing the data on various EV models, manufacturers can identify the features and factors that influence the prices and improve their products accordingly. On the other hand, buyers can use data analysis to compare different EV models and make informed choices based on their needs and preferences.

8.10 Description of the Dataset

The dataset utilized for the analysis includes facts on various EV models, such as their make, model, range, battery capacity, and price. The dataset is obtained from a reliable source [38] and is preprocessed to guarantee that the data are clean and ready for analysis.

8.11 Features and Factors That Influence the Prices of EVs

In this section, we will explore the various features and factors that influence the prices of EVs. We will begin by discussing the range and battery capacity, which are the two most important features that influence the prices of EVs.

Range: EVs with longer ranges (distances) are generally more expensive than those with shorter ranges (distances). This is because longer ranges require larger battery packs, which are more expensive to manufacture.

Battery Capacity: The battery capacity of an EV refers to the amount of energy it can store. EVs with larger battery capacities are generally more expensive than those with smaller battery capacities. This is because larger battery packs require more materials, which increases the manufacturing cost.

Other features that impact the prices of EVs include the make and model, the level of luxury and features, and the government incentives and subsidies. For example, luxury EVs with advanced features such as autonomous driving and premium sound systems are generally more expensive than basic EV models. Similarly, government incentives and subsidies can significantly reduce the prices of EVs, making them more affordable for buyers.

8.12 Price Prediction of EVs

The price prediction of EVs is made using ML algorithms. By training the ML models on the dataset, the prices of EVs can be forecasted based on their features and factors. The ML models can be used to identify the most significant features and factors that affect the prices of EVs and make accurate predictions.

This section highlights the growing trend and interest in EVs due to their lower carbon footprint, energy efficiency, and reduced fuel costs. However, the higher upfront cost of EVs compared to traditional gasoline vehicles is one of the foremost concerns of EV buyers. Therefore, predicting the prices of EVs and identifying the factors that influence the prices are crucial.

The dataset from Kaggle [38] is used for the analysis, which includes information on various EV models, such as their make, model, range, battery capacity, and price as shown in Table 8.1. Data are preprocessed to ensure clean and ready data.

The article includes several graphs that illustrate the relationship between different features and factors that influence the prices of EVs.

The matrix has nine variables/features in total, including:

- KWH: battery capacity in kilowatt-hours
- Acceleration: the time taken by the car to reach from 0 to 100 km/h.
- Top Speed: the maximum speed that a car can achieve.
- Range: the overall distance that an electric powered automobile can travel.
- Efficiency: the energy efficiency of the car.
- Fast Charge Speed: the time taken with the aid of using the automobile to fee as much as 80% of the battery potential.
- Number of Seats: the number of seats in the car.
- Price in Germany: the price of the car in Germany.
- Price in UK: the price of the car in the UK.

The correlation coefficient is a statistical degree that indicates the connection among variables. It tiers from −1 to 1, in which a cost of −1 suggests a sturdy terrible correlation, zero suggests no correlation, and 1 suggests a sturdy fine correlation.

Table 8.1 Information on various EV models.

	KWH	Acceleration	Top speed	Range	Efficiency	Fast charge speed	Number of seats	Price in Germany	Price in UK
KWH	1.000000	-0.622985	0.691415	0.890092	0.356426	0.659743	0.134447	0.598084	0.707831
Acceleration	-0.622985	1.000000	-0821041	-0.712077	0.14-09114	-0.741670	0.318302	-0.501519	-0.571236
Top Speed	0.691415	-0.821041	1000000	0.760319	-0.047383	0.773190	-0.242567	0.676865	0.773805
Range	0.890092	-0.712077	0.760319	1.000000	-0.072273	0.767574	-0.068851	0.569037	0.675256
Efficiency	0.356426	0.14-09114	-0.047383	-0.372273	1.000000	-0.059914	0.602469	0.239221	0.270455
Fast Charge Speed	0.659743	0.741670	0.773190	0.767574	-0.059914	1.000000	-0.119644	0.595609	0.549171
Number of seats	0.134447	0.318302	-0242567	-0.068851	0.602469	-0.119644	1.000000	-0.012455	-0.007433
Price in Germany	0.598084	-0.501519	0.678865	0.569037	0.239221	0.595609	-0.012455	1.000000	0.915390
Price in UK	0.707831	-0.571236	0.773805	0.575256	0.270455	0.649171	-0.007433	0.915390	1.000000

- In this correlation matrix, we can see that the battery capacity (KWH) has a positive correlation with the range, top speed, fast charge speed, and price of the car in both Germany and the UK. This means that the higher the battery capacity of the car, the more range it can travel, the faster it can be charged, and the higher the price it commands.
- Acceleration has a negative correlation with the price of the car in both Germany and the UK, indicating that cars with faster acceleration tend to be less expensive. Similarly, top speed has a negative correlation with battery capacity, indicating that cars with higher battery capacities tend to have lower top speeds.
- Efficiency has a weak positive correlation with the range, fast charge speed, and price in Germany, indicating that more efficient cars can assure longer distances on a single charge, can be charged faster, and tend to be more expensive.
- The number of seats has a weak positive correlation with the acceleration, indicating that cars with more seats may have slower acceleration.

The correlation matrix can be used to recognise the characteristics that influence the prices of EVs and to develop predictive models for forecasting the prices of EVs. Figure 8.2 represents a correlation matrix that suggests the correlation coefficients among different variables or functions of electrical cars.

Figure 8.3 is a plot useful for understanding the distribution of electric car models across different car manufacturers. It can help identify which manufacturers have a larger or smaller share of the electric car market and may provide insights into the relative popularity or competitiveness of different brands.

The connection between KWH capacity and acceleration for the EV models in the dataset can be understood, and the patterns in the data can be found using Figure 8.4. The plot enables us to determine whether there is a positive or negative correlation between two variables and whether electric and hybrid vehicles differ from one another in terms of their KWH capacity and acceleration.

Figure 8.5 is useful for understanding the joint distribution of KWH capacity and range for electric car models in the dataset and for recognizing patterns in the data. The plot allows us to see if there is a connection

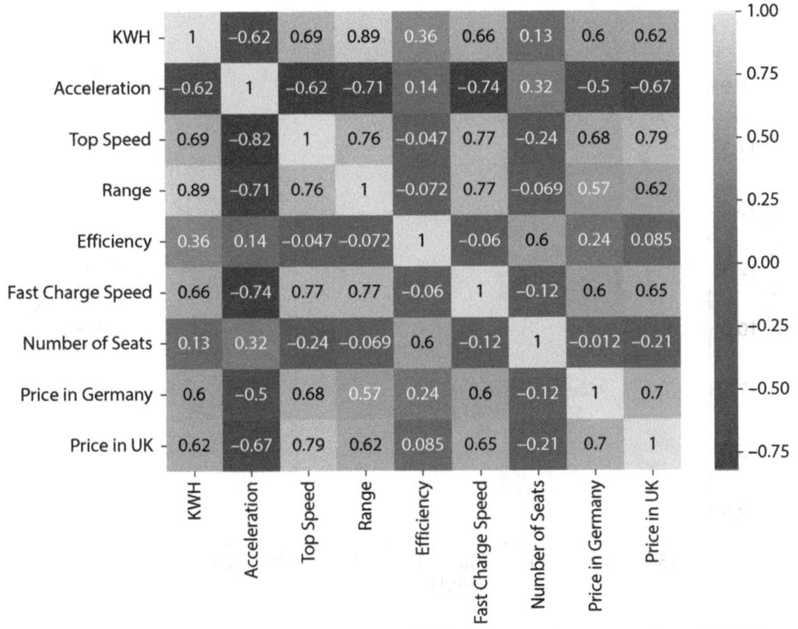

Figure 8.2 Correlation coefficients between different variables of electric vehicle.

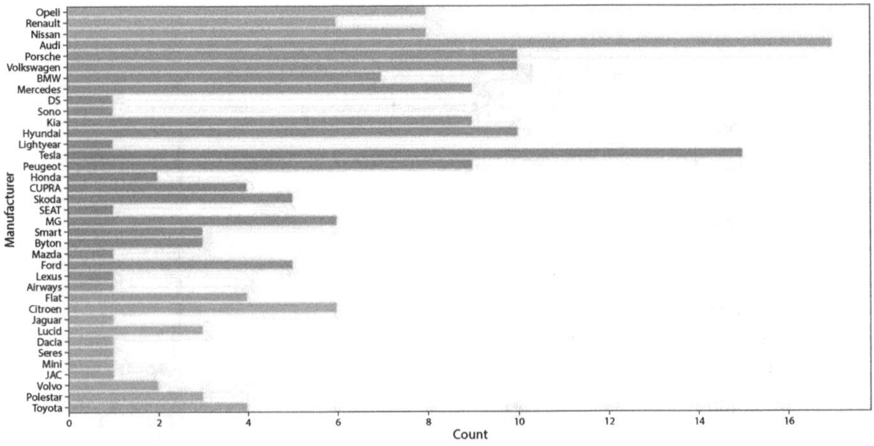

Figure 8.3 Distribution of electric car models across different car manufacturers.

between the two variables and if there are any clusters or patterns in the data. In this plot, the hexagonal bins that are darker in color indicate that there are more data points in those regions, which suggests that there may be a positive correlation between KWH capacity and range for electric cars.

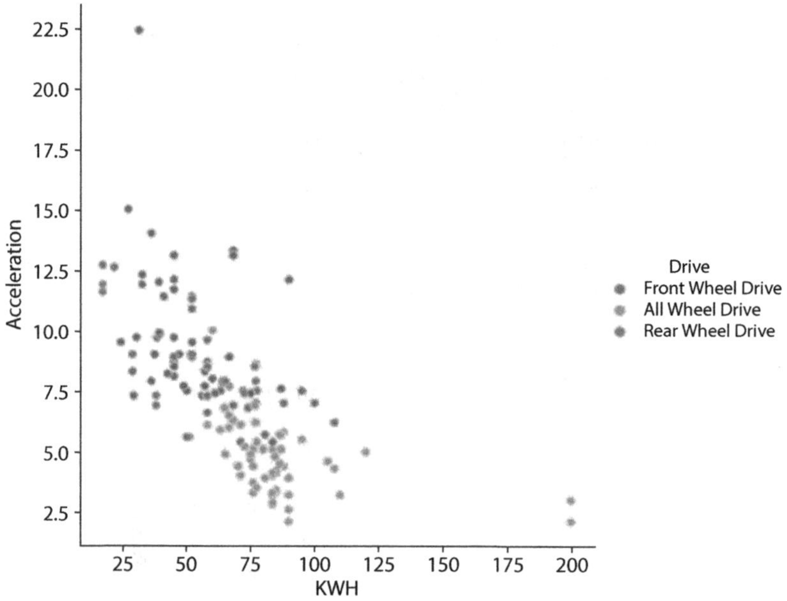

Figure 8.4 The relationship between KWH capacity and acceleration for electric cars.

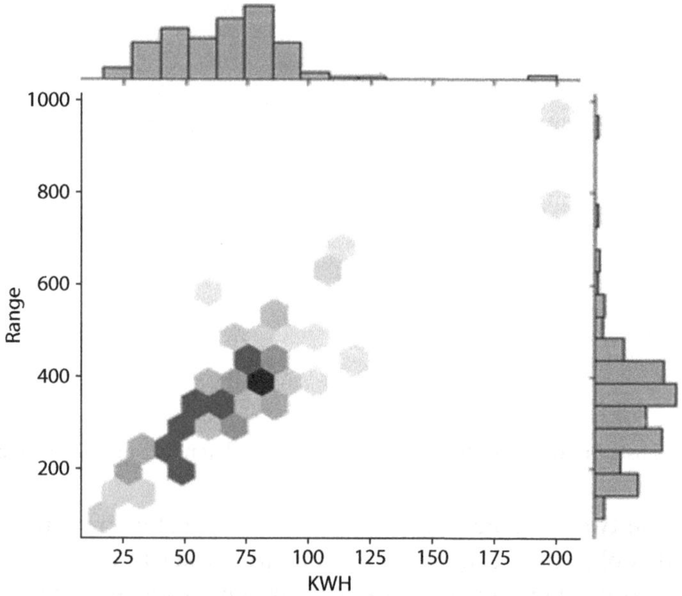

Figure 8.5 The relationship between KWH capacity and range for electric cars.

8.13 Random Forest–Based Price Prediction of Electric Vehicles

The ML part of the article involves building a predictive model using the Random Forest algorithm by considering data from [38]. This includes data preparation, feature selection, hyper-parameter tuning, and cross-validation. Feature importance analysis is also discussed, which helps to identify the most significant attributes in the data that contribute to the prediction of the target variable. The article highlights how ML can be used to analyze data and make predictions about complex systems, such as the market for EVs.

The Random Forest algorithm is a popular ML algorithm used for predictive modeling. It is based on the concept of creating multiple decision trees and combining their results to make a final prediction. In the context of the provided Kaggle notebook, the author has used the Random Forest algorithm to predict the prices of EVs. The dataset used for this analysis contains information about various EVs, such as their make, model, range, battery capacity, and price.

Initial data preprocessing includes steps like removing missing values and changing categorical factors into numerical ones. After that, the data are divided into training and testing groups. The Random Forest algorithm is then trained using the scikit-learn library on the training collection. To enhance the performance of the model, grid search is used to adjust the hyper-parameters of the algorithm, such as the number of trees in the forest and the maximum depth of each tree.

The model is used to generate predictions on the testing set after it has been trained. A variety of metrics, including mean absolute error (MAE) and root mean square error (RMSE), are used to assess the model's success. Evaluation of the Random Forest model's success on the testing set comes after it has been trained on the training set of data. This is done to evaluate the model's ability to generalise to fresh, untested data.

To evaluate the performance of the model, the author of the notebook uses two common regression metrics: MAE and RMSE.

- MAE: It calculates the average amount of errors in a collection of forecasts without taking their direction into account. It is determined by summing the absolute differences between the predicted and actual values across all instances in the testing group. The source data units are used to express the MAE.
- RMSE: It calculates the residuals' standard deviation and accounts for both the size and the direction of the errors. The average of the squared differences between the predicted and

Table 8.2 Comparison of the features and prices of different electric vehicles in the market.

	Name	KWH	Acceleration	Top speed	Range	Efficiency	Fast charge speed	Drive	Number of seats	Price in Germany	Price in UK	Hanufacturer
0	Opel Ampera	58.0	7.3	150	335	173	210	Front Wheel Drive	5	42990	36541.5	Opel
1	Renault Kangoo MaxiZE 33	31.0	22.4	130	160	194	0	Front Wheel Drive	5	0	31680.0	Renault
2	Nissan Leaf	36.0	7.9	144	220	164	230	Front Wheel Drive	5	29990	25995.0	Nissan
3	Audi e-tron Sportback 55 Quattro	86.5	5.7	200	375	231	600	All Wheel Drive	5	0	79900.0	Audi
4	Porsche Taycan Turbo S	83.7	2.8	260	390	215	860	All Wheel Drive	4	186336	138330.0	Porsche

real values is taken into account in its calculation. The initial data units of measurement are also used to express the RMSE.

Better model performance is indicated by reduced MAE and RMSE values. In other terms, a model with reduced MAE and RMSE values makes predictions with smaller errors. The author then determines which features are most crucial for predicting the prices of electric cars by using feature importance scores acquired from the Random Forest algorithm. They come to the conclusion that the price of an electric car is primarily affected by the battery capacity and range. All things considered, the Random Forest algorithm offers a potent tool for estimating the costs of electric cars based on their specifications. The model can be improved to make precise predictions by identifying the most crucial characteristics and optimising the hyper-parameters.

The data are represented in Table 8.2 of EVs with colorful features and specifications. Each row represents a different vehicle, and the columns give information on colorful parameters similar as name, KWH (battery capacity), acceleration, top speed, range, effectiveness, fast charge speed, drive (two-wheel drive or four-wheel drive), number of seats, and price in Germany and the UK. The last column indicates the manufacturer of the vehicle. The table provides a comparison of the features and prices of different EVs in the request.

8.14 Machine Learning Model

Prediction of the price of the EV based on all the features and data is presented in the following section. The price of cars is considered in UK Pound.

```
X= df_train.drop(['Name','PriceinGermany','PriceinUK','Manufacturer'],
axis=1)
X=X.values

Out:
array([ 69905, 30730, 42000, 38845, 110980, 32550, 43248, 28420,
      48000, 45145, 28500, 48824, 49750, 0, 57500, 51531,
      49465, 45507, 41330, 25491, 45000, 88700, 74739, 59990,
      21675, 55577, 29740, 43265, 32150, 87820, 0, 54000,
      30500, 90980, 20495, 32121, 54850, 42075, 31500, 63905,
      28730, 48490, 45900, 91905, 49495, 60600, 0, 30594,
      45000, 87000, 118980, 65720, 30255, 30550, 43900, 41570,
      48195, 70665, 29750, 83580])
```

In:
y_pred

Out:
array([54441.6325 , 31066.27555556, 41380. , 45479.69333333,
114520.10011111, 35399.57666667, 40735.42897222, 32830.70111111,
54001.08666667, 54295.79666667, 31681.26083333, 49643.10804365,
40735.42897222, 35409.45458333, 64010.02277778, 49643.10804365,
46875.94537302, 44265.14 , 47877.97145238, 26695.64938889,
53521.39666667, 71343.52896032, 74522.66555556, 86362.50333333,
31944.27194444, 49637.88687698, 33099.70666667, 50518.21761905,
36664.97611111, 84756.28011111, 26485.26066667, 55484.98333333,
33099.70666667, 81142.91716667, 23535.63061111, 36504.92666667,
49637.88687698, 41719.56333333, 31008.96444444, 71014.14102381,
33094.92033333, 53272.6 , 54491.59777778, 64035.14169048,
49643.10804365, 58952.72694444, 23120.55166667, 37367.93833333,
48409.47016667, 70814.42507143, 127800.99333333, 53660.58111111,
32744.97711111, 28413.25134921, 35301.64230556, 41388.21666667,
40735.42897222, 56676.54561111, 31510.11222222, 76457.32611111])

In:
#Mean Absolute Error(MAE)

```
from sklearn.metrics import mean_absolute_error
print("MAE: ",mean_absolute_error(y_test,y_pred))
```
MAE: 7155.050407804233

In:
#Mean Squared Error(MSE)
```
from sklearn.metrics import mean_squared_error
print("MSE: ",mean_squared_error(y_test,y_pred))
```
MSE: 107902152.72013463

In:
linkcode
#Root Mean Squared Error(RMSE)
```
print("RMSE: ",np.sqrt(mean_squared_error(y_test,y_pred)))
```
RMSE: 10387.596099200942

In:

#R Squared (R2)
R-squared (R2), also known as the measure of determination, is a metric used to measure the virtuousness of fit of a retrogression model.

It represents the proportion of friction in the dependent variable (y) that can be explained by the independent variables (x) included in the model.

```
from sklearn.metrics import r2_score
r2 = r2_score(y_test,y_pred)
print("R2: ",r2)
R2: 0.815987092758551
```

Thus accuracy of the EV price prediction using the Random Forest algorithm is in between 80% and 85%. The error is limited to 10% to 12%.

8.15 Electric Vehicle Usage in India

The adoption of EVs in India has been on the rise in the last 5 years, with the Government of India (GoI) taking various initiatives to promote EVs in the country [39]. The website [40] provides statistics on the sales of EVs in India, collected by the Society of Manufacturers of Electric Vehicles (SMEV) as shown in Table 8.3. Here is a breakdown of the different statistics provided on the website.

Here are some statistics on EVs in India in the last 5 years:

1. The GoI has launched various initiatives to encourage the adoption of EVs in the country. The Faster Adoption and Manufacturing of Hybrid and Electric Vehicles in India (FAME India) scheme was launched in 2015 to promote the manufacturing and adoption of EVs in the country.
2. The GoI has set a target of accomplishing 30% EV penetration in the country by 2030. The government has also

Table 8.3 Statistics on the sales of EVs in India in the last 5 years.

Category	2017–2018	2018–2019	2019–2020	2020–2021	2021–2022	2022–2023 (until 28 March)	Grand total
E-2 Wheelers	1,981	27,478	26,512	44,394	249,615	703,171	1,053,151
E-3 Wheelers	90,411	114,136	140,754	88,497	178,169	3,90,378	1,002,345
E-4 Wheelers	2,433	2,460	2,740	5,952	20,172	44,895	78,652
E-Buses	27	53	363	217	1,066	1,909	3,635
Grand Total	94,852	144,127	170,369	139,060	449,022	1,140,353	2,137,783

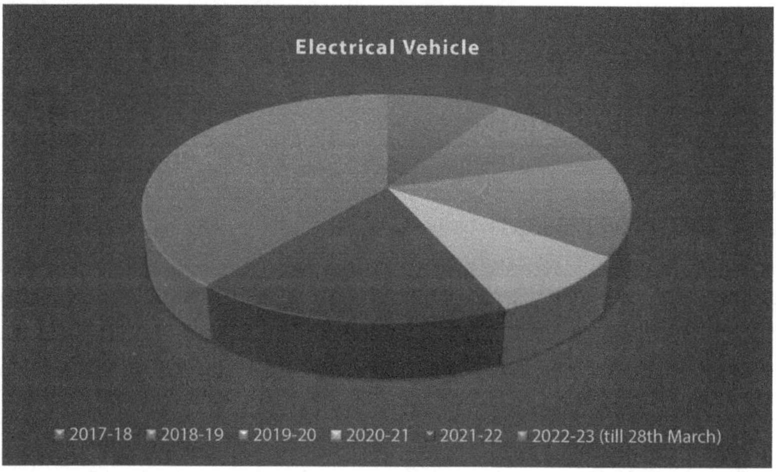

Figure 8.6 The improvement in the EV numbers in the last 5 years.

announced various incentives and subsidies to push the adoption of EVs in the country.

3. EV sales by category: This section provides a breakdown of the number of EVs sold in India by category, including electric two-wheelers, electric three-wheelers, and electric cars.
4. Year-wise EV sales: Figure 8.6 provides a year-wise breakdown of EV sales in India, starting from 2016 to 2017 to the current financial year.
5. Model-wise EV sales: This section provides a model-wise breakdown of EV sales in India, highlighting the top-selling EVs in the country.

The GoI has paid Rs. 3618.06 crores incentives under Faster Adoption and Manufacturing of Hybrid and Electric Vehicles in India (FAME) scheme phase-II until October 2022.

In India, different states' policies have proposed vehicle manufacturing and mobility policy and are available in [41]. There are 6,586 operational public EV charging stations in India [42, 43], and the number may likely increase rapidly in the coming days.

8.16 The Challenges of Adopting EV in India

There are many challenges to adopting EVs in India; there is also significant potential for growth. With the right infrastructure, incentives, and awareness campaigns, India could become a global leader in EV adoption [42]. Adopting EVs in India faces several challenges, including the following:

- Inadequate Charging Infrastructure: One major obstacle in the adoption of EVs in India is the inadequate charging infrastructure. Although some charging stations are present in major cities, the requirement for more widespread and easily accessible charging points remains crucial for facilitating long-distance travel.
- Expensive EVs: The present high cost of EVs is attributed to the expensive batteries and components they require, rendering them unaffordable for numerous Indians who are unable to purchase a new car.
- Limited Range: EVs currently have a limited range, which makes long-distance travel difficult. This is especially problematic in a country like India, where people often travel long distances on a regular basis.
- Insufficient knowledge and awareness about the advantages of EVs and their functioning among the populace in India is a major hurdle for their widespread adoption. This lack of awareness poses a challenge, as people tend to avoid investing in something they do not comprehend.
- Battery Technology: The battery technology used in EVs is still evolving, and this presents a significant challenge for manufacturers. The cost of batteries needs to come down, and their efficiency needs to improve to make EVs more affordable and practical.
- Power Supply: India faces a significant power supply problem, with many areas experiencing frequent power outages. This makes it difficult to charge EVs, especially in areas with limited charging infrastructure.
- Infrastructure Development: Building charging infrastructure and setting up battery recycling facilities requires significant investment, and India's infrastructure is still developing. This limits the government's ability to support EV adoption through incentives and other measures.

8.17 How to Increase Renewable Energy in India to Meet EV Demand

By considering the following steps, India can increase its renewable energy capacity and meet the demand for EVs, reducing its dependence on fossil fuels and mitigating the impact of climate change. To increase renewable energy in India and meet the EV demand, the following steps can be taken:

- Encourage Investment: The Indian government can encourage investment in renewable energy by providing incentives and tax breaks to companies that invest in RES. This will help to create a more conducive environment for renewable energy growth.
- Set Renewable Energy Targets: The Indian government can set renewable energy targets for the country and for individual states. This will help to create a roadmap for renewable energy growth and help to ensure that progress is made in the right direction.
- Increase Awareness: The Indian government can increase awareness about renewable energy and its benefits through public campaigns, school programs, and community outreach initiatives. This will help to instruct people about the importance of RES and encourage them to adopt it.
- Build Infrastructure: The Indian government can build the necessary infrastructure for RES, such as solar and wind farms, transmission lines, and storage facilities. This will help to create a more reliable and efficient renewable energy network.
- Develop Energy Storage Solutions: The development of energy storage solutions, such as batteries and pumped hydro storage, can aid in the storage of surplus renewable energy generated during peak periods, ensuring a consistent energy supply to satisfy the demand for EVs.
- Encourage Innovation: The Indian government can encourage innovation in renewable energy technologies by funding research and development initiatives, setting up incubators and accelerators, and providing support to start-ups.
- Facilitate Private Sector Participation: The Indian government can facilitate private sector participation in renewable energy by creating a more conducive policy environment, providing subsidies and incentives, and promoting public-private partnerships.

Conclusion

As EVs gain popularity, their impact on the smart grid cannot be ignored. However, integrating EVs into the grid necessitates meticulous planning and data analytics to ensure that the grid can handle the additional load. By employing data analytics, we can optimize the charging and discharging of EVs, reduce peak demand, and enhance grid stability. Moreover, it can offer valuable insights into customer behavior and preferences to develop new services and pricing models that encourage EV adoption. Despite obstacles like the lack of charging infrastructure and high EV costs, the advantages of EVs and smart grids are indisputable. By leveraging data analytics and implementing smart grid solutions, we can establish an energy system that is sustainable, efficient, and beneficial for both consumers and the environment.

In the future, the continued development of data analytics and smart grid technology will play a critical role in the integration of EVs into the grid, enabling us to build a cleaner and more sustainable future.

References

1. F. Norouzi, T. Hoppe, L.M. Kamp, C. Manktelow, P. Bauer. (2023). Diagnosis of the implementation of smart grid innovation in The Netherlands and corrective actions. *Renew. Sust. Energy Rev.*, 175. 113185.
2. Ravinder M and Kulkarni V. (2023). Intrusion detection in smart meters data using machine learning algorithms: A research report. *Front. Energy Res.*, 11:1147431.
3. D. J. Kim, B. Kim, C. Yoon, N. -D. Nguyen and Y. I. Lee. (2023). Disturbance Observer-Based Model Predictive Voltage Control for Electric-Vehicle Charging Station in Distribution Networks. *IEEE Trans. Smart Grid*, 14 (1), 545-558.
4. Arora, K.; Kumar, A.; Kamboj, V.K.; Prashar, D.; Shrestha, B.; Joshi, G.P. Impact of Renewable Energy Sources into Multi Area Multi-Source Load Frequency Control of Interrelated Power System. *Mathematics*, 2021, 9, 186.
5. B. Wang. (2023). An IoT-Enabled Stochastic Operation Management Framework for Smart Grids. *IEEE Trans. Intell. Transpo. Syst.*, 24 (1). 1025-1034.
6. Y. Jiang, T. H. Ortmeyer, M. Fan and X. Ai. (2023). Data-Driven Low-Rank Tensor Approximation for Fast Grid Integration of Commercial EV Charging Stations Considering Demand Uncertainties. *IEEE Trans. Smart Grid*, 14 (1). 517-529.
7. Arora, K.; Kumar, A.; Kamboj, V.K.; Prashar, D.; Jha, S.; Shrestha, B.; Joshi, G.P. Optimization Methodologies and Testing on Standard Benchmark Functions of Load Frequency Control for Interconnected Multi Area Power System in Smart Grids. *Mathematics*, 2020, *8*, 980.
8. M.M. Sufyan Beg, Md Muzakkir Hussain, Mohammad Saad Alam, and Shahedul Haque Laskar. (2019). Big Data Analytics Platforms for Electric

Vehicle Integration in Transport Oriented Smart Cities: Computing Platforms for Platforms for Electric Vehicle Integration in Smart Cities. *Int. J. Digit. Crime For.*, 11(3).
9. Sarthak Mohanty, Subhasis Panda, Shubhranshu Mohan Parida, Pravat Kumar Rout, Binod Kumar Sahu, Mohit Bajaj, Hossam M. Zawbaa, Nallapaneni Manoj Kumar, Salah Kamel. (2022). Demand side management of electric vehicles in smart grids: A survey on strategies, challenges, modeling, and optimization. *Energy Rep.*, 8. 12466-12490.
10. Syed Rahman, Irfan Ahmed Khan, Ashraf Ali Khan, Ayan Mallik, Muhammad Faisal Nadeem. (2022). Comprehensive review & impact analysis of integrating projected electric vehicle charging load to the existing low voltage distribution system. *Renew. Sust. Energy Rev.*, 153. 111756.
11. Dulichand Jaraniya, Shailendra Kumar. (2022). Integration of multi voltages, multi electric vehicle spots based three phase photovoltaic array charging station to the modern distribution grid with improved electric vehicle charging capability and power quality. *Int. J. Energy Res.*, 46 (12). 16896-16917.
12. Omid Sadeghian, Arman Oshnoei, Behnam Mohammadi-ivatloo, Vahid Vahidinasab, Amjad Anvari-Moghaddam. (2022). A comprehensive review on electric vehicles smart charging: Solutions, strategies, technologies, and challenges. *J. Energy Storage*, 54. 105241.
13. Zahra Azimi, Rahmat-Allah Hooshmand, Soodabeh Soleymani. (2021). Energy management considering simultaneous presence of demand responses and electric vehicles in smart industrial grids. *Sustain. Energy Technol. Assess.*, 45. 101127.
14. M. Satish Kumar Reddy, K. Selvajyothi. (2021). Mitigation on the Impact of Electric Vehicle Charging Stations by Splitting the Capacity and Optimally Locating on a Reconfigured RDS. *Green Eng. Technol.*, 151-176.
15. Amith Khandakar, Annaufal Rizqullah, Anas Ashraf Abdou Berbar, Mohammad Rafi Ahmed, Atif Iqbal, Muhammad E. H. Chowdhury, S. M. Ashfaq Uz Zaman. (2020). A Case Study to Identify the Hindrances to Widespread Adoption of Electric Vehicles in Qatar. *Energies*, 13(15). 3994.
16. P. Singh, K. Arora, U.C. Rathore, Control Strategies for Improvement of Power Quality in Grid Connected Variable Speed WECS with DFIG – An Overview, *J. Phys. Conf. Ser.*, 2022.
17. S. Tirunagari, M. Gu and L. Meegahapola. (2022). Reaping the Benefits of Smart Electric Vehicle Charging and Vehicle-to-Grid Technologies: Regulatory, Policy and Technical Aspects. *IEEE Access*, 10. 114657-114672.
18. K. Arora *Internet of Things-Based Modernization of Smart Electrical Grid, Smart Electrical Grid System*, CRC Press, 2022.
19. C. Jayapalan, Thangatur Sukumar Hariharan, L.S. Ganesh. (2022). Power supply to electric vehicle charging stations in India: Justification of a framework for a dynamic and adaptive electricity tariff policy. *Electr. J.*, 3 (10).

20. D.R. Bairrão, J. Soares, B. Canizes, F. Lezama, Z. Vale. (2022). Retail Electricity Tariffs for Electric Vehicles in Europe: A Multivariate Analysis in 4 European Countries. *IFAC-Papers OnLine*, 55 (9). 303-308.
21. B. Li, M. C. Kisacikoglu, C. Liu, N. Singh and M. Erol-Kantarci. (2017). Big Data Analytics for Electric Vehicle Integration in Green Smart Cities. *IEEE Commun. Mag.*, 55 (11). 19-25.
22. D. Dallinger, J. Link and M. Büttner. (2014). Smart Grid Agent: Plug-in Electric Vehicle. *Trans. Sustain.*, 5(3). 710-717.
23. Sinan Kahveci, Bugra Alkan, Mus'ab H. Ahmad, Bilal Ahmad, Robert Harrison. (2022). An end-to-end big data analytics platform for IoT-enabled smart factories: A case study of battery module assembly system for electric vehicles. *J. Manuf. Syst.*, 63. 214-223.
24. Jena, R. (2020). An empirical case study on Indian consumers' sentiment towards electric vehicles: A big data analytics approach. *Ind. Mark. Manag.*, 90, 605–616.
25. Murugan, M., & Marisamynathan, S. (2022). Elucidating the Indian customers requirements for electric vehicle adoption: An integrated analytical hierarchy process – Quality function deployment approach. *Stud. Transp. Pol.*, 10(2), 1045–1057.
26. Hossain, M. S., Fang, Y. R., Ma, T., Huang, C., & Dai, H. (2023). The role of electric vehicles in decarbonizing India's road passenger toward carbon neutrality and clean air: A state-level analysis. *Energy*, 273, 127218.
27. Elma, O., Cali, U., & Kuzlu, M. (2022). An overview of bidirectional electric vehicles charging system as a Vehicle to Anything (V2X) under Cyber–Physical Power System (CPPS). *Energy Rep.*, 8, 25–32.
28. Xie, X., Ma, W., Zhang, Q., Liu, Y., & Wang, Y. (2021). Machine Learning Applications for Electric Vehicle Charging and Discharging. *Energies*, 14(16), 4927.
29. Sharma, A., Kumar, R., & Singh, B. (2021). Machine learning for electric vehicle battery management: A review. *J. Energy Storage*, 38, 102260.
30. J. Zhang, Y. Zhang, Y. Song, and Y. Zhou. (2020). Machine Learning for Electric Vehicle Routing Optimization: A Review. *App. Energy*, 276. 115530.
31. Kim, M., Lee, K., Lee, K., & Kim, J. (2021). Machine Learning for Electric Vehicle User Behavior Analysis: A Review. *Energies*. 14(8), 2146.
32. Y. Zhang, Y. Li, Y. Sun, Q. Li, X. Du. (2020). Machine Learning for Electric Vehicle Integration in Smart Grids: A Review. *IEEE Access*. 8. 154824-154840.
33. K. S. Kumbhar, R. V. Kshirsagar, and R. J. Ramteke. (2019). A Review of Machine Learning Techniques for Electric Vehicle Battery Management. *IEEE Trans. Transp. Electrif.*, 5 (2). 412-429.
34. Wang, D., Huang, Y., Liu, Y., & Wen, F. (2019). Machine learning for electric vehicle charging station optimization: A review. *IEEE Trans. Industr. Inf.*, 15(7), 3864-3874.

35. Chen, J., Song, Y., Liao, G., Wang, Y., & Wang, Y. (2021). A Review of Machine Learning Techniques for Electric Vehicle Range Estimation. *IEEE Trans. Intell. Transpo. Syst.*, 22(9), 5481-5499.
36. Chen, R., Li, G., Liu, X., Chen, Y., & Zhang, J. (2021). A Review of Machine Learning Techniques for Electric Vehicle Energy Management. *IEEE Trans. Intell. Transpo. Syst.* 22(3), 1422–1438.
37. Liu, H., Song, Y., Wang, J., & Zhang, J. (2021). Machine Learning for Electric Vehicle Fleet Management: A Review. *IEEE Access*, 9, 121505–121524.
38. R. Sahani, M. Daga, I. Jamatia, P. Kumar and S. L. Tripathi, Design of a self-charging electric vehicle with embedded multiple natural energy sources, *2023 IEEE Devices for Integrated Circuit (DevIC)*, Kalyani, India, 2023, pp. 296-301, doi: 10.1109/DevIC57758.2023.10134774.
39. https://www.kaggle.com/datasets/kkhandekar/cheapest-electric-cars [Open source - Accessed on 30-03-2023]
40. https://www.smev.in/statistics [Accessed on 30-03-2023]
41. https://www.smev.in/govt-notification.php [Accessed on 30-03-2023]
42. https://pib.gov.in/PressReleasePage.aspx?PRID=1910392 [Accessed on 30-03-2023]
43. https://www.iea.org/data-and-statistics/data-tools/global-ev-data-explorer [Accessed on 30-03-2023]

9
Hybrid Electrical Vehicle Designs

T. Meenakshi[1]*, K. Mahendran[1], N. Suthanthira Vanitha[2] and C. Shanmugam[1]

[1]Jansons Institute of Technology, Karumathampatti, Coimbatore, India
[2]Muthayammal Engineering College, Rasipuram, Namakkal, India

Abstract
The hybrid electric vehicle is one of the most preferred modes of transportation in recent years and has had an impact in reducing global warming. It comprises a mixed source of internal combustion engine, ultra-capacitor or battery pack, and an electric motor for propulsion. The plug-in hybrid electric vehicles, battery electric vehicles, and fuel cell electric vehicles have attracted the automobile industry and are improving in their structural design. The upgradation has been done as a means to increase the efficiency of energy conversion and reduce losses in the aerodynamic drag, rolling resistance, and braking losses.

This chapter discusses the different hybrid electric vehicular designs and the system design analysis. The control strategy used in series hybrid drivetrain configuration and its modes of operation is discussed in this chapter. The design of the fuel cell hybrid electric vehicle with fuel economy is also presented with different modes of operation incorporating the fuel economy.

Keywords: Hybrid electric vehicles, drivetrain design, fuel cell hybrid electric vehicle, fuel economy, SoC control

9.1 Introduction

The hybrid electric vehicle (HEV) is the most preferred efficient mode of transportation in recent years and has had a great impact in reducing global warming. It comprises a mixed source of internal combustion engine, ultra-capacitor or battery pack, and an electric motor for propulsion. The HEVs that have a plug-in nature and are battery-operated and

**Corresponding author*: mechand22@gmail.com

Krishan Arora, Suman Lata Tripathi and Himanshu Sharma (eds.) Electric Vehicle Design: Design, Simulation and Applications, (197–214) © 2024 Scrivener Publishing LLC

fuel cell–powered have attracted the automobile industry and are improving in their structural design for efficient power usage, improved efficacy in energy conversion by loss reduction in the aerodynamic drag, rolling resistance, and braking. With the evolution of hybridization of sources, various structural topologies have been devised and have been extensively deployed for a variety of vehicle types based on the performance requirements. This ranges from passenger cars of low-power requirement to transit buses of large power requirement. This chapter discusses the vehicle technology spectrum and provides a comprehensive review of hybrid power trains and the control algorithms deployed for control of electric vehicles (EVs).

9.2 Plug-In Hybrid Electric Vehicles

A plug-in hybrid electric vehicle (PHEV) encompasses different power sources inside the vehicle. A battery bank is applied to provide propulsion power to the electric motor, and fuel like gasoline is used to energize the internal combustion engine (ICE). The battery bank or ultra-capacitors predominantly store energy and the ICE converts fuel into usable energy. The combinations can be a fuel cell with a battery/ultra-capacitor. Figure 9.1 shows the layout of a PHEV.

Depending on the usage of the source the vehicle source transition spectrum travels from ICE to EV as shown in Figure 9.2. The vehicle spectrum elucidates the various grades of use of electrical energy in the HEV.

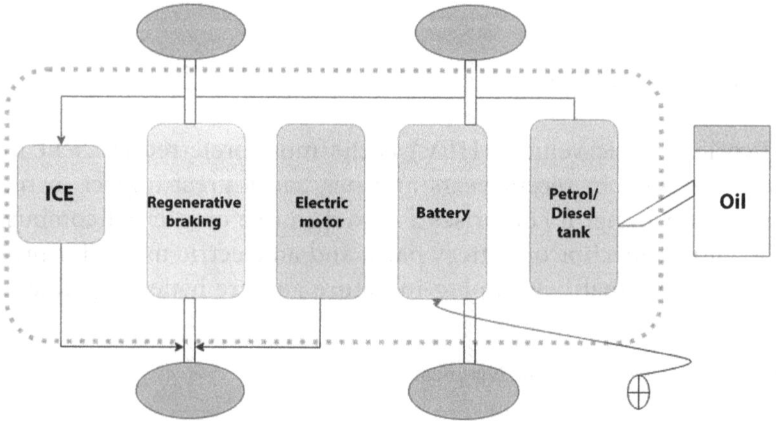

Figure 9.1 Plug-in hybrid electric vehicle layout.

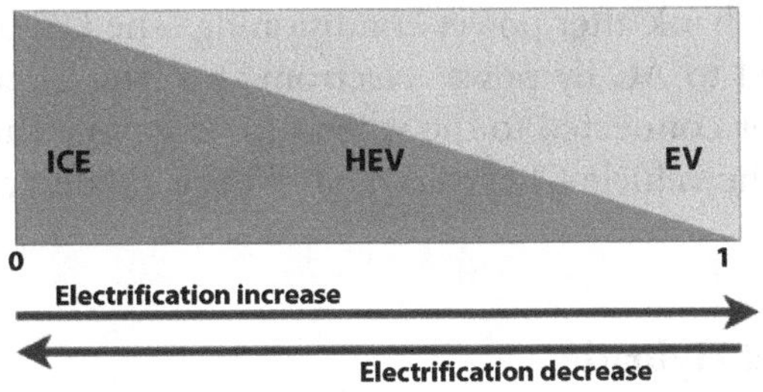

Figure 9.2 Source transition spectrum from ICE to EV.

The grade of electrification falls between a fractional value of 0 and 1. Vehicles falling to zero value are conventional vehicles, and electrical vehicles are those with one. The amount of power delivered in the PHEVs falls between 0 and 1 and is based on the ratio shared by the battery system and the ICE.

9.3 Classification of HEVs

HEVs exhibit different topological connections and are widely classified as series, parallel, and series parallel [1–3]. The classification is duly based on the assembly of the sources and the power flow for propulsion through the vehicle drive system from the fuel and electrical source and other sources integrated in the vehicle.

9.3.1 Series Hybrid

Of the various configurations, the simplest one is the series hybrid configuration, also called as range extender, in which an electric motor is the source for propulsion of the wheel. The ICE generates mechanical power that runs a generator to produce electrical power. The resultant power is

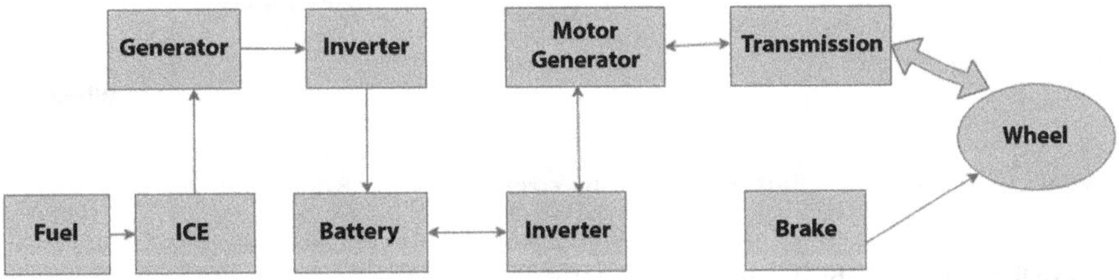

Figure 9.3 Structure of series hybrid HEV.

stored in battery bank after power conditioning. The DC power from battery is converted to AC by power electronic inverter circuit and used to propel the motor connected to the wheels for traction. This simple structure is used in the vehicle Chevrolet Volt. Figure 9.3 depicts the structure of series hybrid.

9.3.2 Parallel Hybrid

Figure 9.4 illustrates the block diagram of parallel hybrid. In a parallel HEV, the sources power the drivetrain concurrently. The ICE is directly connected to the transmission system, and the electric motor powered by the battery-inverter setup delivers power to drive the wheels. Depending on the power requirement, the driving force will be delivered by ICE only, electric motor alone, or both. The Honda Insight vehicle uses parallel hybrid connection.

9.3.3 Series-Parallel Hybrid

The features of the series and parallel connection are integrated in the series-parallel hybrid connection topology; the block diagram is displayed in Figure 9.5. The configuration has a power split circuitry that divides the power from the ICE to the drivetrain and the battery bank. Additional motor generator set and power conditioning circuit are added with the power split circuit. The battery bank is, in turn, connected to the transmission through the power electronic circuit and motor. Regeneration during braking renders power flow in both directions and stores energy in the battery.

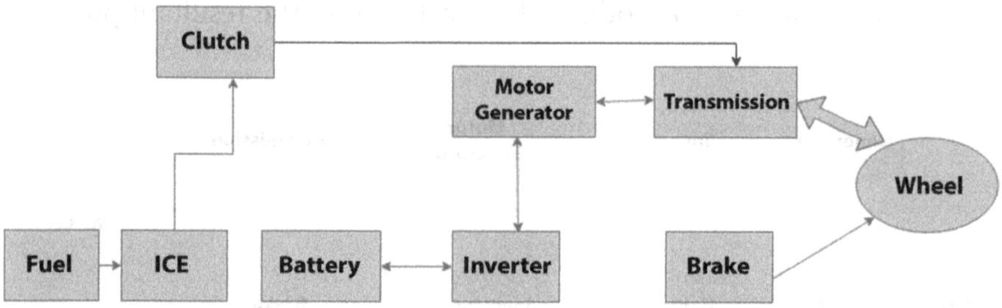

Figure 9.4 Parallel hybrid connection in HEV.

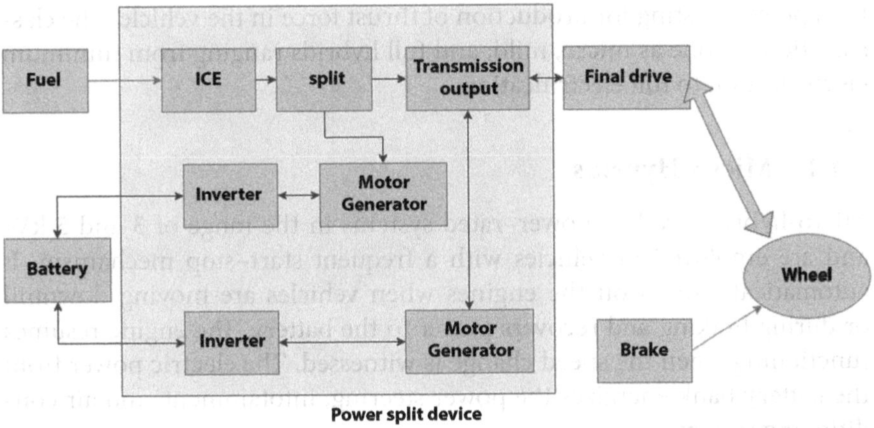

Figure 9.5 Block diagram of series-parallel hybrid HEV.

9.4 Fuel Cell Electric Vehicles (FCEVs)

FCEVs are similar to EVs; they are power-driven by a hydrogen fuel cell, and the storage is assisted by a small battery. The blocks in the FCEV are shown in Figure 9.6. The FCEVs are costlier because of the availability of the source and the storage facility for the fuel in the vehicle as it is of lower density. To overcome it, the fuel is stored as liquid under amplified pressure and at extremely low temperature. The output of the fuel stack is fed to a DC-DC converter and is stored in the battery and also fed to the inverter that runs the motor and propels the wheels.

Apart from topological classification, the HEVs are also categorized on the basis of percentage of the power shared by the electric system to the

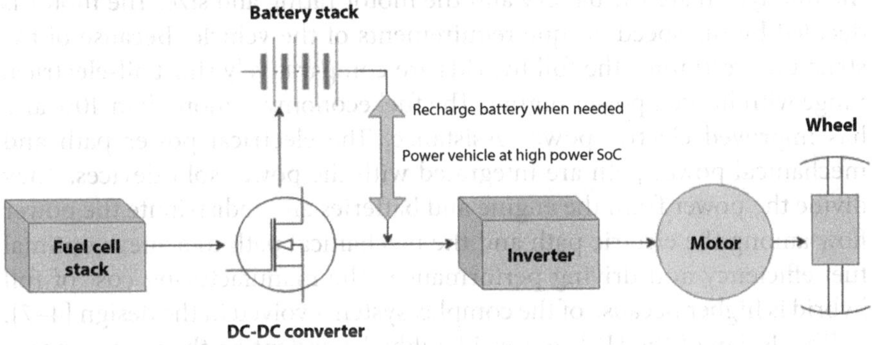

Figure 9.6 Block diagram of FHEV.

total power existing for production of thrust force in the vehicle. The classification is done as micro, mild, and full hybrids ranging from minimum electrification to full electrification.

9.4.1 Micro-Hybrids

Micro-hybrids are low-power–rated systems in the range of 3 and 5 kW and are employed in vehicles with a frequent start–stop mechanism. It automatically turns off the engines when vehicles are moving downhill or during braking and recovers power to the battery. The engine resumes functioning when the speed change is witnessed. The electric power from the battery bank energizes the power steering, infotainment, and air conditioning system.

9.4.2 Mild Hybrids

The HEVs in which the electric power rating is between 7 and 15 kW are classified as mild hybrids. Hence, fuel economy gain is higher with a result of 20% saving in fuel compared to ICE. The use of an electrical motor for propulsion systems between the engine crank and the transmission input shafts provides the start–stop, regenerative braking, and further electric power to drive the electrical systems.

9.4.3 Full Hybrids

Full hybrids exhibit the maximum electric portion with the power rating of 30 kW or more. Full hybrids rely on a complete electric propulsion system and use only electric motors to propel the vehicle and also energize all the internal power loads in the vehicle. The major constraints met by the full hybrid are the battery and the motor rating and size. The motor is decided by the speed–torque requirements of the vehicle. Because of the structural restraints, the full hybrids are comparatively short all-electrical range with limited power output. The fuel economy is more than 40% and has improved electric power assistance. The electrical power path and mechanical power path are integrated with the power split devices. They divide the power from the engine and batteries and redistribute the power flow among the electric path and the mechanical path to achieve optimal fuel efficiency and driving performance. The manufacturing cost of full hybrid is higher because of the complex system evolved in the design [4–7].

The design of the HEV is considerably dependent on the power energy requirement of the vehicle and is depicted in Figure 9.7. The power energy

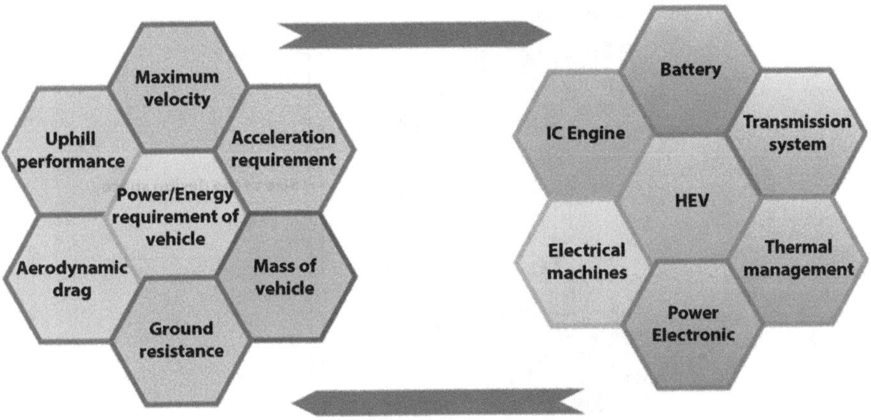

Figure 9.7 Power energy requirement of the vehicle and their influence on HEV design.

requirement is based on velocity, acceleration, aerodynamic drag, mass of the vehicle, and upsloping performance. It directly decides the battery size, motor rating, power electronic system, ICE power and the propel power transmission system.

9.5 Hybrid Electric Vehicle System Design and Analysis

Fuel economy is an important aspect of HEV design. The design should be made optimum such that it consumes minimum fuel and yields maximum efficiency. The fuel economy is obtained as follows:

1. Harnessing power during regenerative braking is one method of optimizing fuel usage. When the driver desires to slow the vehicle down by braking, mechanical power is converted to electric power and is stored in the battery. The friction brake is applied only when needed, and, during other braking conditions, the hybrid power train recovers the energy.
2. Automatically fastening the engine to off during the idle state. This considerably reduces the engine energy loss due to rotation and pumping and the engine idle fuel consumption. The substitution of the electric path during this state or during light load condition results in fuel economy.

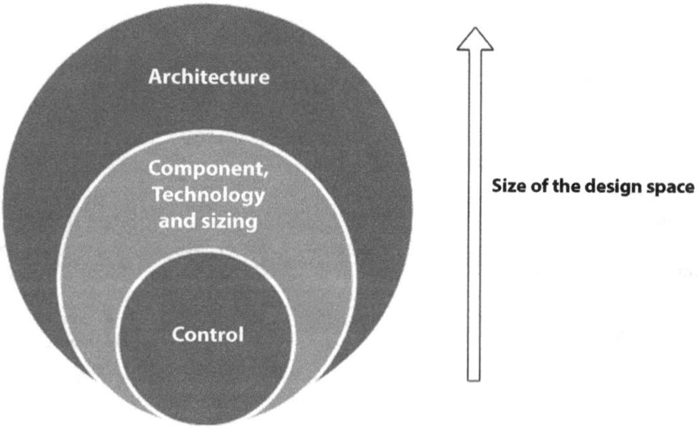

Figure 9.8 Structure level in HEV.

3. The engine is made to operate more efficiently by adjusting the operating points of the engine by controlling engine speed or engine load.
4. The load created by accessories and its management plays a significant role in reducing overall vehicle power consumption. The vehicle accessory load can be reduced by efficient aerodynamic designs and use of lightweight materials.

HEV power trains integrate battery traction with conventional engine traction and evolve diverse HEV power train architectures. Based on the architecture variant, the selection of components, battery rating, and the gear ratio are decided. In the HEV, the design complexity is dependent on numerous variables, and optimization is recommended at every stage to achieve the desired performance. The design aspect is classified into different structures in an HEV and is depicted in different levels as given in Figure 9.8: (1) architectural level, (2) component technology and sizing level, and (3) control level with architecture holding the maximum size of the design space [8].

9.6 Control Strategy in Series Hybrid Drivetrain Configuration

Fuel consumption optimization is a key area in which several researches have been conducted, including automated design process for planetary

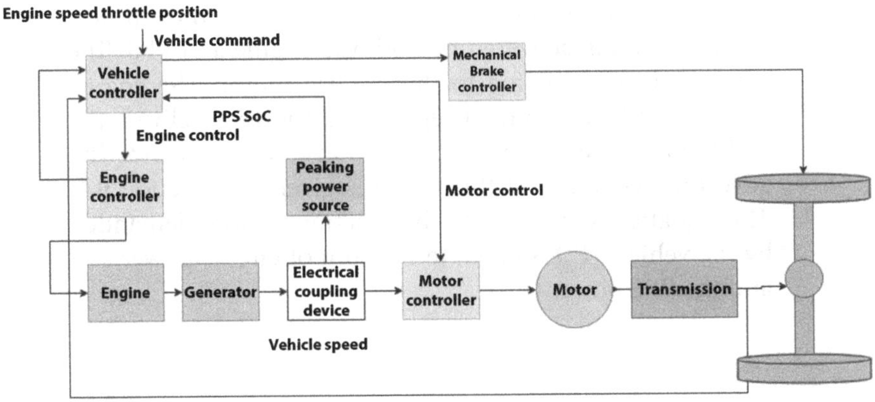

Figure 9.9 Drivetrain of hybrid electric vehicle with control blocks.

gear sets (PGs). The provision of additional gears for speed conversion results in fuel economy and efficiency improvement [9–11].

The vehicle system of the series hybrid has engine, peaking power source (PPS) like a battery to power the vehicle. Various control strategies are implemented through the controllers deployed for vehicle control management, motor control, control for engine, and brake control for optimized fuel consumption and speed control. This is displayed in Figure 9.9.

In the configuration of the series hybrid, no direct connection is established between the ICE and the propelling wheels, and they are mechanically decoupled. Hence, the speed and torque of the engine do not directly impact the vehicle speed and traction torque, and they are independent of each other. This feature simplifies the control of the speed in the speed–torque plane. Furthermore, the fuel consumption of the engine could be minimized by operating the engine in its optimal region. This is established much more simply because of the decoupling of the mechanical system of the engine from the transmission and wheels. However, the control of the fuel consumption is largely influenced by the operating modes and the control scheme of the propulsion drivetrain.

9.6.1 Modes of Operation

Based on the driving conditions, the drivetrain has several operating mechanisms as presented below:

1. Hybrid traction mode: This is the mode where heavy power is demanded by the vehicle. During this mode, more pressure

is applied to the gas pedal by the driver. Because more power is required, it is delivered by both the engine and PPS. The combined power is fed to the electric motor. The engine is operated in its optimal region and the PPS delivers the additional power to meet the traction power demand. The governing equation of this mode is given in Equation 9.1. The equation depicts that the traction power demanded by the vehicle (P_d) is equal to the sum of engine power (P_e) and the PPS power (P_{ps}).

$$P_d = P_e + P_{ps}, \tag{9.1}$$

2. PPS alone traction mode: This mode is established during very low power demanded by the traction. Here, the PPS supplies the entire power demand. The governing equation is given by Equation 9.2.

$$P_d = P_{ps} \tag{9.2}$$

3. Engine alone traction mode: In this operating mechanism, the engine meets the entire power demand by the vehicle: The equation is given by Equation 9.3.

$$P_d = P_e \tag{9.3}$$

4. PPS charge from engine: During the condition of the energy in the PPS decreasing to a minimum value, it should be charged back. The charging of the PPS is initiated in this mode. It is accomplished during regenerative braking or by directly feeding the power from the engine to the PPS. The engine supplies power to propel the vehicle and charge the PPS. The power equation is given by Equation 9.4, and it is observed that the engine power has to be more than the power claimed by the load, and the surplus power of the engine is used to charge the PPS.

$$P_e = P_d + P_{ps} \tag{9.4}$$

5. Regenerative braking mode: In this mode, the charging of PPS happens during the braking condition. When the brake

is applied to a vehicle, the motor acts as a generator and generates electricity by converting the locomotive energy of the vehicle. This is fed to power the PPS, and the regenerative power is effectively utilized.

The vehicle controller present in the control system of the vehicle executes a vital role in yielding efficient performance of the drive. The operation of each component in the system is based on the power requested by the driver; the sensing and feedback signal is produced by each of the components and also relies on the control plan. The vehicle controller is designed to meet the control purposes such as (1) fulfill the power request by the driver, (2) perform each component at its maximum efficiency, (3) utilize braking energy to the fullest extent to power PPS, and (4) maintain the state of charge (SoC) of the PPS.

The vehicle controller is preloaded with the control algorithm that supervises the function of each component. The vehicle controller receives the command of the operation to be performed from the driver. Apart from it, it receives the signals of the speed feedback from the drivetrain, the PPS SoC, and the engine state from the engine controller. This assists the vehicle controller to take appropriate decisions on choosing the proper operational modes.

The control strategy plays an essential role in deciding the performance of the drivetrain. One of the control strategies is maximum SoC of the PPS control strategy

9.6.2 Max. SoC-of-PPS Control Strategy

The focus of the Max. SoC-of-PPS control method is to fulfill the power request of the driver and maintain the SoC of the PPS at a desired high level. In this mode, the engine acts as the primary power source and the PPS as the secondary source. Vehicles that consider speed, acceleration, and performance as the first concern use this control strategy. The vehicles with recurrent stop–go driving patterns and military vehicles use this strategy as they require a high SoC level in the PPS.

Figure 9.10 depicts the illustration of Max. SoC-of-PPS control strategy. The Max. SoC of PPS is separated into four points as A, B, C, and D based on the power required by the driver. The demand of power may be either for traction or during braking.

- At Point A, the driver demands the traction power (Pcommanded), which is greater than the power of the

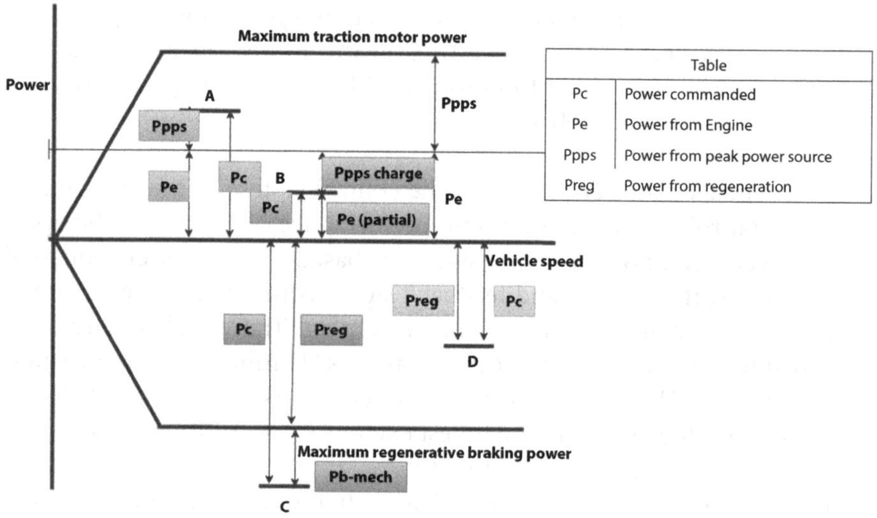

Figure 9.10 Maximum PPS SoC control strategy.

engine (Pe). Because the engine cannot wholly support the demand, the shortage of power is met by the PPS.

- At Point B, the demanded power by the driver is less than the power produced by the engine. Hence, the engine can fully meet the demand. In this case, based on the SoC level of the PPS, two operating modes shall be initiated. The engine is operated at its maximum limit where a portion of its power goes to propulsion based on the driver's demand and the other part goes to charge the PPS as the level of PPS is less than 70% of its capacity. Alternatively, if the SoC of PPS is maximum, then the engine supplies power only for traction such that power produced is equal to the power demanded, and the PPS is kept idle.
- At Point C, the driver side request of the brake power is higher than the brake power of the motor that could be produced during regenerative braking. Because the request cannot be fulfilled, hybrid braking is used, in which the electric motor produces its maximum braking power, and the mechanical braking system produces the remaining braking power.
- At Point D, the driver demands the braking power that is lower than the maximum braking power of the motor.

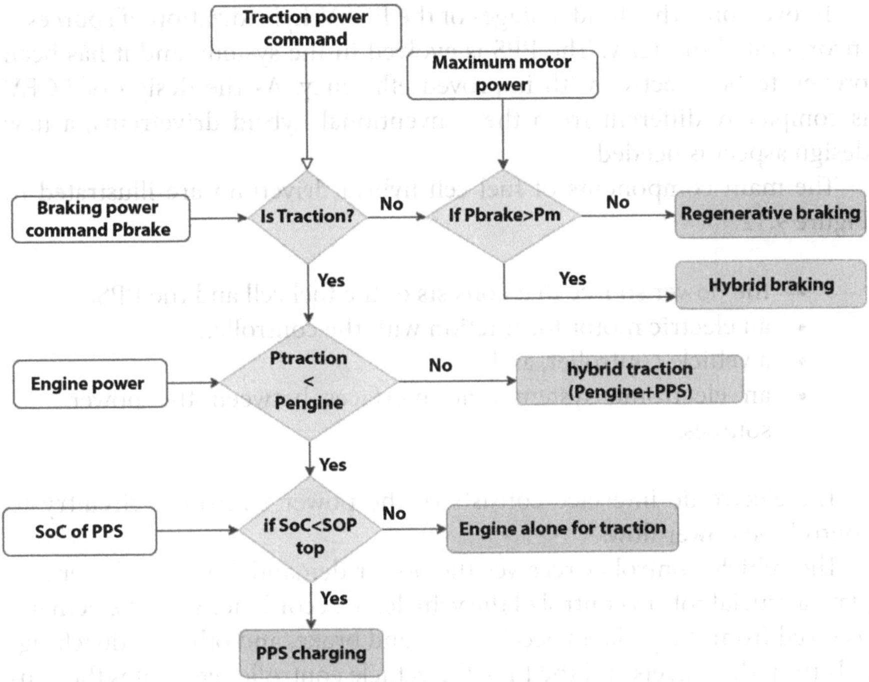

Figure 9.11 Flowchart of maximum SoC of peak power source.

Here, because the power requirement is less, only regenerative braking is used. The flowchart pertaining to the Max. SoC-of-PPS is depicted in Figure 9.11.

9.7 Design of Fuel Cell Electric Vehicles with Fuel Economy

Fuel cells are one of the cutting-edge power sources for transportation applications. They exhibit much less emission and have higher efficiency compared to ICEs. Although the FCEVs have considerable advantages, they exhibit the disadvantage of a bulky power source. This occupies more space, and the FCEV takes a longer time for startup and shows a slow response. The efficiency is further lowered during the dynamic conditions of sharp acceleration and low-speed driving. The production of enormously large power output during quick acceleration and rendering of very low power during low-speed driving lead to a decrease in efficiency [12–15].

To overcome the disadvantages of the FCEV, hybridization of sources is incorporated in FCEV. The PPS is evolved in the system, and it has been proven to be effective with improved efficiency. As the design of FCEV is completely different from the conventional hybrid drivetrains, a new design aspect is needed.

The main components of fuel cell hybrid drivetrain are illustrated in Figure 9.12:

- the power source that consists of the fuel cell and the PPS,
- an electric motor for traction with the controller,
- a vehicle controller, and
- an electronic system that interfaces between the power sources.

The electronic interface consists of the power electronic circuitry to control the power flow.

The vehicle controller receives the power demand from the driver and plays a crucial role in control of the vehicle. In accordance with the demand received from the pedal of acceleration and brake, and other feedback signals from the wheels and the PPS, the vehicle controller generates the control signal. The control signal actuates the motor power output accordingly and reorganizes the energy flows between the sources and the drivetrain.

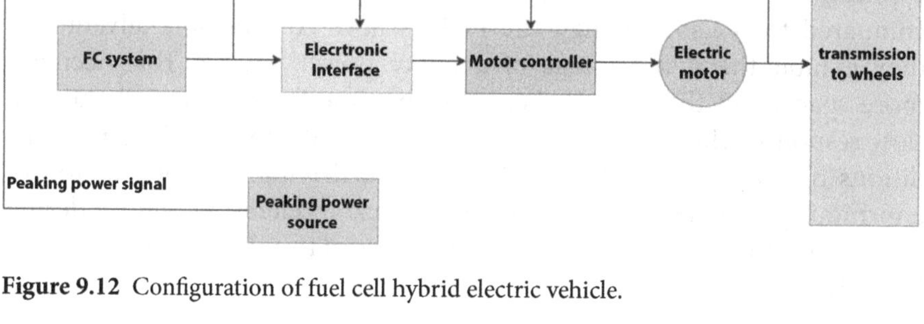

Figure 9.12 Configuration of fuel cell hybrid electric vehicle.

- During sharp acceleration, the vehicle demands peak power and is delivered by the fuel cell and the PPS together.
- During braking, the electric motor harnesses the power by working as generator and the recovered power is stored in PPS.
- The PPS charges from the additional power available at the fuel cell when the load power demanded goes lower than the rated power of the fuel cell.

Thus, by implementing proper design and control strategy, the PPS gets charged within the system.

The vehicle controller implements the control technique in the vehicle, and it controls the power flow between the sources and the drivetrain. It should ensure that the power requested by the driver is always met by the traction motor and that the sources are maintained at its optimum operating region.

The power command P_c is the power that the motor is expected to produce based on the brake command or traction command received from the driver. The $P_{m\text{-}in}$ is the electric power input to the motor to produce the power command P_c. Thus, in traction mode, the electric power input to the motor drive can be expressed as in Equation 9.5.

$$P_{m\text{-}in} = P_c/\eta_m, \tag{9.5}$$

and η_m indicates the efficiency of the motor.

During the regenerative braking mode, the motor recovers power by functioning as a generator, and the electric power output ($P_{m\text{-}out}$) from the motor is expressed as

$$P_{m\text{-}out} = P_{mb\text{-}c} * \eta_m, \tag{9.6}$$

where power requested during braking is given by $P_{mb\text{-}c}$.

The flowchart of control strategy in fuel cell HEVs is showcased in Figure 9.13. The fuel cell and the PPS are controlled to produce the required power based on the power request and other information collected from the vehicle. The drivetrain is operated in various modes with different power control strategy. The modes are standstill mode, braking mode, and traction mode.

Standstill mode: In this mode, the drivetrain is in idle condition, and no power is supplied to the drivetrain.

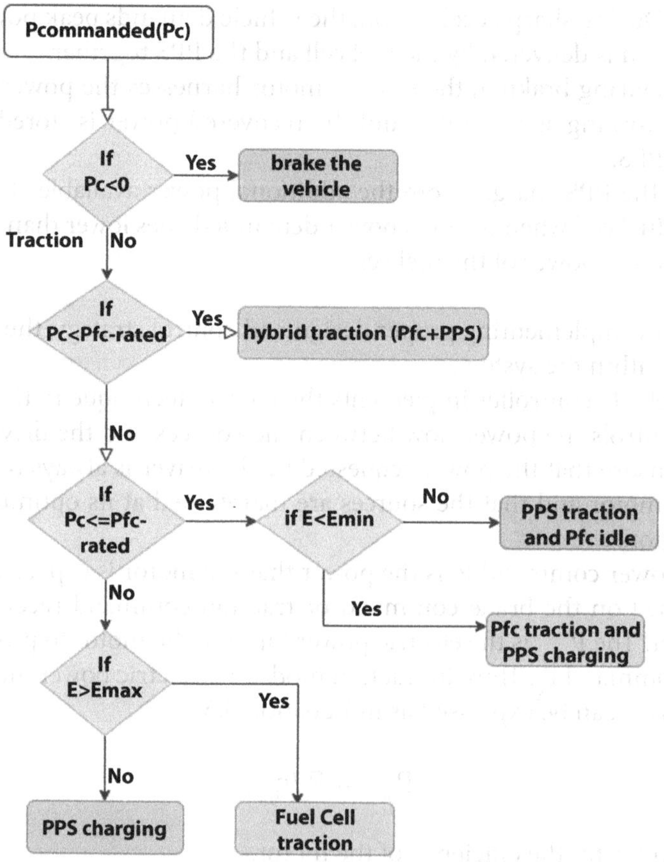

Figure 9.13 Flowchart of control strategy in fuel cell hybrid electric vehicle.

Braking mode: During this mode, regeneration happens and power is recovered by the PPS. The fuel cell system operates at idle condition.

9.7.1 Traction Mode

1. Hybrid traction mode is used if the power requirement of the motor commanded input is greater than the rated power of the fuel cell system. Here, the fuel cell is unable to meet the demand of the vehicle, and the remaining power is supplied by the PPS.
2. If the power requirement of the motor commanded input is less than the minimum power of the fuel cell, it supplies power to the vehicle and powers the PPS depending on the

SoC of the PPS. If PPS is above 70% charged, then it supplies power to the traction system and maintains the fuel cell system idle.

3. In this mode, the load power commanded is less than the rated power of the fuel cell. If PPS is at its optimum charge, then charging is not required for it. Under this condition, the fuel cell system alone drives the vehicle. If PPS needs charging, the fuel cell supplies power to the drivetrain and charges the PPS.

Thus, the various modes are effectively deployed by the vehicle controller and achieve maximum fuel efficiency.

9.8 Conclusion

Hybrid electric vehicles (HEVs) employ two or more power sources and exhibit the benefits of both ICE and EVs, overcoming the disadvantages of a conventional engine. This chapter has described the different configuration design of HEV based on the connection of the sources and the power supplied by the sources to the traction motor. The power demand is based on the requirement of the driver, and the control strategy used by the vehicle controller to govern the power flow in the vehicle for different modes of operation was discussed in detail. A fuel cell has the advantage of high efficiency and low emissions and is used for hybridization with the ICE and the PPS to enhance the performance of the vehicle. A general methodical design and a control scheme for series hybrid power train with optimum SoC of the peak power source control technique and fuel cell hybrid electric drivetrains with fuel efficient operating modes were discussed with the necessary equations. The optimum design of the HEV can efficiently use the sources and render a smooth operation for the entire ride.

References

1. Kabalan, B., Vinot, E., Yuan, C., Trigui, R., Dumand, C., & El Hajji, T., Efficiency improvement of a series–parallel hybrid electric powertrain by topology modification. *IEEE Trans. Veh. Technol.*, 68(12), 11523-11531, 2019.

2. Miller, J. M. Electric Powertrain: Energy Systems, Power Electronics and Drives for Hybrid, Electric and Fuel Cell Vehicles. *IEEE Power Electron. Mag.*, 5(4), 86-87, 2018.
3. Arora, K.; Kumar, A.; Kamboj, V.K.; Prashar, D.; Shrestha, B.; Joshi, G.P. Impact of Renewable Energy Sources into Multi Area Multi-Source Load Frequency Control of Interrelated Power System. *Mathematics*, 2021, 9, 186.
4. Benajes, J., García, A., Monsalve-Serrano, J., & Martínez-Boggio, S., Optimization of the parallel and mild hybrid vehicle platforms operating under conventional and advanced combustion modes, *Energy Convers. Manage.*, 190, 73-90, 2019.
5. Liu, Wei. *Hybrid electric vehicle system modeling and control.* John Wiley & Sons, 2017.
6. Sabri, M. F. M., Danapalasingam, K. A., & Rahmat, M. F., A review on hybrid electric vehicles architecture and energy management strategies, *Renew. Sust. Energ. Rev.*, 53, 1433-1442, 2016.
7. Arora, K.; Kumar, A.; Kamboj, V.K.; Prashar, D.; Jha, S.; Shrestha, B.; Joshi, G.P. Optimization Methodologies and Testing on Standard Benchmark Functions of Load Frequency Control for Interconnected Multi Area Power System in Smart Grids. *Mathematics*, 2020, *8*, 980.
8. Dagci, O. H., Peng, H., & Grizzle, J. W., Hybrid electric powertrain design methodology with planetary gear sets for performance and fuel economy, *IEEE Access*, 6, 9585-9602, 2018.
9. Tran, M. K., Akinsanya, M., Panchal, S., Fraser, R., & Fowler, M., Design of a hybrid electric vehicle powertrain for performance optimization considering various powertrain components and configurations, *Vehicles*, 3(1), 20-32, 2020.
10. Tribioli, L., Fumarola, A., & Martini, F., Methodology procedure for hybrid electric vehicles design (No. 2011-24-0071). SAE Technical Paper, 2011.
11. P. Singh, K. Arora, U.C. Rathore, Control Strategies for Improvement of Power Quality in Grid Connected Variable Speed WECS with DFIG – An Overview, *J. Phys. Conf. Ser.*, 2022.
12. Ehsani, M., Gao, Y., Longo, S., & Ebrahimi, K., *Modern electric, hybrid electric, and fuel cell vehicles*, CRC Press, 2018.
13. Hayes, J. G., & Goodarzi, G. A., Electric powertrain: energy systems, power electronics and drives for hybrid, *Electric and Fuel Cell Vehicles*, 2018.
14. Nikowitz, M., *Advanced Hybrid and Electric Vehicles: System Optimization and Vehicle Integration*, Springer, 2016.
15. R. Sahani, M. Daga, I. Jamatia, P. Kumar and S. L. Tripathi, Design of a self-charging electric vehicle with embedded multiple natural energy sources, *2023 IEEE Devices for Integrated Circuit (DevIC)*, Kalyani, India, 2023, pp. 296-301, doi: 10.1109/DevIC57758.2023.10134774.

10

EV Battery Charging System

Balamurugan M.[1]*, Narendiran S.[2], Sarat Kumar Sahoo[3] and Fernando Yanine[4]

[1]Department of EEE, Dayananda Sagar College of Engineering, Bangalore, Karnataka, India
[2]Department of EEE, Sri Krishna College of Technology, Coimbatore, Tamil Nadu, India
[3]Department of Electrical Engineering, Parala Maharaja Engineering College, Behrampur, Odisha, India
[4]Faculty of Engineering, Universidad Finnis Terrae, Santiago, Chile

Abstract
Electric vehicles (EVs) have been recognized as one of the capable technologies of the future. Electric vehicles, including fuel cell, hybrid electric, and plug-in hybrid EVs, have the potential to improve fuel economy and decrease gas emissions when compared to ordinary vehicles. One of the vital challenges within the progression of EVs is to create a control technique that meets the control requirement of the vehicles. The battery is a primary energy depository device in an EV. As of late, large projects are underway for in-battery improvement such as Li particle, which is being presently utilized by the new age of EVs. Because distinctive sorts of motors show diverse qualities, it is vital to assess motors according to some fundamental guidelines in order to pick a specific type of motor for an EV. Electric motors utilized in an EV ought to possess properties as in easy plan, high vitality, low upkeep cost, and great control. One of the major challenges of electric-drive vehicles is the advancement of a productive control methodology. The control methodology is an algorithm that decides when and at what control level to run the vehicle's Internal Combustion Engine (ICE) that receives control request at the wheels, the state of charge of the battery, and the current control level of the ICE. The main objective of this chapter is to examine various control strategies for charging EVs and its real-time challenges.

***Keywords*:** Battery, grid, electric vehicle, motor, internal combustion engine

**Corresponding author*: balamurugan-eee@dayanandasagar.edu

Krishan Arora, Suman Lata Tripathi and Himanshu Sharma (eds.) Electric Vehicle Design: Design, Simulation and Applications, (215–234) © 2024 Scrivener Publishing LLC

10.1 Introduction

Design of an electric vehicle (EV) is an uphill task. The battery is the heart of an EV. Batteries must be developed to meet the necessities of the engine and charging methodology used by the vehicle. These include physical limitations such as competent packing within the car body to make the most of capacity. As a major weight factor for EVs, designers must also contemplate replacement of battery within the vehicle. This is because it can affect the vehicle's energy efficiency and drivability [1, 2].

EV charging requirements are based on battery specifications and required performance. The battery is supplied with the correct voltage and current chargeable level [3]. Typical power and voltage ratings vary on the basis of the number of EV batteries as shown in Table 10.1.

Here, the outline of the specifications, safety requirements, and management arrangements used to design EV batteries have been provided.

Electric vehicle batteries often consist of tiny discrete cells organized in a series/parallel arrangement to reach the anticipated voltage and power in the final package. A typical package consists of a block of 18 to 30 parallel cells connected in series to attain the anticipated voltage. For example, 96 series blocks are arranged in a nominal 400-V package.

Typical pack voltage ratings for today's vehicles are 100 V to 200 V for hybrid or plug-in hybrid vehicles and 400 V to 800 V or higher for pure EVs. This is because the higher the voltage, the more power the equivalent diameter of copper wire can carry with less loss. Disadvantages of high voltage include the need for components with high-voltage ratings throughout the system. Also, the low voltage prevents the possibility of using a DC fast-charging station without integrating some kind of DC-DC step-up converter in the car charger [4].

Table 10.1 Battery capacity and voltage ratings for various EV segments.

Vehicle type	Capacity of battery	Battery voltage
Electric Two-Wheelers	1.2 kWh to 3.3 kWh	48 V to 72 V
Electric Three-Wheelers	3.6 kWh to 8 kWh	48 V to 60 V
Electric Cars (first generation)	21 kWh	72 V
Electric Cars (second generation)	30 kWh to 80 kWh	350 V to 500V

The battery blocks must be monitored and balanced, and dedicated circuit boards are involved in a package to accomplish this task. These boards must contain a remote communication interface, as every board's ground reference differs by several hundred volts and with the foremost battery management system.

These boards measure the temperature and voltage of every block and also measure temperature between the blocks. It also comprises tiny groups of resistors to complete balancing duties. Cells in the package must be reserved a few millivolts apart to permit maximum energy transfer in and out of the pack. Due to normal variances in cell production, some cells will discharge or charge slightly faster when compared to others. To counter this, at the time of charging, equalization has been done to draw a reduced amount of power from the higher voltage blocks to bring them closer when compared to other blocks [5, 6].

These block monitoring boards will include an extra packet security feature, allowing very precise temperature monitoring of cells and connection points in the package. In the case of damaged cells, for example, this indicates that a fault may be reported earlier, and, thus, serious impairment or the possibility of a fire may be avoided.

10.2 Electric Vehicle Charging Infrastructure

The electric vehicle supply equipment (EVSE) is the fundamental unit of the EV charging structure. EVSE gets power from sources and uses controller systems and wired connections to securely charge EVs. The EVSE control system permits a variety of purposes such as user validation, charging permission, information measuring and network management exchange, privacy, and safety [7, 8]. We recommend using an EVSE with minimum fundamental control and managing capabilities for all charging purposes. The power ratings of various EVSE segments are shown in Table 10.2.

There are currently three models of EV chargers: stage 1, stage 2, and stage 3. Stage 3 has been generally referred to as DCFC (direct current fast charging). In stages 1 and 2, the EV is coupled to a 120-V/240-V AC power source, and the EV's battery charger converts AC power to DC power required to charge the battery and also to control the charging method. DCFC alters AC power to DC power, which bypasses on-board battery charger and goes directly to the EV's battery. This allows the DCFC to unswervingly charge the EV battery. Stage 1 and 2 charging is usually restricted by accessible AC power and the magnitude of the EV's on-board battery charger. DCFC is restricted by DCFC device rating and availability

Table 10.2 Power ratings of EVSE.

Charging type	Power rating	Type of current	EV compatibility
Normal Power Charging	P ≤ 7 kW	AC and DC	Electric two-wheelers, three-wheelers, and cars and light commercial vehicles to 1 ton
	7 kW < P ≤ 2 kW	AC and DC	
High Power Charging	22 kW < P ≤ 50 kW	DC	Electric cars, light commercial vehicles, and medium commercial vehicles (1–6 tons)
	50 kW < P < 200 kW	DC	

of power from grid or mains power source. Charging rate, EV range, and how long EV can charge (dwell time) combine to govern the finest sort of EV charging system desired for your application [9, 10].

Stage 1

Stage 1 charging is the modest EV charging, connecting the EV to a typical 120-V AC outlet utilizing a proprietary electrical cable with suitable plugs on respective end. This type of charging is usually restricted by the quantity of current that can be provided by the outlet. Typically, it is no more than 12–16 A (1.44–1.92 kW). This is based on an EV rated at 3 kWh every 5.8 miles, equivalent to 1 h of load. After 10 h of charging overnight, the battery only lasts 58 miles. Stage 1 charging is only beneficial for EVs with restricted mileage, low daily mileage, or if the EV is only used a few days apart.

Stage 1 chargers are cheap, are readily accessible, and can be provided with your EV.

Stage 2

Stage 2 chargers permit the EV to be coupled to a 240-V outlet, such as that utilized for electric stoves/tumble dryers. Stage 2 chargers are now accessible up to about 20 kW and also add 60 miles per hour charged at 20 kW. Most of the stage 2 chargers are available in the range of 7 kW to 10 kW. Long-distance cars or delivery trucks can have a battery rating of

100 kWh and can be recharged in just 10 h using a 10-kW level 2 charger that takes into account system losses. Higher-capacity stage 2 chargers are fixed, but lower-rating portable chargers are accessible. However, finding a 240-V outlet to plug in can be significantly more difficult than a stage 1 120-V outlet.

Stage 3

A stage 3 charger uses DC current to charge EV batteries instead of going through the built-in AC battery charger. This stage of charge permits the system to charge the battery with a greater capacity. DCFC chargers are typically restricted to viable use, at commercial fast-charging stations. The 100-kW DCFC can charge an EV with a 100-kWh battery in around an hour [11–13].

DCFC chargers are suggestively more affluent than stage 1 or stage 2 chargers and need a three-phase 480-V power supply. This mainly restricts them to viable EV charger connections.

10.3 Power Electronics Converters Used for Charging System

In this section, the power electronic converters for electrical vehicles are discussed. Several configurations of EV power supply require at least one DC-DC converter to interface with the DC supply source (fuel cell, renewable, and supercapacitor) or battery to the DC link [14].

DC-DC converters are utilized to convert fixed DC to variable DC voltage source in one direction from input to output. For electrical vehicles, due to the capability of the regenerative braking, bidirectional DC-DC converters are used to transmit power in either direction. Two major types of DC-DC converters used in EVs are unidirectional and bidirectional converters. The application of unidirectional converters is provided to several onboard loads like controls, sensors, utility, entertainment, and protection equipment. It is mostly utilized in tractions and DC motor drives. Bidirectional converters are mostly used for regenerative braking, charging the battery, and power backup. It facilitates safe operation of EV when both the drives and engine fail to drive motor. Both bidirectional and unidirectional DC-DC converters are isolated to offer the protection for loading devices. Most of the DC-DC converters include a high-frequency transformer.

The presence of a transformer in the DC-DC converter will lead to subsequent problems, as follows:

- The high-voltage stress over diodes and switches of the converter due to the ringing effect brought by the inclusion of leakage inductance and the semiconductor yield capacitance.
- It will increase converter weight, volume, Electromagnetic Interference (EMI), area and cost.

The main objective of designing DC-DC converters is to provide cost-effective and highly efficient converters. Bidirectional converters are in high demand because the power will flow in both directions. To reduce the high current or voltage stresses because the energy was kept in the transformer leakage inductance, active clamping and solutions of soft-switching can be amalgamated [15, 16].

DC-DC converters take up a significant part in allocating EV power well. By providing more attention for power electronic converters in EVs, it might be logical that DC-DC converters will keep on playing a vital part in EV developments in the future.

In the motor drive system, the converter is generally used to commutate the motor phases. The objective of power converter utilized in the drive system not only provides energy to an electron device but also controls the current to run into the necessity of the particular device and should also have the competency of increasing and reducing the phase current. The routine in which the energy is controlled provides a way to several exceptional converter configurations for switched reluctance motor (SRM) drives. A decent number of converters have been established for SRM. All the converter configurations except the two-stage power converter topology adopt that a DC voltage source has been accessible for their contributions. The DC source is generally obtained from rectified AC supply with filter or from a battery to the power converter. The converter configurations provided for the SRM can be broadly arranged as q, q+1, 1.5q, and 2q switch topology configurations where q denotes the quantity of the machine phases.

To excite the phases of the motor, asymmetric converters have been used, and it has three probable voltage states. Figure 10.1 displays the asymmetrical converter topology for four-phase SRM drive. The probable combination of switching states required for the circuit is 3. The particular phase of motor is excited by turning on the switches S_1 and S_2. Ignoring the drop of the devices, a voltage of V_{dc} is provided to the phase winding. If the winding current surpasses the reference current, then either S_1 or S_2 is turned off, which indicates that the voltage through the winding is

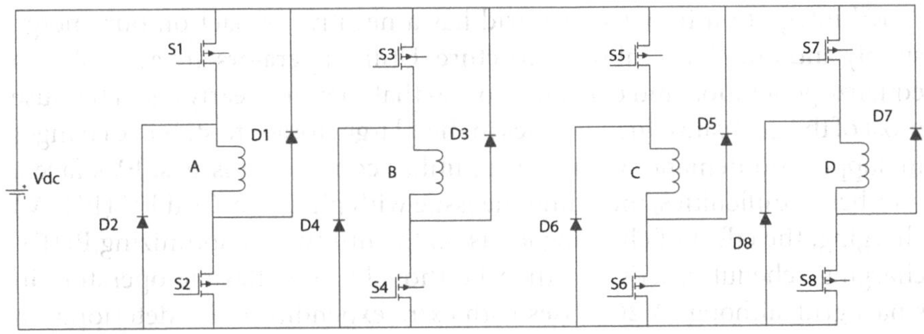

Figure 10.1 Circuit diagram for asymmetric converter.

0 and that the current will flow through the freewheeling diode through the diodes D1 or D2 [17, 18].

The winding current is commutated by turning off together the phase switches, and the voltage equal to $-V_{dc}$ is provided to the winding as long as current flows through the winding. The zero voltage is provided to the winding until it must be turned on once again. Throughout the current control phase, the upper switch is kept on as long as the phase is to be on and the lower switch is turned on and off.

The merits of the converter are as follows:

1. Complete autonomy to apply $+V_{dc}$, $-V_{dc}$, or zero to the phase.
2. Two or even three phases can be turned on irrespective of the current in the other phases, by this means permitting the negative torque to be immediately produced for charging the direction of rotation.
3. If one of the switches is damaged, then the drive can still operate at reduced power levels with the remaining switches without any operational interruption.

10.4 Control Strategies of EV Charging System

EVs are regarded as movable power sources that may be connected to any substructure, including a power grid, in various geographic positions to create a dynamically altering system. Future smart grids are expected to include vehicle-to-grid (V2G) and grid-to-vehicle (G2V) connections, which can significantly improve the system's performance in several ways. The power system's performance is significantly improved by the control method for integrating EVs.

EV integration into a smart grid has a negative impact on our energy supply and our electrical infrastructure. Utility operators are adaptable to equalize generation and consumption so that they are nearly equal because most of the grid causes deficiencies in their large storage to stabilize changes in supply and demand. Considering utility connection issues, EVs face a number of difficulties, including the issue with plug-in hybrid EV (PHEV) charging, the effect of charging loads on the utility, and optimizing PHEV charging schedules. EVs can then be thought of as flexible operators in smart grid, although V2G arises with extra expenditures for deterioration of battery. By supporting electricity grid, V2Gs are potential options for boosting future smart grid resilience. Furthermore, it can be utilized as power sources to supply electricity for essential loads in harsh weather circumstances wherever there are incidences of power breakages and in some facilities like hospitals, cell towers, and various emergency scenarios where there is a higher requirement for power sources [19].

i. Multi-Agent Control Strategies

In Low Voltage (LV) distribution networks, EV charging operations are managed using the multi-agent control (MAC) technique. It is made up of two or more intelligent agents, each having a local purpose that enables the achievement of a global goal in the lack of principle control. Autonomy, minimal maintenance, extensibility, and fault tolerance are among MAC's core benefits. Flexibility, which is used to operate under diversely distributed generation and EV dispersion levels, is one of the advantages of MAC methods. The other is scalability, which proves the capacity to function well in both single-liable and multiple-liable scenarios. It involves service restoration for testing systems of all sizes, both minimal and maximum. In order to control power system issues such as outage administration and service refurbishment, control of microgrid and virtual power plant control is essential. Figure 10.2 depicts the decentralized MAC-based techniques that employ three-layer designs.

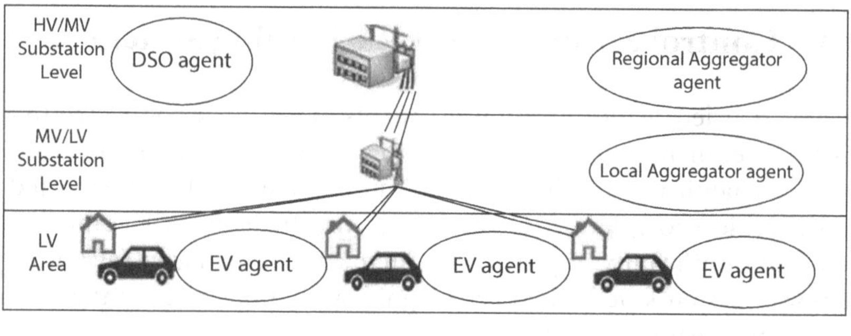

Figure 10.2 Hierarchy in the MAC method.

It was determined that MAC techniques outperform centralized ones in terms of computing effectiveness and scalability. The architecture of this MAC is composed of three layers. The MAC can further be utilized as a self-healing system, auto-detection capability, and functionality recovery where it meets casualties, regardless of how little or serious they are. The word refers to actions like prompt problem detection, promotion of efforts to mitigate any severe consequences brought on by casualties, and prompt restoration of the process' circumstances to a stable operational state in the context of a power system. The reactive phase and the restorative phase are the two main phases of the self-healing process. Emerging structures where dynamic reorganization is essential; evaluating systems to evaluate complex problems that cannot be resolved by one agent alone; relating to such problems that comprise numerous problem-solving approaches and require several types of proficiency and knowledge or where there are further lookouts—these are the tasks were MAC claims applications. Utmost of the advantages of MAC expertise are resilience, flexibility, quick computation, extensibility, development- and maintenance-friendly, affordability, and reusability.

Regarding MAC's challenges, they include a variety of issues with their design and implementation, including their effective and functional relating conventions, planning their task and its issues, fragmentation, assigning each agent a task and synthesizing their subtasks, agent identification, survey and open system position, continual and comprehensible model behavior when avoiding unfavourable connections, and representing the agents. There are two types of MACs: closed MAC and open MAC. Static designs, pre-assigned modules, and their functions are all parts of a closed MAC. The shared vernacular, the ability to industrialize any agent into a skilled and helpful agent, and the ability for multiple developers to work on the system simultaneously are all previously recognized aspects of such systems and should be borne in mind when recommending MAC to a company, for instance. In addition to the already mentioned advantages of load distribution, expertise, comprehensibility, likelihood, as well as an identification of interaction conventions and language supported the mediators and the sharing of architecture and software among them. It has the following drawbacks: it is difficult to work with other systems, maintenance costs can be very high, and there is little tolerance for malfunctions. Open MAC is a different sort of MAC in which there is only one agent present and no previous static design. Agents in this situation are probably unaware of others; hence, a mechanism for finding, evaluating, and outlining others is required.

ii. Aggregated Control Strategies

It announces many EVs as well as an aggregator storage strategy that asserts that the EVs' sustainable operation is confirmed through modelling of driving statistics. In this sort of control method, the scheduled charging is automatically completed, and the centralized V2G and G2V system successfully completes universal energy and power management distributed to various EVs. It is imperative to have a logical or control relationship while speaking with an aggregator. Presumptuously equal to renewable energy generation and gathered electrical loads in the instance of distribution feeders, the collector is responsible for the appeal of the secondary regulator market as long as the regulator power is symmetrical and the communication with TSO, on one hand, and with EV, on the other, is permitted. The policy of combined storage for the collector and various EVs enables for the incorporation of not only individual EVs but also renewable energy sources into the power system while reducing inspirations on the charging procedure of the vehicle to that of distribution arrangement [20].

The independent planned charging system is used in such a scheme to support each vehicle with its charge request and battery management, which are essential for vehicle application and usage.

An aggregator also reports control signals in the event of global energy and power management units related toward idling EV is shown in Figure 10.3. In addition to these factors, there are other details concerning compromising examination in terms of losses in energy and its challenging

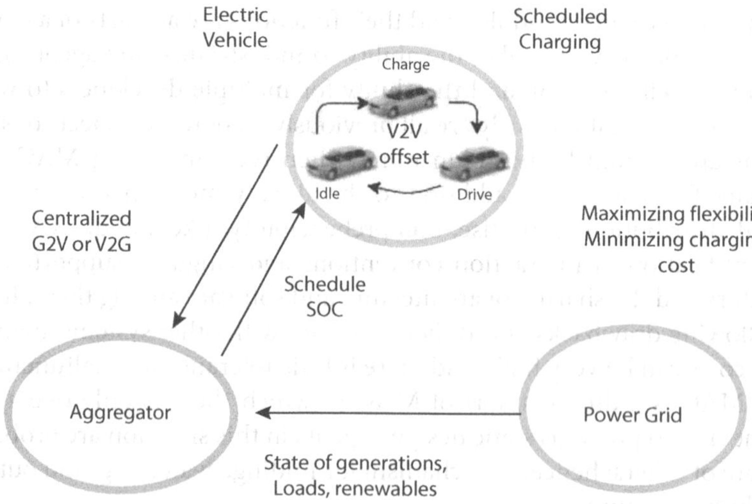

Figure 10.3 Aggregated incorporation of EVs with utility.

conformations like renewable energy resources, which are regarded as one of the valuable data for the operational statistics of EV collector. For instance, only 38.2% of the battery's total net capacity from a fleet of combined PHEVs relating to several automobiles must be obtained for supply of static control of positive or negative standby power at any given period in that stated scenario. Through the aid of power assets of the combined type and unmatched transmission from their provided pool of available capacity in accordance with the needs of the grid, the manager directly utilizes the shared load-frequency control (LFC) [21]. Combined regulator standby power can be flawlessly communicated from a set of serviceable regulator reserve capacity that has been selected in accordance with grid requirements, as well as a reduction in actuator utilization charges and any associated technical limitations like the increasing and fluctuating obtainability of actuators. PHEVs are engaged into consideration for establishment of added controller reserve power when it consists of united battery size from fleets of PHEVs as additional regulator reserve volume at exclusion. In this scenario, this dynamically rapid retorting battery volume can also be utilized for leading relatively minor and fluctuating frequency conflicts.

iii. Load-Frequency Control (LFC) of EV Connections

In order to balance the provided frequency at its target rate, active power must stay generated and consumed in a balanced manner in order to preserve the generation and load side's stability. This is done by introducing the concepts of load frequency and moving closer to common features of frequency control services. To perform this control, a specified quantity of active power that has been too tied to frequency controller reserve is maintained available. Three control structures are presented and categorized as primary, secondary, and tertiary for frequency reserves. In general, these sorts of controller in specified extents for each of their respective control areas, where an upsurge in generation side results in an equivalent reduction in load, where as reduction in the generation side results in an equivalent rise in load. The amount depends on the control area's size and generation set. It is possible to achieve the LFC through V2G by properly controlling the charging and discharging procedures for EVs in accordance with the LFC signal. It is dispatched by the fundamental load dispatching center and is thought to be sent to the EVs via the locally managed centers. The strategy includes a thermal power model, wind power and generation design, solar generation design, load design, EV design, and LFC system design, as shown in Figure 10.4. The LFC signal is produced by the LFC model [22].

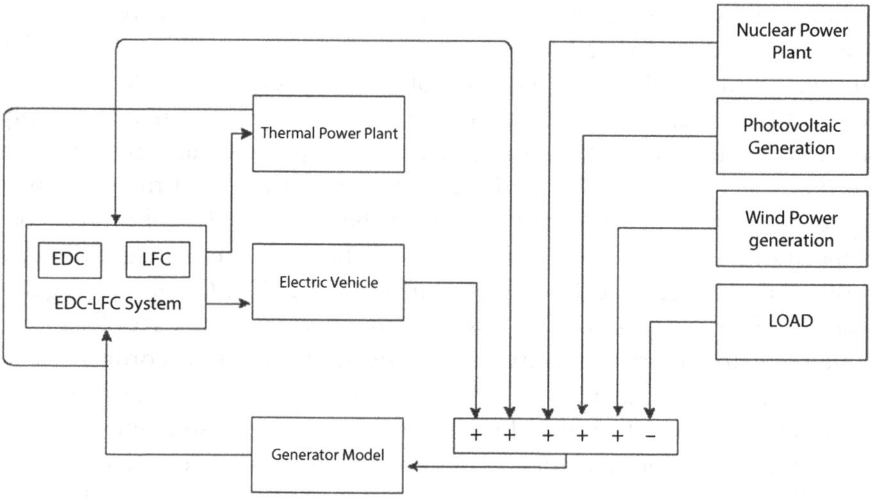

Figure 10.4 LFC strategy for incorporation of EVs with utility grid.

10.5 Various Modes of Charging System

An electric car's battery pack can be charged in a diversity of methods. It can be a little daunting at first to have to choose between the slow and fast charging techniques and multiple connector kinds. It really boils down to two decisions: where you choose to charge and how quickly you choose to charge. The speed of charging will differ based on type of EV, the size of battery, and the charging station type used. These factors are interconnected [23, 24].

Charging modes:
The EV conductive charging system IEC 61851-1 Committee has subsequently established four charging modes. These modes are the ones that are frequently used in various E-Mobility technical documents, studies, and marketing brochures in Europe.

1. **Mode 1** (Schuko mode) describes the charging of a domestic outlet using a straightforward extender cable without the use of any safety precautions. Although having a fuse to protect it, the household outlet's response time is so poor that Mode 1 charging is exceedingly risky. Also, it is prohibited in many areas of the world (yet it is in use in some parts of Europe). Schuko charging is something that we highly advise against.

2. Another term for charging from a standard outlet is **Mode 2**, but, this time, a control and safety device is put into the charging cable. Although this method of charging is far safer than the Schuko, the maximum charging capacity is restricted by the outlet's maximum rating, which is normally 10 A at 230 V (or a maximum of 2.3 kW).
3. The EVSE with suitable control and protection is referred to as a dedicated charging point in **Mode 3**. In Europe, this can range from 3.8-kW to 22-kW AC charging, making it the most used charging method globally.
4. DC charging is covered by **Mode 4**. The onboard charger for EVs is bypassed, and the charging station delivers DC energy to the battery directly through a DC connector.

10.6 Real-Time Challenges of EV Charging Infrastructure

EV charging structure must provide with mandate from automakers presenting novel models as the EV revolution gains traction. Also, they must stay current with the quick-changing skills and charging methodologies utilized by contemporary EVs. Nonetheless, the efficacy of EVs has been frequently checked. Every brand faces network problem as an outcome of the influx of competing quick or home charger models onto the market. There are situations when the charging point makers are not at liability [25].

Below, we go over each of these problems in further detail to help you understand what challenges manufacturers encounter while setting up and managing their charging network. In addition to being less prevalent than conventional fuelling stations, charging networks frequently require connection to the electrical grid. This involves balancing the demand during peak hours and managing an increase in electricity supply for individuals who require up to 350 kW of rapid charging. Here are the top eight issues that EV charging networks confront, and how they affect overall electric car efficiency.

i. Poor Charging Infrastructure

Behind the scenes, there are several agreements involving landowners, charging device makers, and grid operators. As a result, installing a charging network is costly, costing somewhere from $2,500 for a slow charger to $35,800 for a fast charger. It could be challenging to comprehend the charging networks that are most appropriate for use at home or in a place of business. For instance, it may be challenging to determine the optimal electric charging infrastructure to employ in a densely populated location. The ease of finding a charging station is a crucial consideration for owners considering purchasing EVs. Some owners worry about how the network is set up. Infrastructure for charging is also lacking in rural regions. To make EV charging accessible to impoverished populations, significant investment is required in this area. A step in the right direction may be to install overnight and midway charges.

ii. Vehicle-to-Grid Interoperability

There are currently few agreements, logistics, and channels of communication between the grid, utilities, and charging manufacturers. The inability to reach consensus on a broad-scale deployment of an inexpensive, dependable, and secure EV charging network for drivers has limited the situation. All these partners are important for the necessary procedures and technologies that the V2G infrastructure has made possible. This frequently occurs as a result of information being withheld by the government or utility companies that is otherwise not provided to the charging network providers. This frequently lengthens the development process and reduces the project's profitability.

iii. Station Performance and Downtime

Driving anxiety has a psychological component that affects the thinking of the driver. Even if most EVs currently have a range of more than 300 km, this notion still needs to be overcome. Finding a working charger is the next challenge if an owner reaches a charger with a battery range of less than 10 miles. The unreliability of the electric charging infrastructure is frequently criticized on social media and in the media. About 60% of EV owners experience these difficulties, with the remaining owners reporting no problems at all. The manufacturer error check might be the first step in this process before, during, and after each charger is used. When charging providers are alerted right away to a problem, they can fix it right away and spare users any additional time or stress throughout the charging process.

iv. Payment Method Integration

You can pay for charging your automobile in a variety of ways, including apps, contactless cards, QR Codes, and RFID cards. If you could employ all these techniques at once, it would not be a problem, but that is not usually the case. A single app or method to pay for power while a device is charging would be the obvious solution, but all brands would need to work together to make this happen. Otherwise, those who are not comfortable using technology can become confused and be opposed to the EV charging network, which would prevent them from buying an electric car. There is a recent law enacted in Germany, under which companies are compelled to make it simpler for EV drivers to simply show up and charge by just using a debit or credit card. By July 2023, this will be the law, with most networks having already made the necessary adjustments. One business that is assisting with this is Hubert, which provides EV drivers with a backup payment option, removing the requirement to download and sign up for multiple apps in order to charge your car.

v. Power Issues While Charging

Rapid-charging users frequently require them for on-the-go needs at gas stations. This indicates that they are traveling farther distances and require a charging pace that is consistently rapid and uses a lot of power. One of the biggest issues for electricity companies is managing the grid during peak periods, and installing chargers just makes things worse. What is the optimum circumstance? That is, arriving when you need a power boost and plugging in for no more than 10 to 20 min while families on the road have coffee before continuing their journey. The precise charging network, however, is what determines how much electricity is given. While some new EVs can charge at 300 kW, most chargers do not use this speed to its full potential. Sometimes, the grid cannot supply power, and, other times, it could happen that the charger overheats if the peak power is used for an extended period.

vi. Charging Station to Network

Electric charging infrastructure can only interact using protocols that are now available from a single network provider. If the charger must be upgraded or if they want to switch networks, then this is a problem. A whole new universe of options would become available if the protocols were open for switching between networks. A change in network service providers promotes competition and adds a level of upgradeability to the current network, reducing the cost of future improvements. Without switching network providers, more charging stations can be established. The Open Charge Point Protocol (OCPP) is useful in this situation. Europe employs the open networking standard known as OCPP. It is also gaining popularity in the US. Although the configuration of the current versions is limited, most networks accept them, and their growth may be crucial for network integration [26].

vii. Finding Balance Between Rapid Charge Points and On-Street Units
The delicate balance between having too many chargers in gas stations and not enough at home is a controversial topic. According to studies, homeowners frequently charge their devices overnight at their homes. Nevertheless, not every residence can accommodate a charger. By 2035, England's key A-roads and motorways should have at least 6,000 high-powered charging stations installed there, according to a £950 million fund made available by the UK government. To demonstrate their dedication to the current EV issues, they recently emphasized that more than 250 chargers had opened in just 30 days. Richard Bruce, a government representative from the DfT, noted that "if you need to open up the grid connection to it, it is extremely expensive" when discussing concerns like power distribution in rural places.

viii. Ease of Use While Charging
The user experience of recharging an EV has, in the words of Richard Bruce of the DfT, been "very unsatisfactory for too long." For drivers, the unreliability and the unknown of whether a charger will be accessible at the destination are serious problems. In addition to the difficulties using numerous applications and cards, there is also the general annoyance of paying for EV charging. Users will eventually be able to view the entire network on a single app, according to long-term plans. To get the most out of their experience, find out which chargers are available and working as well as their prices. The UK government has pledged to improve accessibility for disabled drivers so they can charge their vehicles swiftly and conveniently.

Things must alter in order to monopolize the market for all charging networks with a single technique or app, making the procedure considerably simpler. Customers worry about not being able to find a charging station when their range is short, preventing them from purchasing an EV, according to consumer surveys. But, according to other consumer polls, 40% of EV owners report no problems with charging. If the money, infrastructure, and communication between the grid, charge point manufacturers, and local governments are present, then these difficulties can be overcome with the appropriate combination of home and highway service station chargers.

Conclusion

In this chapter, the battery charging system and its infrastructure for charging electrical vehicles have been briefly discussed. Electrical vehicles are on a path toward green mobility and also protect the environment from pollution. The major challenge in implementation of EVs is the range, cost, and infrastructure. Then, the various power electronics converters and their control strategies have been included. The various modes of the charging system and its ratings for electric two-wheelers, three-wheelers, and cars have been briefly discussed. The researchers have been focussed on improving the energy source and development of efficient battery charging system for vehicle technology. The various modes of EV battery charging system and their challenges are presented in this chapter.

References

1. B. Sakhdari, N. Azad and L. (2015). An Optimal Energy Management System for Battery Electric Vehicles. *IFAC-Papers Online*, Vol. 48, Issue 15, 2015, pp. 86-92.
2. P. Singh, K. Arora, U.C. Rathore, Control Strategies for Improvement of Power Quality in Grid Connected Variable Speed WECS with DFIG - An Overview, *J. Phys. Conf. Ser.*, 2022.
3. P. Sivaraman and C. Sharmeela (2021). Power quality problems associated with electric vehicle charging infrastructure. *Power Quality in Modern Power Systems*, 151-161.
4. C. Suarez and W. Martinez (2019). Fast and ultra-fast charging for battery electric vehicles—A review. *Proc. IEEE Energy Convers. Congr. Expo. (ECCE)*, 569-575.
5. Y. Santha and Pomya Kumari (2014). Electrical Vehicle with Reduced Voltage Induction Motor Drive using MLI. *Int. J. Electr. Electron. Res.*, 2(3), 149-157.
6. H. Fathabadi (2020). Novel stand-alone completely autonomous and renewable energy-based charging station for charging plug-in hybrid electric vehicles (PHEVs). *Appl. Energy*, 260, 194-197.
7. M.K Hassan, T.A.T Mohd, Ishak Aris, C.S Azura and B.S.K.K Ibrahim (2015). A Review of Energy Management System in Battery Electric Vehicle with Hybrid Electrical Energy Source. *International Conference on Engineering Education and Research*, 1-6.
8. Arora, K.; Kumar, A.; Kamboj, V.K.; Prashar, D.; Shrestha, B.; Joshi, G.P. Impact of Renewable Energy Sources into Multi Area Multi-Source Load Frequency Control of Interrelated Power System. *Mathematics*, 2021, 9, 186.
9. T. S. Bryden, G. Hilton, B. Dimitrov, C. Ponce de León and A. Cruden (2019). Rating a stationary energy storage system within a fast electric vehicle charging station considering user waiting times. *IEEE Trans. Transport. Electrific.*, 5(4), 879-889.

10. Lamprecht, S. Narayanaswamy and S. Steinhorst (2018). Improving fast charging efficiency of reconfigurable battery packs. *Proc. IEEE Des. Autom. Test Europe Conf. Exhibit.*, 585-588.
11. Kumar, K., Gupta, S., & Nema, S. (2021). A review of dynamic charging of electric vehicles. In *2021 7th International Conference on Electrical Energy Systems (ICEES),* 162-165.
12. Niculae, D., Iordache, M., Stanculescu, M., Bobaru, M. L., & Deleanu, S. (2019). A review of electric vehicles charging technologies stationary and dynamic. *11th IEEE International Symposium on Advanced Topics in Electrical Engineering (ATEE),* 1-4.
13. Meintz (2017). Enabling fast charging—Vehicle considerations. *J. Power Sources*, 367, 216-227.
14. Arora, K.; Kumar, A.; Kamboj, V.K.; Prashar, D.; Jha, S.; Shrestha, B.; Joshi, G.P. Optimization Methodologies and Testing on Standard Benchmark Functions of Load Frequency Control for Interconnected Multi Area Power System in Smart Grids. *Mathematics,* 2020, *8*, 980.
15. Dakshina M. Bellur, Marian K. Kazimierczuk. (2007). DC-DC converters for electric vehicle applications. *Electrical Insulation Conference and Electrical Manufacturing Expo*, 1-6.
16. T. Porselvi, M. K Srihariharan, J Ashok and S. Ajith Kumar (2017). Selection of Power Rating of an Electric Motor for Electric Vehicles. *International Journal of Engineering Science and Computing (IJESC)*, 7(4), 6469-6472.
17. Ahmed A. Abd Elhafez, Majed A. Aldalbehia, Naif F. Aldalbehia, (2017). Comparative Study for Machine Candidates for High Speed Traction Applications. *Int. J. Electr. Eng.,* 10, 1-20.
18. M. E. Kabir, C. Assi, M. H. K. Tushar and J. Yan (2020). Optimal scheduling of EV charging at a solar power-based charging station. *IEEE Syst. J.*, 14(3), 4221-4231.
19. Z. Gong *et al.* (2020). Distributed control of active cell balancing and low-voltage bus regulation in electric vehicles using hierarchical model-predictive control. *IEEE Trans. Ind. Electron.*, 67(12), 10464-10473.
20. Mi, C. C., Buja, G., Choi, S. Y., & Rim, C. T. (2016). Modern advances in wireless power transfer systems for roadway powered electric vehicles. *IEEE Trans. Ind. Electron.*, 63(10), 6533–6545.
21. Balamurugan, M., Raghu, N., Kamala, N., Nandikolmath, T. V., & Sahoo, S. K. (2021). Solar Powered Electric Vehicle Through Wireless Power Transfer. In: *Electric Vehicles and the Future of Energy Efficient Transportation.* IGI Global publishers 219-24.
22. Global EV Outlook 2021, 2021, [online] Available: https://www.iea.org/reports/global-ev-outlook-2021.
23. M. Yilmaz and P. T. Krein (2013). Review of battery charger topologies charging power levels and infrastructure for plug-in electric and hybrid vehicles. *IEEE Trans. Power Electron.*, vol. 28(5), 2151-2169.

24. Viana, S. Semsar, M. Pathmanathan and P. W. Lehn (2022). Integrated transformer less EV charger with symmetrical modulation. *IEEE Trans. Ind. Electron.*, 69(12), 12506-12516.
25. M. R. Ahmed and A. K. Karmaker (2019). Challenges for electric vehicle adoption in Bangladesh. *Proc. Int. Conf. Electr. Comput. Commun. Eng. (ECCE)*, 1-6.
26. S. Pareek, A. Sujil, S. Ratra and R. Kumar (2020). Electric vehicle charging station challenges and opportunities: A future perspective. *Proc. Int. Conf. Emerg. Trends Commun. Control Comput. (ICONC3)*, 1-6.

11

Optimization Algorithms and Computing Techniques for Electric Vehicles: Advancements in Computing and Algorithms

S. Arun Mozhi[1]*, J. Nishanthy[1], S. Charles Raja[1] and J. Jeslin Drusila Nesamalar[2]

[1]*Department of Electrical and Electronics Engineering, Thiagarajar College of Engineering, Madurai, India*
[2]*Department of EEE, Tamil Nadu Government Polytechnic College, Madurai, Tamil Nadu, India*

Abstract

Recently, the running down of fossil fuels has led to the improvement of the electric vehicle (EV) as it has lower emission and increased energy efficiency compared to a conventional vehicle. Yet, there are number of difficulties regarding EV implementation such as reduced driving range, expensive batteries, and the need of significant charging infrastructure. Vehicle scheduling is the predominant problem in charging infrastructure. Driving range, battery cost, scheduling cost, energy consumptions in grid, vehicle cost, and vehicle and battery performance are some of the objectives need to be optimized. To overcome such difficulties, various optimization algorithms are employed. This chapter defines various fundamental optimization algorithms and computing techniques to enhance the performance, efficiency, reliability, and sustainability of EVs. Furthermore, the energy management between vehicle to grid and grid to vehicle are also included to mitigate the grid stress.

***Keywords*:** Electric vehicles, optimization algorithms, objective function, constraints, linear programming, genetic algorithm, particle-swarm optimization, dynamic programming

**Corresponding author*: arunmozhi49@gmail.com

Krishan Arora, Suman Lata Tripathi and Himanshu Sharma (eds.) Electric Vehicle Design: Design, Simulation and Applications, (235–264) © 2024 Scrivener Publishing LLC

11.1 Introduction

Knowing some basic optimization tactics can help in finding the most practical answer to a particular issue. Also, these techniques are widely used in various fields, including mathematics, engineering, economics, and computer science. In addition, optimization enhances efficiency, reduces costs, and maximizes output in complex systems, with meta-heuristic optimization techniques being widely used to address engineering design issues.

Additionally, this paper uses six major keywords to gather information and present cutting-edge techniques in a variety of categories, including fundamental, modified, and hybrid techniques. To determine which version of optimization methods works better in solving the studied problems, the results of these methods are also given. Additionally, this work helps researchers and practitioners to choose the most efficient method for solving their specific design problems [1]. The objective function is examined in a generalized form that allows for the computation of partial derivatives with respect to both x and y as well as the addition of both explicit and implicit variables. Then, we present the distributed quasi-newton method, which is used to resolve optimization threads over a large High-Performance-Computing (HPC) cluster node. Overall, the article offers a promising approach for resolving challenging optimization problems with high-performance computing tools [2].

Twenty-three objective functions of various unimodal and multimodal types are used to assess the efficacy of the Pelican Optimization Algorithm (POA). When compared to the eight well-known metaheuristic algorithms, the optimization results indicate that the POA is highly effective in exploitation, leading to a better and more competitive performance [3]. In this research, a new robust soft computing technique is used to predict solar insolation for two stations in China's southeast. To create the better hyperparameters for the Least Square Support Vector Machine (LSSVM) model compared to Genetic Algorithm (GE), Gray Wolf Optimization (GWO), Sine-Cosine Algorithm (CSA), and Multi-Verse Algorithm original version, the method also uses the improved version of the Multi-Verse Optimizer with integration of LSSVM [4].

In order to handle project improvement in the construction industry, multi-objective optimization (MOO) has been thoroughly reviewed in this paper. In order to enhance projects, this study aims to comprehend the state of MOO today and, in the future, identify research trends and pinpoint obstacles to its widespread adoption. Therefore, the research suggests three potential directions—adoptive optimization, interpretable mapping, and interactive optimization—to address these challenges [5]. In order to produce more accurate and reliable forecasts for time-varying patterns in economics, meteorology, and transportation, the paper suggests a new ensemble forecasting method called Bayesian optimization–based dynamic ensemble that incorporates statistical methods, machine learning models, and deep neural networks. Moreover, this method adjusts the weights of the models based on their past performance and uses a Bayesian optimization algorithm for hyperparameter tuning to improve its adaptability and generalization performance [6].

This study examines how the particle swarm optimization (PSO) method can be used to increase a sewage treatment system's energy efficiency. The work focuses on the activated sludge sewage treatment method and uses experimental analysis to demonstrate the effectiveness of the PSO in improving energy efficiency. Consequently, the result demonstrates that the algorithm is crucial to obtaining effective sewage treatment and enhancing the process' fiscal and social benefits [7]. This study examines the application of chaotic optimization methods to five different chaotic maps: the Chebyshev, Cubic, ICMIC, Neuron, and Sine maps. To discover the fewest problems, a novel algorithm called the Hybrid Chaotic Pattern Search Algorithm (HCPSA) combines with five chaotic maps patterns. The effectiveness of HCPSA for higher dimensional non-linear functions is demonstrated by the numerical findings over 12 benchmarks [8].

The paper explains that, by taking six different nature-inspired optimization strategies—GA, GWO, Jaya optimization, Whale optimization, PSO, and simulated annealing algorithm, it is possible to optimize the cutting constraints of plasma arc cutting (PAC) procedures for cutting stainless steel (AISI 304). This gives practitioners insight into the strengths and weaknesses of each approach, allowing them to select the strategy that is best suited to their particular application [9]. The case study used in this paper is based on the Harvard Dataverse's keratoconus dataset. As a result, each algorithm's performance—including batch, stochastic, mini-batch gradient descent, and the normal equation algorithm—is assessed on the basis of how well it can reduce the cost function of a linear regression model that has been trained on the dataset using the mean squared error as the cost function. Additionally, it is demonstrated that these algorithms produce large-scale suitable datasets with greater efficiency in terms of computational time and memory utilization [10].

11.2 Fundamental Optimization Techniques

An optimization is a collection of mathematical techniques and algorithms that find the most advantageous resolution to an issue given a particular set of constraints. Moreover, an optimization is defined as a fundamental aspect of machine learning, as it involves finding the optimal set of parameters that can accurately represent the underlying data distribution and make accurate predictions. Moreover, it plays a vital role in aerospace engineering by enabling engineers to design more efficient aircraft and aerospace systems. Thereby, in aircraft design, the objective function is to regularly reduce the overall weight of the aircraft when reaching for specific performance and safety requirements. Also, this needs to solving complex optimization issues involving multiple variables and constraints. Thus, the basic procedure of optimization for any problem involves several steps, as shown in Figure 11.1.

11.3 Problem Formulation of Optimal Solution

Consider that the optimal design optimization is an important task in industrial and engineering activities where the best design solution is selected from a set of alternatives solutions. The target is to achieve the best possible design while satisfying certain constraints and objectives. An optimization algorithm is used to find the best answer after building a mathematical model to accomplish this. Additionally, the optimization problem formulation involves constraints, objective functions, variable bounds, and other relevant parameters. Therefore, it is important to ensure that the design solution remains at limit and does not violate any constraints and to

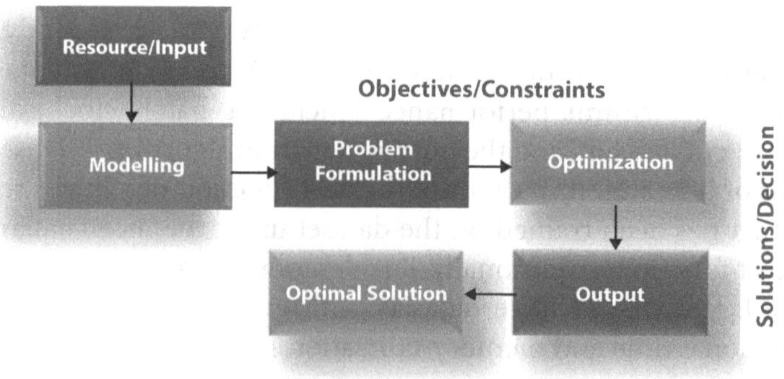

Figure 11.1 Process map to get optimal solution.

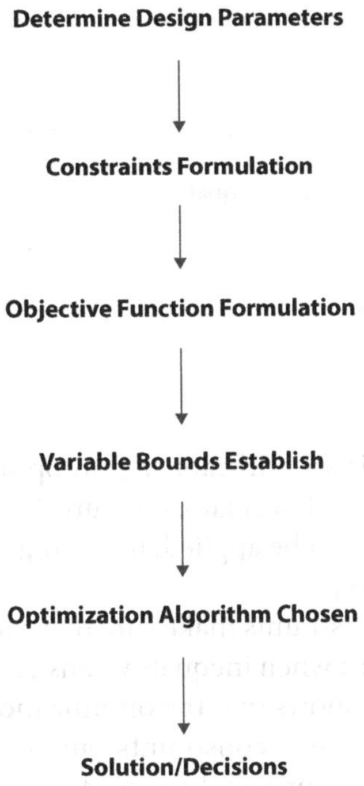

Figure 11.2 Workflow diagram for design optimal technique.

determine the best design solution for optimizing the objective function. Thus, the optimization process involved in optimal design typically carries several steps as shown in Figure 11.2.

11.3.1 Design Parameters

The initial step of selecting design parameters in an optimization problem formulation is critical to achieve the optimization process. Moreover, design variables are the parameters that can be varied due to the optimization process to attain the optimal solution [11]. Additionally, the number of design variables is important to identify the effect of optimization algorithms' efficiency, which are crucial factors to mention in the optimization functions. Furthermore, a few design parameters are having complex optimization problems, which take more time to recover. Thus, it is most required to manage a balance between the number of design parameters and the efficiency of optimization algorithms.

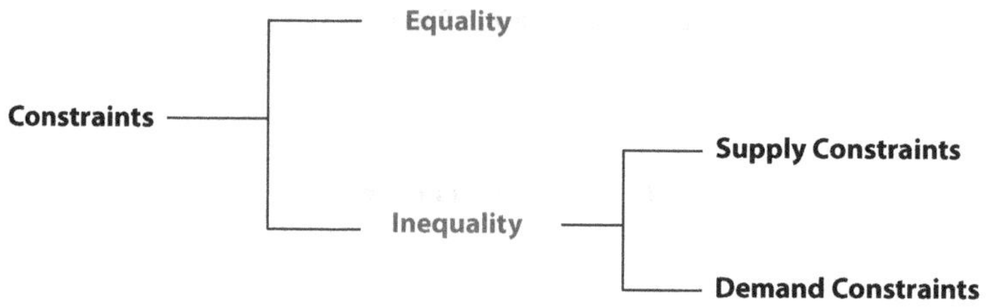

Figure 11.3 Constraint types.

11.3.2 Constraints

Identifying constraints is a significant step in optimization problems as it reveals the limitations and boundaries. Figure 11.3 illustrates the various kinds of constraints that can be applied, including equality constraints and inequality constraints [11].

Moreover, equality constraints make sure that the optimization problem satisfies a set of equations; when inequality constraints limit the values that the variables can take, it shows that the optimization problem must satisfy a set of inequalities. Inequality constraints can be further categorized into greater than, smaller than, or equal to constraints with a resource value. In some cases, constraints can be easily formulated by existing rules or mathematical expressions, whereas, in others, numerical methods such as simulation may be necessary to compute the constraints. Therefore, it is important to understand the simulation procedures of inputs and outputs required for such methods.

11.3.3 Objective Function

The significant role in optimization is an objective function formulation problem. It maps the design variables and other problem parameters to a scalar value that represents the performance or quality of the solution. It can be either a single objective having one objective function that needs to be minimized and expressed in different forms, such as linear, nonlinear, continuous, or discrete, or more objectives that is to be minimized in MOO problems [11]. By utilizing sophisticated optimization methods such as are evolutionary algorithms and swarm intelligence algorithms, there are various ways to solve multi-objective functions into single-objective functions.

In some cases, an approximating mathematical expression can be used to represent the objective function. Therefore, formulating an appropriate objective function is a critical step in solving an optimization problem, and it requires a thorough understanding of the problem and the available data. Thus, the type of maximization and minimization objective function is selected. Thereby, the decision variables' range of possible values is usually constrained by the objective function. The aim of an optimization problem is to determine the decision variables values that meet the constraints and maximize the function of objective in Equation 11.1.

$$Z_{Max} = f(x) = C_1x_1 + C_2x_2 + C_3x_3 + \ldots + C_nx_n \tag{11.1}$$

where $x_1, x_2, x_3, \ldots, x_n$ are the decision variables of the problem; $C_1, C_2, C_3, \ldots, C_n$ are the coefficients of the variables representing the contribution to the objective function; and Z_{Max} is the maximum value of the objective process that need to achieve.

When the data issue is specified in terms of costs, an objective function of the minimization type is chosen. Finding the values of the decision variables to minimize the objective function while meeting the constraints is the goal of an optimization issue in Equation 11.2.

$$Z_{Min} = f(x) = c_1x_1 + c_2x_2 + c_3x_3 + \ldots + c_nx_n \tag{11.2}$$

where $c_1, c_2, c_3, \ldots, c_n$ are the coefficients of the variables representing the cost of the objective function.

Moreover, an expression of minimization is defined as a maximization function by multiplying with (−1) for both objective functions. Hence, both Equations 11.1 and 11.2 should be mentioned below like this in Equations 11.3 and 11.4.

The maximize function is as

$$Z_{Max} = f(x) = -c_1x_1 - c_2x_2 - c_3x_3 - \cdots - c_nx_n \tag{11.3}$$

The minimize function is as

$$Z_{Min} = f(x) = -C_1x_1 - C_2x_2 - C_3x_3 - \cdots - C_nx_n \tag{11.4}$$

11.4 Optimization Techniques [11]

The goal of an optimization issue is to maximize or minimize a function under specific constraints. Optimization techniques are methods and

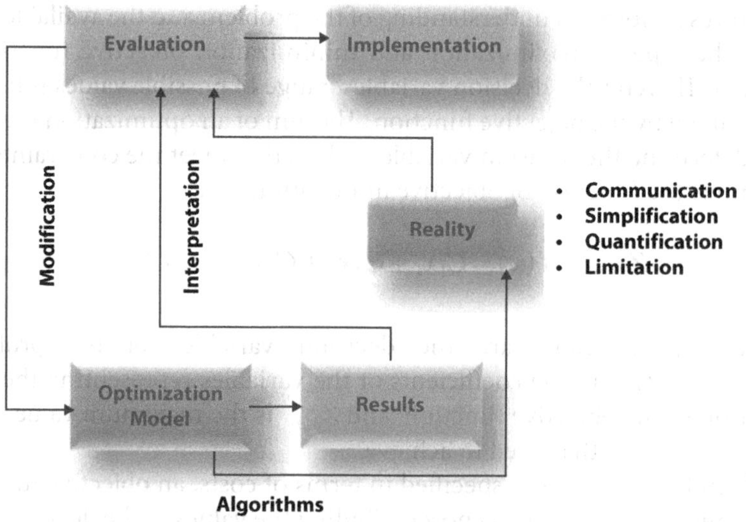

Figure 11.4 Flowchart structure for modeling process.

algorithms used to discover the best solution. Thus, the functioning of the modeling process for optimization fundamental technique structure is shown in Figure 11.4.

Moreover, traditional techniques, such as integer programming, dynamic programming, geometric programming, and linear programming, are used to solve optimization problems. Furthermore, advanced methods like simulated annealing, artificial neural networks, PSO, and genetic algorithms are listed below.

11.4.1 Linear Programming

It is a powerful tool for optimizing resource, and the process of modeling a physical problem into a mathematical formulation that will be solved by linear programming techniques involves determining the decision parameters, formulating the objective function to be optimized, and imposing constraints that limit values of the decision variables. Depending on the size and complexity of the issue, there are different approaches to solving it once it has been formulated as a linear program.

The graphical method is a simple technique that can be used to solve small two-variable problems graphically by finding the intersection of the constraint lines. The simplex method is a more efficient and general-purpose technique that can handle larger and more complex problems with any number of decision variables and constraints.

A. Problem Formulation

As the formulation of the problem is one of the essential procedures for the LPP, it should be done employing the following solutions:

1. Identify the decision variables about constraints and objectives whether which are <, > or = form.
2. Identify the problem data using parameters and constants with the decision variables are controlled but not the data set.
3. Identiffy the constraints into linear equality or inequality in decision variables and ensuring the values which satisfy the constraints for a feasible solution.
4. Identify the objective function should be maximized or minimized and expressed as linear functions of decision variables.

11.4.2 Integer Programming

A particular type of linear programming where the decision factors can only have integer values is integer linear programming. The decision variables represent quantities that are essentially discrete, such as number of workers, batches, and units, in many practical problems, such as production planning, scheduling, and resource allocation. Integer programming issues come in a variety of forms, and each has unique traits and approaches for solving them. The following are some examples of these types:

Programming with

- Binary integers
- Mixed integers
- Integer quadratic equations

Integer programming issues can also take the form of random integer programming or multi-objective integer programming. These need particular modeling and solution techniques and select crucial type for obtaining an accurate and efficient solution.

11.4.3 Dynamic Programming

For problems that can be broken down into smaller sub-problems, where the solution for the overall problem is obtained by integrating the solutions to the sub-problems, dynamic programming is especially helpful. It can be applied to numerous varieties of optimization issues, such as linear programming, integer programming, and nonlinear programming.

There are two types of dynamic programming commonly used in fundamental optimization: forward and backward dynamic programming.

- Forward dynamic programming helps to solve subsequent problem. This approach is particularly useful when the problem has a natural sequential structure, and the optimal solution will be obtained by incorporating the solutions to sub-problems.
- Backward dynamic programming helps to solve preceding sub-problems. This is useful specifically when the problem has a natural structure, and the optimal solution can be obtained by working backwards from the final state.

11.4.4 Geometric Programming

One mathematical optimization method that works well for solving issues where the goal function and constraints are represented by posynomials is called geometric programming (GP). Posynomials are functions that are the sum of monomials, where each monomial is the product of positive variables raised to constant exponents. Because it can manage non-linear constraints and objective functions, GP is frequently quicker and more accurate than other optimization techniques like linear programming or non-linear programming. Also, it is useful for problems involving multiple objectives, where the target is optimizing several objectives simultaneously.

There are several types of GP problems including the following:

- Unconstrained GP: No constraints, objective function is posynomial
- Signomial GP: Both positive and negative terms
- Generalized GP: Only allows for constraints, no objective functions but expressed as posynomials
- Multi-objective GP: Have multiple conflicting objectives

11.5 Electric Vehicle

11.5.1 An Introduction to the Electric Vehicle

An electric car, also known as an EV, utilizes electric motors and battery packs to propel itself, rather than relying on a conventional internal

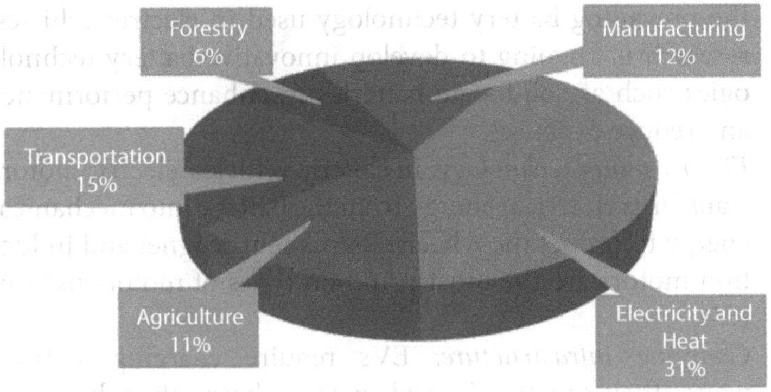

Figure 11.5 CO_2 emission.

combustion engine. EVs come in many forms, including cars, trucks, buses, and even bicycles. Battery electric vehicles (BEVs) and plug-in hybrid electric vehicles (PHEVs) are the two primary categories of electric vehicles available. BEVs rely entirely on electricity and can recharge their batteries by connecting to an external power source, whereas PHEVs have both an electric motor and a conventional gasoline engine. They can run on electricity alone for a limited range before the gasoline engine takes over to extend their driving range. Figure 11.5 shows the emission of carbon-dioxide in the environment. The prominent emission of CO_2 is due to electricity and heat.

Electric vehicles provide numerous advantages compared to conventional gasoline-powered cars. One of the benefits of electric vehicles is that they produce no tailpipe emissions, which is better not only for the environment but also for human health. Furthermore, EVs typically have lower operating costs because electricity is often cheaper than gasoline, and they necessitate less maintenance due to fewer moving components [12]. Additionally, EVs can offer a quieter and smoother driving experience, with instant torque and acceleration. As technology advances and battery costs continue to decrease, it is expected that EVs will become more common and more affordable in the coming years.

11.5.2 Vehicle Technology

Electric vehicle technology refers to the systems and components that enable electric vehicles to operate. Here are some of the key technologies used in electric vehicles:

- *Battery technology:* The most crucial component of an EV is the battery, which stores the energy needed to power the electric motor. While lithium-ion batteries are currently the prevailing battery technology used in electric vehicles, research is ongoing to develop innovative battery technologies such as solid-state batteries to enhance performance and reduce expenses.

- *Electric motor technology:* In electric vehicles, electric motors transform electrical energy from the battery into mechanical energy to propel the wheels. Permanent magnet and induction motors are the most common types of motors used in EVs.
- *Charging infrastructure:* EVs require charging stations where they can be plugged in to recharge their batteries. Public charging stations are becoming more common, and many EV owners also install home charging stations for convenience.
- *Regenerative braking:* Regenerative braking technology in electric vehicles allows the battery to recharge by converting kinetic energy into electrical energy when the vehicle decelerates, during braking.
- *Thermal management:* Battery temperature management is important in EVs to ensure optimal performance and longevity of the battery. Cooling and heating systems are used to maintain the battery's temperature in extreme weather conditions.

The vehicle technology has three types of transportation: conventional vehicle technology, hybrid vehicle technology, and electric vehicle technology, as shown in Figure 11.6 [13].

Conventional Vehicles: Conventional vehicles are the traditional vehicles motorized with gasoline or diesel as fuel. The engine powers the wheels, and the vehicle runs on fossil fuels. These vehicles are widely used and can be found on roads across the world. However, conventional vehicles have several drawbacks such as pollution, high fuel costs, and dependence on oil.
Hybrid Vehicles: A hybrid vehicle is a type of vehicle that incorporates both an internal combustion engine and an electric motor with a battery pack. The electric motor assists the engine during acceleration and deceleration, improving fuel efficiency and reducing emissions. The battery is recharged through regenerative braking and the engine. Hybrid vehicles are a good compromise between conventional and electric vehicles, as they offer better fuel economy and reduced emissions while still being able to operate on gasoline or diesel when needed.

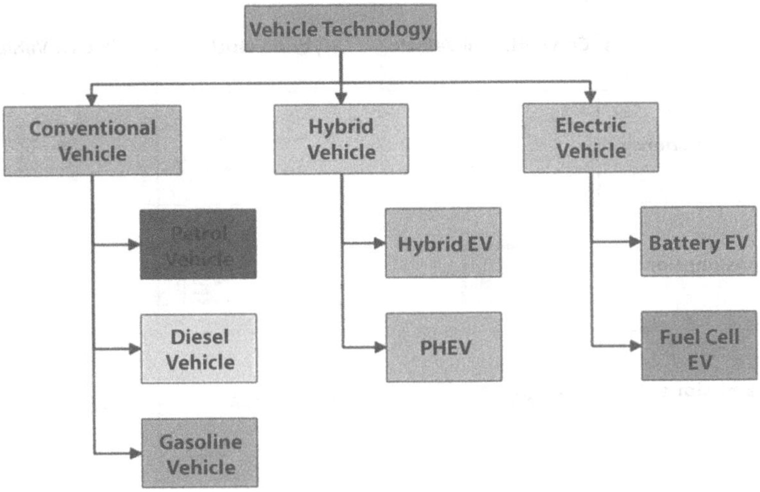

Figure 11.6 Vehicle technology.

Electric Vehicles: These types of vehicles are motorized with both electric motor and rechargeable batteries. They are emission-free and have a much lower operating cost than conventional vehicles. With the help of a charging station, the batteries get recharged. Due to the improvement in battery technology such as increase in range and decrease in charging time, EVs are gaining popularity. However, the infrastructure for charging electric vehicles is still developing in many areas, and EVs can be more expensive to purchase initially than conventional or hybrid vehicles.

In Figure 11.7, the sources of energy, consumption, and emission of conventional, hybrid, and electric vehicle are shown. It shows that the emission of an electric vehicle is less when compared with conventional vehicles. An electric vehicle utilizes electricity as a source of energy for running. A hybrid vehicle uses both conventional fuel and electricity to operate.

11.5.3 Charging Technology

Electric vehicle charging technology refers to the methods and infrastructure used to charge electric vehicles. As electric vehicles become more popular, the advancement of charging facility is becoming increasingly important to support the widespread adoption of these vehicles. The three main types of electric vehicle charging include battery swapping, wireless

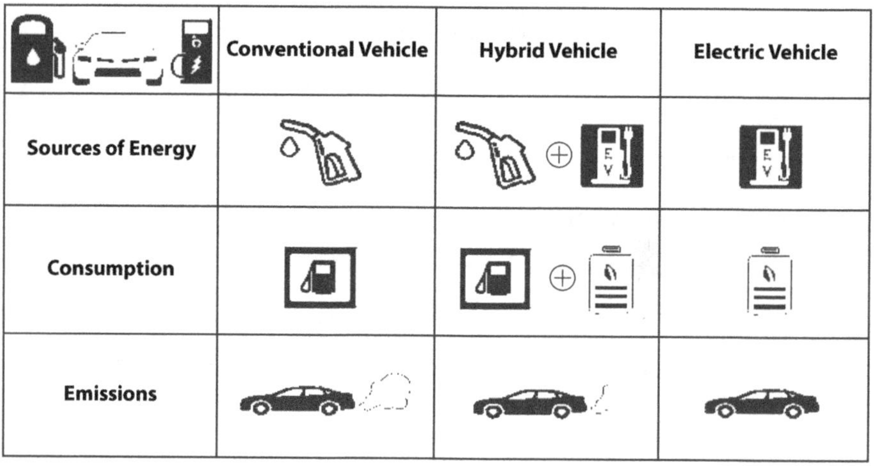

Figure 11.7 Sources and effect of vehicles.

charging, and wired charging. They are categorized on the basis of how the charging process takes place.

- *Battery Swapping*: Battery swapping involves removing an exhausted battery and replacing it with a fully charged battery. This method can be quicker than traditional charging methods and may be useful for EV owners who have limited access to charging stations [29]. However, battery swapping requires specialized infrastructure and standardized battery design.
- *Wireless/conductive Charging:* This type of charging uses a pad on the ground to transfer energy wirelessly to a coil in the EV's undercarriage. This technology requires the vehicle to be parked over the pad, and the charging process is initiated automatically [14]. While this method is convenient, its efficiency is much lower than the wired charging.
- *Wired Charging:* Wired charging involves physically connecting a charging cable from a charging station to the EV's onboard charging port. It consists of level 1 charging, level 2 charging, both of which are applicable in the home itself, and DC fast charging which is done in a public charging station [26]. In the former charging type, the outlet is standard like household and it is the slowest charging type. But in level 2, the outlet utilized is 240 V from the home. DC fast charging

provides a very rapid charge but requires specialized equipment and infrastructure.

The two modes of charging to be used in electronic devices, such as smart phones, tablets, and laptops includes constant current (CC) mode in which the current is maintained as constant and the next mode is the constant voltage (CV) mode [22].

In CC mode, the charging circuitry maintains a CC flow to the battery, which gradually increases the battery's voltage. This mode is used during the initial stages of charging when the battery is depleted and needs to be charged quickly. During this stage, the battery's internal resistance is low, allowing it to accept a higher current flow.

In CV mode, the charging circuitry maintains a constant voltage across the battery terminals while gradually reducing the charging current. This mode is used during the later stages of charging when the battery's voltage has reached a certain level. During this stage, the battery's internal resistance is higher, and it can only accept a lower current flow to avoid overcharging.

Most charging circuits in electronic devices use a combination of both CC and CV modes to optimize charging time and prevent overcharging or overheating of the battery. The charging circuitry switches from CC mode to CV mode once the battery reaches a certain voltage level and switches back to CC mode if the battery's voltage drops too low. This ensures that the battery is charged quickly and efficiently while protecting it from damage due to overcharging or overheating.

11.5.4 Energy Sources for EV Charging

There are several energy sources for electric vehicle charging, including the following:

- Grid electricity: This is the most common source of energy for electric vehicle charging. The electricity is supplied through the electrical grid and can be accessed using public or private charging stations.
- Solar power: Solar panels can be used to generate electricity to charge electric vehicles. This can be done through home-based solar charging stations or public solar-powered charging stations [21].
- Wind power: The electricity generated by the wind turbine is also utilized for EV charging. Similar to solar power, this

can be done through home-based or public wind-powered charging stations.
- Hydro power: Hydroelectric power plants can generate electricity that can be used for electric vehicle charging. This can be done through public hydro-powered charging stations.
- Geothermal power: Geothermal power plants can generate electricity that can be used for electric vehicle charging. This can be done through public geothermal-powered charging stations.
- Biomass power: Biomass power plants can generate electricity that can be used for electric vehicle charging. This can be done through public biomass-powered charging stations.
- Battery storage: Batteries can store energy from any of the above sources and be used for electric vehicle charging. This can be done through home-based battery storage or public battery-powered charging stations.

Overall, the choice of energy source for electric vehicle charging depends on factors such as availability, cost, and environmental impact.

11.6 Challenges in EV Implementation

While electric vehicles have many benefits, their widespread implementation also presents several challenges. Some of these challenges include the following:

- Limited driving range: The driving limit of electric vehicles is typically slightly less than traditional gasoline-powered vehicles. While battery technology is improving, it is still expensive, and EVs with longer ranges are often more expensive.
- Charging infrastructure: Unlike traditional gas stations, electric charging stations are still not widely available in many regions, particularly in rural areas. This can make long-distance travel in EVs challenging.
- Cost: In comparison to gasoline-powered vehicles, electric vehicles are usually higher in price, although the cost is decreasing as battery technology improves and economies of scale are realized.

- Charging time: Even with fast charging technology, EVs can take longer to charge than filling up a gas tank, which can be inconvenient for drivers who are used to the speed and convenience of traditional gasoline-powered vehicles.
- Battery recycling: Lithium-ion batteries used in EVs are not easily recyclable, and their disposal can have negative environmental impacts.
- Public awareness: There is still a lack of awareness among the general public regarding the benefits of electric vehicles and the availability of charging facility.
- Electric grid capacity: As more EVs are adopted, there may be strains on the electric grid to provide the increased power capacity required for charging.

Overall, these challenges need to be addressed for the widespread adoption of EVs to become a reality. This can be achieved through government policies that incentivize EV adoption, investment in charging infrastructure, and continued advancements in battery technology.

Electric vehicle scheduling problems refer to the challenges of coordinating the charging of a fleet of EVs to meet operational requirements and minimize costs. Figure 11.8 describes the challenges to be faced while implementing the electric vehicle [18].

The following are some common EV scheduling problems:

- *Vehicle routing and scheduling problem:* The challenge at hand entails determining the optimal routes [15, 20, 25]

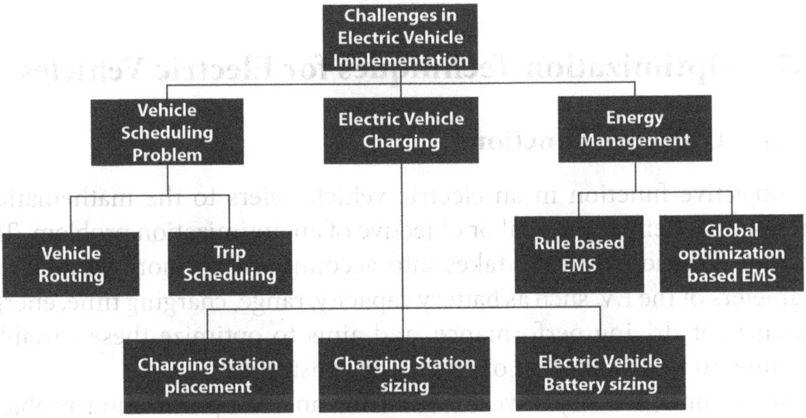

Figure 11.8 Challenges in EV implementation.

and schedules [23] of electric vehicles to cater to a specific customer base, all the while keeping costs to a minimum and adhering to delivery at timeframes. It involves finding the optimal sequence of deliveries and pickups, taking into account vehicle capacity, charging requirements, and other constraints.

- *Electric vehicle charging:* It includes the placement of charging station, sizing of charging station, and the battery sizing for electric vehicle. The problem in charging station assignment involves determining which charging stations should be used to charge a fleet of EVs to meet operational requirements while minimizing costs [19]. It involves deciding which charging stations to use, how many EVs can be charged at each station, and the charging time required for each EV. The problem involved in charging station incorporation is balancing the load of charging stations to avoid overloading the electric grid while meeting operational requirements. It involves managing the distribution of charging demands across different charging stations, considering the availability of charging stations and the charging requirements of each EV.
- *Energy management problem:* This problem involves managing the energy consumption of a fleet of EVs to optimize energy usage, minimize costs, and avoid overloading the electric grid [27]. It involves scheduling the charging of EVs during periods of low energy demand and using renewable energy sources to power charging stations [28].

11.7 Optimization Techniques for Electric Vehicles

11.7.1 Objective Function

The objective function in an electric vehicle refers to the mathematical function that defines the goal or objective of an optimization problem. The objective function typically takes into account one or more variables or parameters of the EV, such as battery capacity, range, charging time, energy efficiency, or driving performance, and aims to optimize these variables according to a particular set of criteria or constraints.

For example, the objective function for an EV optimization problem may be to minimize the overall cost of ownership over the vehicle's lifetime, taking into account factors such as initial purchase price, maintenance costs, fuel or electricity costs, and resale value. Alternatively, the objective function may be to maximize the range of the EV while minimizing the time required for charging, subject to constraints such as battery capacity and charging infrastructure availability.

The specific form and parameters of the objective function in an EV optimization problem depend on the specific goals and constraints of the problem, as well as the methods used for optimization, such as linear programming, nonlinear optimization, or heuristic algorithms [17]. In general, the objective function in an EV optimization problem plays a critical role in defining the problem and guiding the search for optimal solutions.

11.7.2 Objectives

The objectives in electric vehicles are related to various aspects of their performance, efficiency, safety, and user experience. Some common objectives in EVs include the following:

- Range: Maximizing the range of the vehicle on a single charge is a key objective for most EVs, as it determines the practicality and convenience of the vehicle for daily use.
- Energy efficiency: Improving the energy efficiency of the EV, which is typically measured in miles per kWh (kilowatt-hour), can help to extend its range and reduce the overall energy consumption and operating costs of the vehicle.
- Performance: Enhancing the performance of the EV, including its acceleration, speed, and handling, can improve the driving experience and make the vehicle more appealing to consumers.
- Safety: Ensuring the safety of the EV and its occupants is a critical objective and involves designing the vehicle to meet rigorous safety standards and incorporating features such as collision avoidance systems, airbags, and structural reinforcements.
- Charging time: Reducing the time requisite to charge the EV can improve its convenience and practicality and may involve optimizing the charging infrastructure and battery management system.
- Cost: Reducing the cost of the EV, both in terms of initial purchase price and ongoing operating costs, has the

potential to broaden the consumer base and make it more inclusive and accelerate the transition to electric mobility.

The objectives in EVs are closely related to the technical, economic, and social factors that influence the adoption and use of electric vehicles, and will continue to evolve as the technology and market for EVs mature [18].

11.7.3 Constraints

Constraints in electric vehicles refer to the limitations or conditions that restrict their design, operation, or performance. The constraints are listed below:

- Battery capacity: The capacity of the EV's battery is a critical constraint that determines its range, weight, and cost, and must be optimized to balance these factors.
- Powertrain efficiency: The efficiency of the EV's powertrain, which includes the motor, controller, and inverter, is another key constraint that affects the vehicle's range, performance, and energy consumption.
- Charging infrastructure: The charging infrastructure's accessibility and capacity, encompassing public and private charging stations, can limit the practicality and convenience of EVs, especially for long-distance travel.
- Vehicle weight and size: The weight and size of the EV must be balanced against factors, such as aerodynamics, performance, and safety, and must comply with regulatory requirements for vehicle size and weight limits.
- Cost: The cost of EVs, including both the initial purchase price and ongoing operating costs, is a critical constraint that affects their adoption and market penetration and must be balanced against factors such as range, performance, and features.
- Environmental regulations: EVs must comply with environmental regulations for emissions, noise, and other factors, which can affect their design, operation, and performance.

Constraints must be carefully balanced and optimized to achieve the optimal trade-off between performance, cost, and environmental impact.

11.7.4 Categorization of Energy Management in Electric Vehicle

➢ *Rule-based energy management*

- *Deterministic rule-based:* Deterministic rule-based energy management for electric vehicles is an approach that uses predetermined rules or algorithms to manage the energy usage of an EV. The rules are programmed into the vehicle's control system and govern factors such as acceleration, speed, and power usage to maximize efficiency while ensuring that the battery is not drained completely [18]. This approach can be effective for fleet operations or in predictable driving scenarios, but it may not be as flexible or adaptable to unexpected situations. Overall, deterministic rule-based energy management can help optimize the energy usage of an EV and improve its overall efficiency.
- *Fuzzy logic rule-based:* This approach enhances the energy usage of EV with fuzzy logic control algorithm. Fuzzy logic systems use linguistic variables, such as "high," "medium," and "low," to describe different driving conditions and inputs. The control algorithm then uses these variables to make decisions about how to manage the energy usage of the vehicle. Fuzzy logic rule-based energy management can adapt to changing driving conditions and can be customized for different driving scenarios and individual driving styles, which can further improve efficiency [18]. However, fuzzy logic systems can be more complex to develop and implement than deterministic rule-based systems. Conventional, adaptive, and predictive fuzzy rule-based energy management are three approaches to managing the energy usage of electric vehicles using fuzzy logic control algorithms:

i) Conventional fuzzy rule-based energy management involves setting fixed rules that govern how the vehicle uses its battery power based on predetermined conditions. For example, the vehicle may be programmed to maintain a certain speed or acceleration rate to optimize energy efficiency.
ii) Adaptive fuzzy rule-based energy management is more flexible than conventional systems, as it allows the

system to adapt to changing driving conditions in real-time. The system can learn from the driver's behavior and adjust its rules accordingly, which can optimize energy usage and improve efficiency.

iii) Predictive fuzzy rule-based energy management uses predictive algorithms to anticipate future driving conditions and adjust energy usage accordingly. The system can analyze data such as the vehicle's location, traffic patterns, and weather conditions to optimize energy usage in advance, which can further improve efficiency.

All three approaches to fuzzy rule-based energy management have their advantages and disadvantages. Conventional systems are simple to implement but may not be optimized for individual driving styles, whereas adaptive and predictive systems are more complex but can provide greater flexibility and efficiency. Overall, fuzzy rule-based energy management can be an effective approach to optimizing energy usage in electric vehicles, as it allows for greater flexibility and adaptability to changing driving conditions.

➢ *Global optimization-based energy management*

The energy management of electric vehicle with global optimization [13] includes linear programming, dynamic programming, stochastic dynamic programming, genetic algorithm, game theory, convex optimization, and PSO, as shown in Figure 11.9.

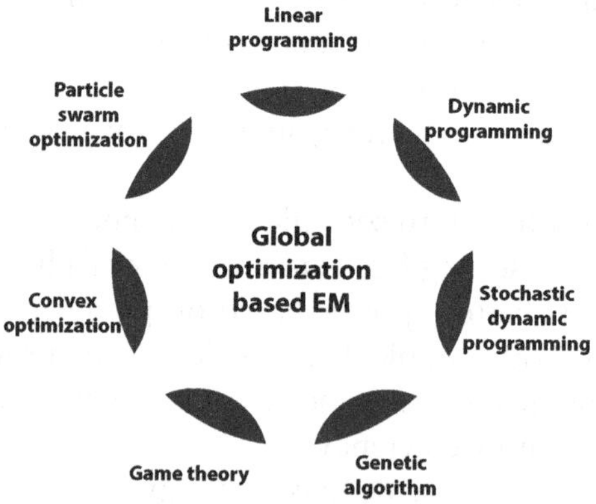

Figure 11.9 Global optimization-based energy management for EV.

- *Linear programming optimization of EV:* The mathematical modeling of linear programming for energy management is a set of equations and constraints. Let $x(t)$ be the state of charge (SoC) of the battery at time t, $p(t)$ be the power demanded by the motor and auxiliary systems at time t, and $q(t)$ be the power generated by the battery and regenerative braking system at time t. The LP model is devised as follows:

$$\text{Minimize: } \sum p(t)\Delta t \tag{11.5}$$

The constraints included are given in Equations 11.6 to 11.9:

$$x(t) - x(t-1) = \Delta t(q(t) - p(t))/C \tag{11.6}$$

$$x\_min \le x(t) \le x\_max \tag{11.7}$$

$$q(t) = k(p(t) - r) \tag{11.8}$$

$$-q\_max \le q(t) \le q\_max \tag{11.9}$$

where Δt is the time step, C denotes the capacity of battery, the battery SoC levels are given in minimum and maximum levels as *x_max* and *x_min*, *k* is the battery efficiency, *r* is the power losses in the battery and motor system, and *q_max* is the utmost power that can be generated by the regenerative braking system.

The first constraint ensures that the battery SoC at time *t* is equal to the battery SoC at time *t – 1* plus the change in SoC due to charging or discharging, whereas the second and third constraints ensure that the battery SoC remains within the allowable limits. The fourth constraint relates the power generated by the battery to the power demanded by the motor and auxiliary systems, whereas the fifth and sixth constraints ensure that the generated power by regenerative braking system remains within the allowable limits.

By solving the LP model using optimization software, it provides the optimal allocation of power and energy for the given set of constraints. Thereby, the efficacy and performance can be enhanced. The LP model can be updated in real time to adjust the allocation of power and energy based on changing driving conditions, making it a powerful tool for energy management in EVs.

• *Dynamic programming optimization of EVs:* In energy management for EVs, the mathematical modeling of dynamic programming for electric vehicles involves developing a set of equations and constraints to optimize the allocation of power and energy over a longer time horizon compared to linear programming. Dynamic programming considers the future states of the battery and driving cycle to determine the optimal control strategy.

The dynamic programming algorithm solves a recursive optimization problem that considers all possible state transitions and control actions. The algorithm finds the optimal control sequence that minimizes the objective function, which is typically the total energy consumption or a weighted combination of energy consumption and driving comfort.

The mathematical model for dynamic programming is devised as follows:

$$J(x(t),t) = min[\Sigma u(t)H(x(t),u(t),t) + J(x(t+1),t+1)] \quad (11.10)$$

The constraints include the following:

$$x(t+1) = f(x(t),u(t),d(t)) \quad (11.11)$$

where $J(x(t),t)$ is the optimal cost-to-go function at time t, the control input is shown as $u(t)$, $H(x(t),u(t),t)$ is the instantaneous cost function, the state of system for time (t) is $x(t)$, $d(t)$ is the driving cycle for time t, and the state transition function given in Equation 11.10 as $f(x(t),u(t),d(t))$ is the state transition function.

The state of the system $x(t)$ includes the battery SoC, the motor and auxiliary system power demands, and the vehicle speed. Charging of battery, discharging of battery, and power balance with both motor and battery are the control inputs. The driving cycle $d(t)$ includes information about the vehicle speed, acceleration, and road grade.

The objective function seeks to mitigate the energy consumed by fulfilling the constraints, which may include battery capacity, power balance, speed, and acceleration constraints. The state transition function $f(x(t),u(t),d(t))$ relates the state of the system at time $t + 1$ to the state of the system at time t, the control input at time t, and the driving cycle at time t. Solving the dynamic programming algorithm provides the feasible control sequence for the given driving cycle and set of constraints. The algorithm can be run offline to generate a lookup table or online to update the control sequence in real-time based on changing driving conditions. Dynamic programming offers a more accurate and comprehensive optimization approach compared to linear programming, but it requires more computational resources and is more complex to implement.

• *Genetic algorithm:* This is a computational technique inspired by the process of nature in biology. It is used to solve complex problems by mimicking the evolutionary processes of nature with selection, mutation, and reproduction. The flowchart of this type is given in Figure 11.10. The algorithm begins by creating a random population of potential solutions to the problem. Each member of the population embodies a prospective resolution to the current problem [24]. Then, the algorithm measures the competence of every individual by employing a fitness function that evaluates their ability to solve the problem.

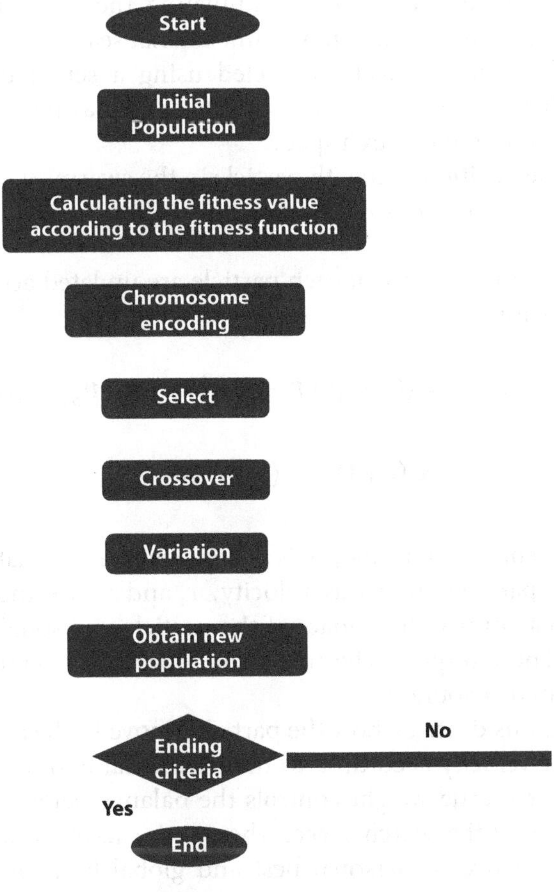

Figure 11.10 Process diagram of genetic algorithm.

Individuals possessing superior fitness have a greater probability of being chosen for reproduction, which entails the merging of selected individuals' components through crossover and mutation operations to generate new offspring. Crossover necessitates the swapping of components between chosen individuals to create novel solutions, whereas mutation involves randomly modifying components of chosen individuals to maintain population diversity. The process continues with the assessment of the new individuals until a satisfactory outcome is reached or a pre-established termination criterion is fulfilled. Genetic algorithms [30] have been efficaciously employed to address various optimization challenges such as function optimization, scheduling dilemmas, and machine learning.

- *Particle swarm optimization:* This is a computational optimization technique motivated by the social behavior of bird flocking. It is used to solve complex optimization problems by simulating the social behavior of a swarm of particles moving in a multidimensional search space [16].

PSO can be mathematically modeled using a set of equations that describe the behavior of the particles in the swarm as they search for the optimum solution in the search space.

Let x_i be the position of the ith particle in the swarm, v_i be its velocity, and p_i be its best position so far. Let p_g be the best position of the swarm so far.

The position and velocity of each particle are updated according to the following equations:

$$v_i(t+1) = wv_i(t) + c_1r_1(p_i - x_i(t)) + c_2r_2(pg - x_i(t)) \quad (11.12)$$

$$x_i(t+1) = x_i(t) + v_i(t+1) \quad (11.13)$$

where t is the current iteration; w is the inertia weight that controls the impact of the particle's previous velocity; c_1 and c_2 are the acceleration coefficients that control the impact of the particle's personal best and the swarm's global best, respectively; and r_1 and r_2 are random numbers drawn from a uniform distribution.

These equations describe how the particles move in the search space by adjusting their velocity according to their personal best and the swarm's global best. The inertia weight controls the balance between exploration and exploitation of the search space, whereas the acceleration coefficients control the influence of personal best and global best on the particle's movement.

The PSO algorithm terminates when a stopping criterion is met, such as reaching a maximum number of iterations or finding a satisfactory solution.

In summary, the mathematical model of PSO describes how the particles in the swarm interact with each other and explore the search space to find the optimum solution to an optimization problem.

11.8 Conclusion

This paper presented a comprehensive technical overview of the wide-ranging and diverse research carried out on the crucial subject of optimization techniques. This topic has gained significant attention in recent times due to several initiatives that encourage novel and cutting-edge research in this field. In conclusion, optimizing electric vehicle performance is critical to ensuring their widespread adoption as a sustainable alternative to traditional combustion engine vehicles. Fundamental optimization algorithms and computing techniques play a crucial role in achieving this goal. Some of the key optimization algorithms and techniques include battery management systems, power electronics control, and energy management systems. These systems help improve electric vehicle efficiency, extend battery life, and enhance overall vehicle performance. Overall, continued research and development in fundamental optimization algorithms and computing techniques are essential to advancing the electric vehicle industry and ensuring a sustainable future for transportation. This paper presents the fundamental optimization algorithms with the objective functions, constraints, and the design parameters. Finally, the optimization algorithms such as linear programming, dynamic programming, genetic algorithm, and PSO algorithms are discussed in brief.

References

[1] Abualigah Laith, Mohamed Abd Elaziz, Ahmad M. Khasawneh, Mohammad Alshinwan, Rehab Ali Ibrahim, Mohammed AA Al-Qaness, Seyedali Mirjalili, Putra Sumari, & Amir H. Gandomi. (2022). Meta-heuristic optimization algorithms for solving real-world mechanical engineering design problems: a comprehensive survey, applications, comparative analysis, and results. *Neural Comput. Appl.*, 1-30.

[2] Gao, Guohua, Yixuan Wang, Jeroen C. Vink, Terence J. Wells, and Fredrik JFE Saaf. (2022). Distributed quasi-Newton derivative-free optimization

method for optimization problems with multiple local optima. *Comput. Geosci.*, 26, 4, 847-863.

[3] Trojovský Pavel, & Mohammad Dehghani. (2022). Pelican optimization algorithm: A novel nature-inspired algorithm for engineering applications. *Sensors*, 22 (3), 855.

[4] Ikram Rana Muhammad Adnan, Hong-Liang Dai, Ahmed A. Ewees, Jalal Shiri, Ozgur Kisi, & Mohammad Zounemat-Kermani. (2022). Application of improved version of multi verse optimizer algorithm for modeling solar radiation. *Energy Rep.*, 8, 12063-12080.

[5] Guo Kai, & Limao Zhang. (2022). Multi-objective optimization for improved project management: Current status and future directions. *Autom. Constr.*, 139, 104-256.

[6] Du Liang, Ruobin Gao, Ponnuthurai Nagaratnam Suganthan, & David ZW Wang. (2022). Bayesian optimization based dynamic ensemble for time series forecasting. *Inf. Sci.*, 591, 155-175.

[7] Arora, K.; Kumar, A.; Kamboj, V.K.; Prashar, D.; Shrestha, B.; Joshi, G.P. Impact of Renewable Energy Sources into Multi Area Multi-Source Load Frequency Control of Interrelated Power System. *Mathematics*, 2021, *9*, 186.

[8] Rani G Sandhya, Sarada Jayan, & Bilal Alatas. (2023). Analysis of Chaotic maps for Global Optimization and a Hybrid Chaotic Pattern Search Algorithm for optimizing the reliability of a bank. *IEEE Access*, 11, 24497-24510.

[9] Mangaraj Soumya Ranjan, Dilip Kumar Bagal, NirmalyaParhi, Surya Narayan Panda, Abhishek Barua, & Siddharth Jeet. (2022). Experimental study of a portable plasma arc cutting system using hybrid RSM-nature inspired optimization technique. *Mater. Today Proc.*, 50, 867-878.

[10] Mustapha Aatila, Lachgar Mohamed, & Kartit Ali. (2020). An overview of gradient descent algorithm optimization in machine learning: Application in the ophthalmology field. In: *Smart Applications and Data Analysis: Third International Conference, SADASC, Marrakesh, Morocco*, Springer International Publishing, 349-359.

[11] Sukanta Nayak. (2021). Fundamental Optimization with Algorithm. Elsevier. https://doi.org/10.1016/C2019-1-02539-9

[12] Arif, Syed Muhammad, Tek Tjing Lie, Boon Chong Seet, Soumia Ayyadi, and Kristian Jensen. (2021). Review of electric vehicle technologies, charging methods, standards and optimization techniques. *Electronics*, 10(16), 1910.

[13] Zhang, Fengqi, Lihua Wang, Serdar Coskun, Hui Pang, Yahui Cui, and Junqiang Xi. (2020). Energy management strategies for hybrid electric vehicles: Review, classification, comparison, and outlook. *Energies*, 13(13). 3352.

[14] Subudhi, Partha Sarathi, Sanjeevikumar Padmanaban, Frede Blaabjerg, and Dwarkadas Prahladadas Kothari. (2022). Design and Implementation of a PV-Fed Grid-Integrated Wireless Electric Vehicle Battery Charger Present in a Residential Environment. *IEEE J. Emerg. Sel. Top. Ind. Electron.*, 4(1), 78-86.

[15] Barco, John, Andres Guerra, Luis Munoz, and Nicanor Quijano. (2017). Optimal routing and scheduling of charge for electric vehicles: A case study. *Math. Probl. Eng.*, 2017.

[16] Dai, Qiongjie, Jicheng Liu, and Qiushuang Wei. (2019). Optimal photovoltaic/battery energy storage/electric vehicle charging station design based on multi-agent particle swarm optimization algorithm. *Sustainability*, 11(7), 1973.

[17] Devan, P., Fawnizu Azmadi Hussin, Rosdiazli B. Ibrahim, Kishore Bingi, M. Nagarajapandian, and Maher Assaad. (2022). An arithmetic-trigonometric optimization algorithm with application for control of real-time pressure process plant. *Sensors*, 22(2), 617.

[18] Kalakanti, Arun Kumar, and Shrisha Rao. (2023). Computational Challenges and Approaches for Electric Vehicles. *ACM Comput. Surv.*, Vol. 55, Issue 14s, pp. 1-35.

[19] Babu, K., Pratyush Chakraborty, and Mayukha Pal. (2023). Planning of Fast Charging Infrastructure for Electric Vehicles in a Distribution System and Prediction of Dynamic Price. arXiv preprint arXiv: *2301.06807*. https://doi.org/10.48550/arXiv.2301.06807

[20] Jia, Ya-Hui; Yi Mei, and Mengjie Zhang. (2021). A bilevel ant colony optimization algorithm for capacitated electric vehicle routing problem. *IEEE Trans. Cybern.*, 52(10), 10855-10868.

[21] Fachrizal, Reza, Mahmoud Shepero, Magnus Åberg, and Joakim Munkhammar. (2022). Optimal PV-EV sizing at solar powered workplace charging stations with smart charging schemes considering self-consumption and self-sufficiency balance. *Appl. Energy*, 307, 118139.

[22] Nasr Esfahani, Fatemeh, Ahmed Darwish, and Barry W. Williams. (2022). Power converter topologies for grid-tied solar photovoltaic (PV) powered electric vehicles (EVs)—A comprehensive review. *Energies*, 15(13), 4648.

[23] Titus, Femin, Sudhakar Babu Thanikanti, Sanchari Deb, and Nallapaneni Manoj Kumar. (2022). Charge scheduling optimization of plug-in electric vehicle in a PV powered grid-connected charging station based on day-ahead solar energy forecasting in Australia. *Sustainability*, 14(6), 3498.

[24] Yao, Enjian, Tong Liu, Tianwei Lu, and Yang Yang. (2020). Optimization of electric vehicle scheduling with multiple vehicle types in public transport. *Sustain. Cities Soc.*, 52, 101862.

[25] Li, Boda, Ying Chen, Wei Wei, Shaowei Huang, Yufeng Xiong, Shengwei Mei, and Yunhe Hou. (2021). Routing and scheduling of electric buses for resilient restoration of distribution system. *IEEE Trans. Transp. Electrif.*, 7(4), 2414-2428.

[26] Alwesabi, Yaseen, Zhaocai Liu, Soongeol Kwon, and Yong Wang. A novel integration of scheduling and dynamic wireless charging planning models of battery electric buses. *Energy*, 230. 120806.

[27] Ji, Jinhua, Yiming Bie, and Linhong Wang. (2023). Optimal electric bus fleet scheduling for a route with charging facility sharing. *Transp. Res. Part C Emerg. Technol.*, 147, 104010.

[28] Wu, Yuankai, Huachun Tan, Jiankun Peng, Hailong Zhang, and Hongwen He. (2019). Deep reinforcement learning of energy management with continuous control strategy and traffic information for a series-parallel plug-in hybrid electric bus. *Appl. Energy*, 247, 454-466.

[29] Zhan, Weipeng, Zhenpo Wang, Lei Zhang, Peng Liu, Dingsong Cui, and David G. Dorrell. (2022). A review of siting, sizing, optimal scheduling, and cost-benefit analysis for battery swapping stations. *Energy*, 124723.

[30] Abdullah-Al-Nahid, Syed, Tafsir Ahmed Khan, Md Abu Taseen, Taskin Jamal, and Tareq Aziz. (2022). A novel consumer-friendly electric vehicle charging scheme with vehicle to grid provision supported by genetic algorithm based optimization. *J. Energy Storage*, 50, 104655.

12

Economic Load Dispatch Solutions at Small, Medium, and Large Scales Utilizing Chaotic Spotted Hyena Optimization

Tanuj Mishra[1]*, Amit Kumar Singh[1] and Vikram Kumar Kamboj[1,2]

[1]*School of Electronics and Electrical Engineering, Lovely Professional University, Phagwara, India*

[2]*Department of Electrical and Computer Engineering, Schulich School of Engineering, University of Calgary, Alberta, Canada*

Abstract

Incorporating ideas from both spotted hyenas and chaotic functions, the newly developed chaotic spotted hyena optimization (CSHO) is a meta-heuristic search algorithm that takes its cues from the natural hierarchy and hunting strategy of hyenas. Prey is either (a) actively sought after and pursued, (b) deemed too difficult to approach, (c) causing problems and encircled, or (d) rendered immobilized before being attacked. In this work, we demonstrate how the CSHO method may be used to solve the economic load dispatch problem (ELDP) for a power grid that is both non-convex and subject to dynamic changes. Comparisons are made to other algorithms like particle swarm optimization, ant lion optimizer, complex algorithm (CM), enhanced swift converging simulated annealing (ESCSA), hybrid many-objective particle swarm optimization (HMAPSO), parallel particle swarm optimization (PPSO), multi objective particle swarm optimization (MPPSO), discrete particle swarm optimization (DPSO), adaptive phasor particle swarm optimization (APPSO), k-Logic, and spherical vector-based particle swarm optimization (SPSO) to validate the findings, which show that CSHO performs well for ELDP in systems of various sizes. In both the absence and presence of transmission losses, the CSHO and traditional approaches are used to find optimal solutions to the economic load dispatch issue for the 3-, 6-, 10-, 13-, 38-, 40- (including valve point loading), and 140-unit systems, respectively. The CSHO algorithm is shown to be capable of producing competitive outcomes in a variety of benchmarking

**Corresponding author*: tanuj.12948@lpu.co.in

Krishan Arora, Suman Lata Tripathi and Himanshu Sharma (eds.) Electric Vehicle Design: Design, Simulation and Applications, (265–294) © 2024 Scrivener Publishing LLC

exercises, against the backdrop of various well-known traditional, heuristic, and meta-heuristic search algorithms.

***Keywords*:** Spotted hyena optimization, chaotic spotted hyena optimization, economic load dispatch problem, metaheuristic, unit commitment

12.1 Introduction

The production and load satisfaction of electric power is a challenging task for every country. Meeting the ever-growing demand for electricity while ensuring reliable and affordable power supply is a complex endeavor that requires careful planning, management, and investment in various aspects of the power sector.

One of the main challenges in electric power production is the need to balance supply and demand in real time. Electricity demand fluctuates throughout the day, seasonally, and based on various economic and social factors. Power plants must be able to generate enough electricity to meet the demand at all times to avoid blackouts or brownouts, which can have significant economic and social impacts.

Another challenge is the need to transition toward sustainable and low-carbon power generation sources to mitigate the effects of climate change. Many countries are striving to reduce their dependence on fossil fuels and increase the share of renewable energy sources such as solar, wind, hydropower, and geothermal in their power generation mix. However, integrating intermittent renewable energy sources into the power grid can be challenging due to their variable nature and the need for energy storage solutions to ensure a consistent power supply.

The power generation's economic load dispatch is essential because of the rising demand for electric energy and the limited availability of energy resources, making the electric power sector the most difficult and complicated. Growth in energy demand, together with limitations on traditional energy sources, necessitates the use of renewable energy as the transportation sector electrifies. Scheduling committed generating units to meet the system load demand at the lowest possible operating cost while adhering to physical and operational (equality and inequality) limitations is the primary goal of the economic load dispatch problem (ELDP) in an electricity generating scheme. Researchers have been inspired to look into renewable energy options by the ever-increasing need for electricity. Further study is also needed to lessen the effects of global warming and ecological decline.

12.2 Unit Commitment and Economic Dispatch Process

Commitment [1] and dispatch of power-generating units are important processes in the operation of a power system to ensure reliable and efficient electricity generation. Let us discuss each of them in detail:

1. Commitment of Power-Generating Units
The commitment of power-generating units, also known as unit commitment or economic dispatch, is the process of determining which power-generating units should be online and generating electricity at any given time, based on factors such as demand for electricity, availability of generation resources, and operational constraints. This process is typically done by power system operators or energy management systems (EMSs) using mathematical algorithms and optimization techniques.

The commitment of power-generating units takes into consideration various factors, including the cost of generation, environmental constraints, system stability, and reliability requirements. The goal is to minimize the overall cost of generating electricity while ensuring that there is enough capacity to meet the demand for electricity at all times. This process is usually performed over a time horizon ranging from hours to days, depending on the operational characteristics of the power system.

2. Dispatch of Power-Generating Units
Once the power-generating units are committed, the dispatch process determines the amount of electricity that each generating unit should produce in order to meet the real-time demand for electricity. Dispatch decisions are typically made in real time or near real time based on the actual demand for electricity, the availability of generation resources, and system operational constraints.

Dispatch decisions take into account various factors, such as unit operating characteristics, fuel availability, transmission constraints, environmental requirements, and system stability. The goal of the dispatch process is to optimize the operation of the generating units in real time, ensuring reliable and efficient electricity generation while adhering to operational constraints and regulatory requirements.

The dispatch process is typically performed by EMSs or power system operators, who continuously monitor the operation of the power system; collect data on generation, demand, and other relevant parameters; and make real-time dispatch decisions to ensure reliable and efficient electricity supply.

In summary, commitment and dispatch of power-generating units are critical processes in the operation of a power system, involving the determination of which units should be online and generating electricity, and how much electricity each unit should produce, respectively. These processes are aimed at ensuring reliable and efficient electricity generation while adhering to operational constraints and regulatory requirements.

12.3 Optimization in Power System

Optimization plays a crucial role in power systems to ensure efficient and reliable operation. Optimization techniques are used to make decisions and find optimal solutions to various operational challenges in power system planning, scheduling, dispatch, and control. Some common areas where optimization is applied in power systems include the following:

Economic Dispatch: Economic dispatch is the process of determining the optimal generation levels of power-generating units to meet the demand for electricity at the least cost. It involves optimizing the allocation of generation resources based on their costs, operating constraints, and demand requirements. Optimization techniques, such as linear programming, mixed-integer programming, and heuristic algorithms, are commonly used in economic dispatch to minimize the total generation cost while considering various operational constraints, such as unit ramping limits, transmission constraints, and environmental requirements.

Unit Commitment: Unit commitment is the process of determining which power-generating units should be online and generating electricity at any given time. It involves deciding which units to start or stop based on the forecasted demand, unit availability, and other operational constraints. Unit commitment is a complex optimization problem that considers various factors, including generation costs, start-up costs, shut-down costs, minimum up/down time limits, and environmental constraints. Mathematical optimization techniques, such as mixed-integer programming, dynamic programming, and genetic algorithms (GAs), are commonly used in unit commitment to find the optimal scheduling of generating units.

Grid Expansion Planning: Grid expansion planning involves determining the optimal expansion and upgrade of the power grid to meet future electricity demand while ensuring reliable and cost-effective operation. Optimization techniques are used to identify the optimal locations for new transmission lines, substations, and other grid components, considering factors such as load growth, generation capacity, transmission losses, and investment costs. Techniques such as linear programming, mixed-integer programming, and power flow optimization are commonly used in grid expansion planning to find the optimal grid expansion strategies.

Renewable Energy Integration: Renewable energy sources, such as wind and solar, are variable and intermittent, which introduces challenges in their integration into the power grid. Optimization techniques are used to manage the variability and uncertainty of renewable energy sources and optimize their integration into the grid. Techniques such as stochastic programming, probabilistic modelling, and advanced forecasting methods are commonly used in renewable energy integration to optimize scheduling, forecasting, and dispatch of renewable energy sources, considering their variable nature and grid operational constraints.

Load Dispatch and Energy Management: Load dispatch and energy management involve real-time or near-real-time optimization of power system operation to ensure reliable and efficient electricity generation. Optimization techniques, such as optimal power flow, model predictive control, and other advanced algorithms, are used to make real-time decisions on generation scheduling, load shedding, and other control actions to optimize power system operation and maintain system stability, while considering factors such as demand variations, generation constraints, and equipment limitations.

These are just a few examples of how optimization techniques are applied in power systems to ensure efficient and reliable operation. Optimization plays a critical role in power system planning, scheduling, dispatch, and control, helping to optimize the allocation of resources, minimize costs, ensure reliable operation, and meet environmental and regulatory requirements.

Some of the recent developed algorithms are as follows: ant colony optimization (ACO) [2], binary gravitational search algorithm [3], bat algorithm [4, 5], elephant herding optimization [6], adaptive gbest-guided search algorithm [7], firefly algorithm [8], grey wolf optimizer [9], krill herd algorithm [10], moth-flame optimization [11], adaptive cuckoo search algorithm [12], runner-root algorithm [13], search group algorithm [14], virus colony search [15], chaotic krill herd algorithm [16], ant lion optimizer (ALO) [17], dragonfly algorithm [18], electromagnetic field optimization [19], cuckoo search algorithm [20], water wave optimization [21], gravitational search algorithm [22], optics inspired optimization [23], animal migration optimization [24], earthworm optimization algorithm [25], biogeography-based optimization [26], forest optimization algorithm [27], grasshopper optimization algorithm [28], exchange market algorithm [29], flower pollination algorithm [30], binary bat algorithm [31], multi-verse optimizer [32], fireworks algorithm [33], invasive weed optimization [34], bacterial foraging optimization algorithm [35], league championship algorithm [36], colliding bodies optimization [37], lightning search algorithm [38], collective animal behavior optimization [39], shuffled frog-leaping algorithm [40], symbiotic organisms search [41], sine cosine algorithm [42], whale optimization algorithm [43], bird swarm algorithm [44], stochastic fractal search [45], seeker optimization algorithm [46], salp swarm algorithm [47], water cycle algorithm [48], backtracking search optimization [49], mine blast algorithm [50], monarch butterfly optimization [51], wind-driven optimization [52], imperialist competitive algorithm [53], human group optimizer [54], interior search algorithm [55], teaching-learning-based optimization [56], weighted superposition attraction [57], etc. The "No Free Lunch" [58, 59] theorem is a concept from machine learning and optimization theory that suggests that there is no single algorithm or approach that can universally perform well on all possible problem domains or tasks. In other words, there is no one-size-fits-all solution for all optimization problems. This theorem implies that different algorithms or methods may be more suitable for specific types of problems or domains and that the effectiveness of an algorithm depends on the particular characteristics of the problem at hand. For example, a machine learning algorithm that works well for image recognition tasks may not perform as well for speech recognition tasks, and vice versa. Furthermore, the theorem also implies that there are inherent limitations to what any algorithm or computation can achieve in terms of solving all possible optimization problems efficiently or optimally. This means that no single algorithm can be universally skilled at handling all types of optimization problems with maximum efficiency or effectiveness. Therefore, when dealing with a wide range of optimization problems, it is important to carefully select or design algorithms or methods that are specifically tailored to the characteristics of the problem domain at hand. This may involve using different approaches or combinations of techniques depending on the specific problem being addressed. It is essential to understand the inherent limitations of algorithms and choose appropriate methods based on the problem's requirements, characteristics, and constraints.

Metaheuristic algorithms are optimization algorithms that are inspired by natural and/or artificial systems and are capable of finding approximate solutions to complex problems. They do not guarantee finding the optimal solution, but they are often able to find good solutions in a reasonable amount of time. Metaheuristic algorithms can be categorized into several categories based on their underlying principles or characteristics.

Some of the common categories of metaheuristic algorithms are shown in Figure 12.1 [61], and a few of them are elaborated below.

Evolutionary Algorithms: These algorithms are inspired by the process of natural selection and evolution. They use genetic operators, such as mutation and crossover, to generate new candidate solutions and evolve them over generations to optimize the objective function.

Examples: Genetic algorithm (GA), genetic programming, and evolutionary strategies.

Swarm Intelligence Algorithms: These algorithms are inspired by the collective behavior of social insects, such as ants or bees, where a group of individuals work together to find a solution. They use local interactions and global communication to explore the solution space and optimize the objective function.

Examples: Particle swarm optimization (PSO), ant colony optimization (ACO), and artificial bee colony.

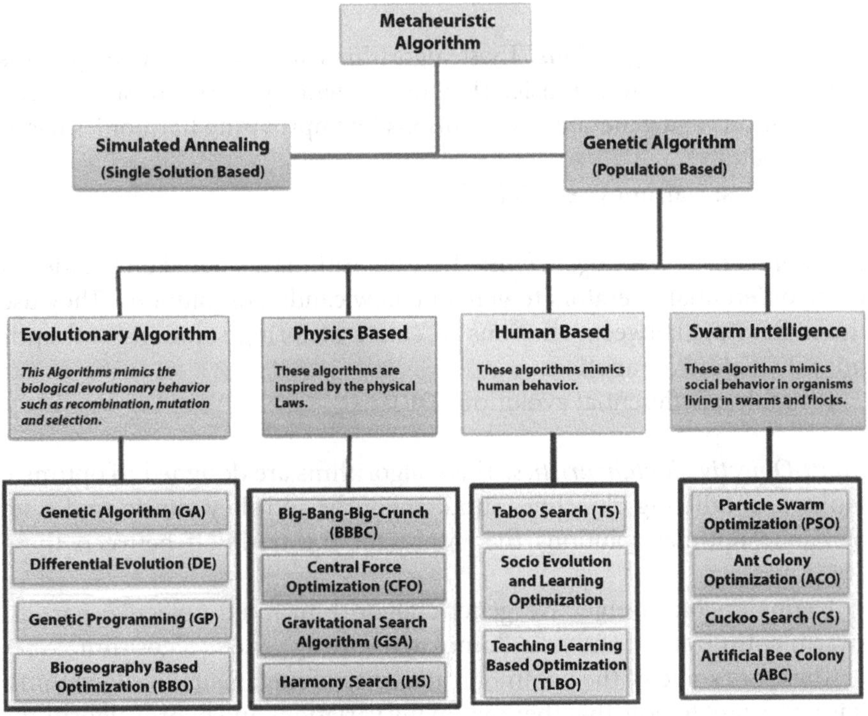

Figure 12.1 Different categories of metaheuristic algorithms.

Simulated Annealing Algorithms: These algorithms are inspired by the annealing process in metallurgy, where a material is heated and slowly cooled to reduce its defects and improve its properties. They use a probabilistic acceptance criterion to accept or reject new candidate solutions and explore the solution space.

Examples: Simulated annealing (SA) and quantum annealing.

Tabu Search Algorithms: These algorithms are based on the idea of using a memory-based mechanism to escape local optima. They use a tabu list to keep track of previously visited solutions and prohibit revisiting them, allowing for more exploration of the solution space.

Examples: Tabu search.

Particle-Based Algorithms: These algorithms are based on the idea of using particles or agents to represent candidate solutions and move through the solution space. They use the movement of particles or agents to explore and optimize the objective function.

Examples: Particle swarm optimization (PSO), and firefly algorithm (FA).

Harmony Search Algorithms: These algorithms are inspired by the process of creating harmonious music. They use a memory matrix to store previous solutions and generate new solutions by improvising harmonies based on the memory.

Examples: Harmony search (HS).

Differential Evolution Algorithms: These algorithms are based on the idea of using differential operators to generate new candidate solutions. They use the differences between solutions to create offspring solutions and optimize the objective function.

Examples: Differential evolution (DE).

Multi-Objective Metaheuristics: These algorithms are designed to optimize multiple conflicting objectives simultaneously, and they aim to find a set of Pareto-optimal solutions that represent the trade-off between these objectives.

Examples: Multi-objective genetic algorithm, multi-objective particle swarm optimization, and non-dominated sorting genetic algorithm.

These are some of the common categories of metaheuristic algorithms. It is important to note that there are many variations and hybrid algorithms that combine different principles from multiple categories, and new algorithms are constantly being developed as the field of optimization continues to evolve.

12.4 Hybrid Optimization Algorithm

A hybrid optimization algorithm is a technique that combines multiple optimization algorithms or approaches to solve complex optimization problems. These problems typically involve finding the optimal solution within a large search space and may have multiple objectives, constraints, or other complexities.

Hybrid optimization algorithms are designed to leverage the strengths of different optimization methods and combine them in a way that can improve the overall performance of the optimization process. The combination of different algorithms can lead to improved efficiency, robustness, and solution quality compared to using a single optimization algorithm.

There are several ways in which hybrid optimization algorithms can be implemented, including the following:

Combination of different algorithms: Hybrid optimization algorithms can combine two or more different optimization algorithms, such as GAs, PSO, and SA. The algorithms may be used sequentially or in parallel, with each algorithm contributing to the search process based on its strengths.

Integration of heuristics or metaheuristics: Hybrid optimization algorithms can incorporate heuristics or metaheuristics, which are problem-specific or problem-independent strategies for guiding the search process. For example, a hybrid algorithm may use a GA to explore the search space and incorporate a local search heuristic to refine the solutions.

Hybridization of deterministic and stochastic methods: Hybrid optimization algorithms can combine deterministic and stochastic methods, where deterministic methods use rules or deterministic procedures to guide the search, while stochastic methods use randomness to explore the search space. The combination of both types of methods can lead to improved performance in terms of both convergence and solution quality.

Multi-objective optimization: Hybrid optimization algorithms can also be used for multi-objective optimization problems, where the goal is to find a set of optimal solutions that represent trade-offs between multiple conflicting objectives. Different optimization algorithms can be combined to explore the trade-off surface and identify a diverse set of Pareto-optimal solutions.

Overall, hybrid optimization algorithms provide a flexible and powerful approach for solving complex optimization problems by leveraging the strengths of different optimization methods. They can be customized to specific problem domains and are often used in fields such as engineering, computer science, finance, and operations research.

12.5 Chaotic Spotted Hyena Optimization Technique

The chaotic spotted hyena optimization (CSHO) [60] is a nature-inspired optimization algorithm that is inspired by the behavior of spotted hyenas in the wild. It is a population-based optimization algorithm that mimics the social hunting behavior of spotted hyenas to search for optimal solutions in complex optimization problems.

The CSHO algorithm incorporates chaotic maps into the standard spotted hyena optimization (SHO) [61] algorithm, which adds a randomizing element to the search process. Chaotic maps are mathematical functions that exhibit chaotic behavior, characterized by sensitivity to initial conditions and random-like, unpredictable outputs. By introducing chaotic maps into the SHO algorithm, CSHO aims to enhance the exploration and exploitation capabilities of the algorithm.

The CSHO algorithm typically starts with an initial population of hyenas, representing potential solutions to the optimization problem. These hyenas move and search for better solutions in the solution space by imitating the social behavior of real hyenas, such as scouting, encircling, and attacking prey. Chaotic maps are used to modify the positions of the hyenas, introducing randomness and unpredictability into their movements, which can help the algorithm escape local optima and explore the solution space more effectively.

Over iterations, the CSHO algorithm refines the population of hyenas by updating their positions based on their fitness, which is a measure of their quality as solutions to the optimization problem. The algorithm continues to iteratively search for better solutions until a stopping criterion is met, such as a maximum number of iterations or reaching a satisfactory solution.

CSHO has been applied to various optimization problems, including but not limited to function optimization, image processing, feature selection, and machine learning, among others. As with any optimization algorithm, the performance of CSHO may vary depending on the specific problem being addressed, the parameter settings, and other factors. Proper parameter tuning and benchmarking against other optimization algorithms are important to ensure its effectiveness and efficiency for a particular problem domain. Flowchart of the CSHO technique is shown in Figure 12.2.

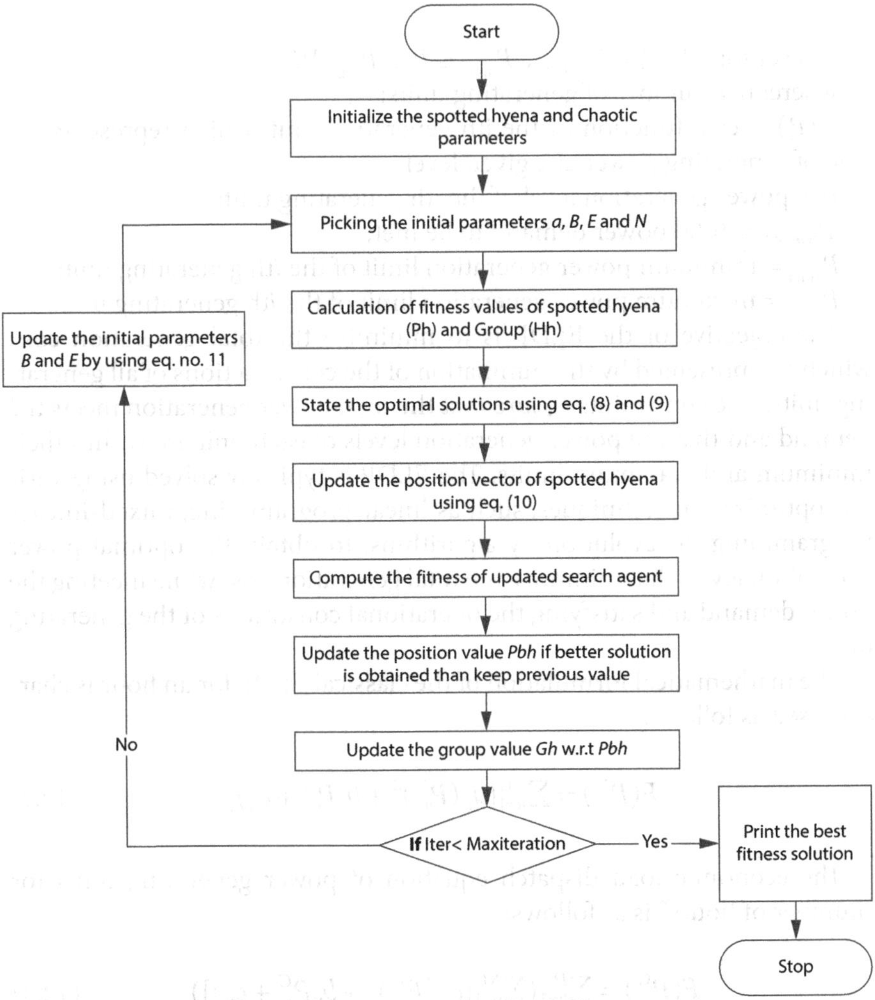

Figure 12.2 CSHO flowchart.

12.6 Economic Load Dispatch Problem

The classical ELDP is a mathematical optimization problem in electrical power systems that aims to determine the optimal power generation levels for a set of power generating units in order to meet the power demand while minimizing the total generation cost. The ELDP is typically formulated as follows:

$$\text{Minimize:} \quad \sum_{i=1}^{n} c_i(P_i) \tag{12.1}$$

subject to: $\sum_{i=1}^{n} P_i = P_{demand}$, $P_{\min,i} \le P_i \le P_{\max,i}$ $\forall i$
where: n = number of generating units;

$c_i(P_i) =$ cost function of the ith generating unit, which represents the cost of generating power at a given level;

P_i = power generation level of the ith generating unit;

P_{demand} = total power demand to be met;

$P_{\min,i}$ = minimum power generation limit of the ith generating unit;

$P_{\max,i}$ = maximum power generation limit of the ith generating unit.

The objective of the ELDP is to minimize the total generation cost, which is represented by the summation of the cost functions of all generating units. The constraints ensure that the total power generation meets the demand and that the power generation levels of each unit are within their minimum and maximum limits. The ELDP is typically solved using various optimization techniques, such as linear programming, mixed-integer programming, or evolutionary algorithms, to obtain the optimal power generation levels that minimize the total generation cost while meeting the power demand and satisfying the operational constraints of the generating units.

The mathematical formulation of the classical ELDP for an hour is characterised as follows:

$$F(P^G) = \sum_{n=1}^{NG}[a_n(P_n^G)^2 + b_n P_n^G + c_n)] \tag{12.2}$$

The economic load dispatch equation of power generating units for "number of hours" is as follows:

$$F(P^G) = \sum_{h=1}^{H}(\sum_{n=1}^{NG}[a_n(P_n^G)^2 + b_n P_n^G + c_n)]) \tag{12.3}$$

12.7 Power Balance Equality Constraints

In power systems, power balance constraints [62] refer to the requirement that the total power generated must be equal to the total power consumed at any given time. This principle is known as power balance, and it is a fundamental constraint that must be satisfied for the stable and reliable operation of an electric power grid.

Power balance constraints are typically expressed mathematically as equations that represent the power flow into and out of a power system. These equations are used in power system analysis and optimization models to ensure that the generation and consumption of electrical power are in equilibrium. Power balance constraints are important for maintaining grid stability, preventing overloads or voltage violations and ensuring that the electricity supply matches the demand in real time.

The power balance equations are based on the principle of conservation of energy and can be formulated in different ways, depending on the level of detail and complexity of the power system model. In a simplified form, the power balance constraint can be expressed as follows:

Total power generation = Total power consumption

$$\sum_{n=1}^{NG} P_n^G = P^{Demand} + P^{Loss} \tag{12.4}$$

where P^{Loss} is the power transmission loss may be represented as

$$P^{Loss} = \sum_{n=1}^{NG} \sum_{m=1}^{NG} P_n^G B_{nm} P_m^G \tag{12.5}$$

and if B_{i0}, B_{00} are loss coefficients, then the P^{Loss} equation can be modified as:

$$P^{Loss} = P_n^G B_{nm} P_m^G + \sum_{n=1}^{NG} P_n^G \times B_{i0} + B_{00} \tag{12.6}$$

The rewritten equation above may look like this:

$$P^{Loss} = [P_1 \quad P_2 \quad \cdots \quad P_{NG}] \begin{bmatrix} B_{11} & B_{12} & \cdots & B_{1n} \\ B_{21} & B_{22} & \cdots & B_{2n} \\ \vdots & \vdots & \ddots & \vdots \\ B_{n1} & B_{n2} & \cdots & B_{nn} \end{bmatrix} \begin{bmatrix} P_1 \\ P_2 \\ \vdots \\ P_{NG} \end{bmatrix} + [P_1 \quad P_2 \quad \cdots \quad P_{NG}] \begin{bmatrix} B_{01} \\ B_{02} \\ \vdots \\ B_{0NG} \end{bmatrix} B_{00} \tag{12.7}$$

This equation ensures that the total power generated by all the power generation units in the system matches the total power consumed by all the loads in the system. Power generation includes all the electricity generated by various sources, such as thermal power plants, hydroelectric power plants, wind farms, and solar power plants, whereas power consumption includes all the electrical loads, such as residential, commercial, and industrial loads, connected to the power grid.

Power balance constraints are a critical aspect of power system operation and planning, and they are considered in various power system analysis and optimization studies to ensure reliable and efficient operation of the electric power grid.

12.8 Generator Power Limit Inequality Constraints

Every power plant's output must be within its minimum and maximum allowable ranges. The generator power limit inequality constraint is a mathematical constraint that restricts the power output of a generator to be within a certain limit. In mathematical notation, it can be expressed as follows:

$P^G_{n(min)} \leq P^G_n \leq P^G_{n(max)}$ where P^G_n is the power output of the generator, $P^G_{n(max)}$ is the maximum power output limit of the generator, and $P^G_{n(min)}$ is the minimum power output limit of the generator. This constraint ensures that the generator does not exceed its maximum power output capacity during operation, which could result in equipment damage, instability in the power system, or violation of operational guidelines.

12.9 Test Systems Results and Discussion

The suggested study actively considers many trial schemes.

Case Study 1: We take into account transmission loss, ramp rate constraints, and no-go zones in a three-unit system. Test results are obtained for three different load demands using CSHO technique. The input data for various case studies are taken from [63, 64]. Figure 12.3 showcases the convergence curve for a three-unit system. The test results using conventional, PSO and CSHO technique is given in Table 12.1. It shows the power output of three generators for three different demands.

Case Study 2: Consider a six-unit system where there is transmission loss, restricted ramp rates and no-go zones. In this analysis, we take into account load requirements of 600 MW, 700 W, and 800 W. After 500 iterations, the relative costs of various techniques of creation are compared in

Table 12.1 Power output for three-generator system including losses.

P_{demand} (MW)	Method	P1 (MW)	P2 (MW)	P3 (MW)	P_{out} (MW)	P_{loss} (MW)	Fuel cost ($/h)
275	**CSHO**	194.838	73.7284	15	283.5664	8.5664	3332.5
	Conv.	193.82	74.78	15	283.6	8.61	3,333.14
	PSO	189.31	79.13	15	283.44	8.44	3,332.69
300	**CSHO**	206.788	88.1127	15	309.9007	9.9007	3,619.8
	Conv.	207.68	87.4	15	310.08	10.08	3,621.53
	PSO	212.56	82.48	15	310.04	10.04	3,620.09
350	**CSHO**	233.6082	114.3868	15	362.995	12.9951	4,210.3
	Conv.	235.58	112.89	15	363.47	13.47	4,215.18
	PSO	228.15	119.73	15	362.88	12.88	4,211.08

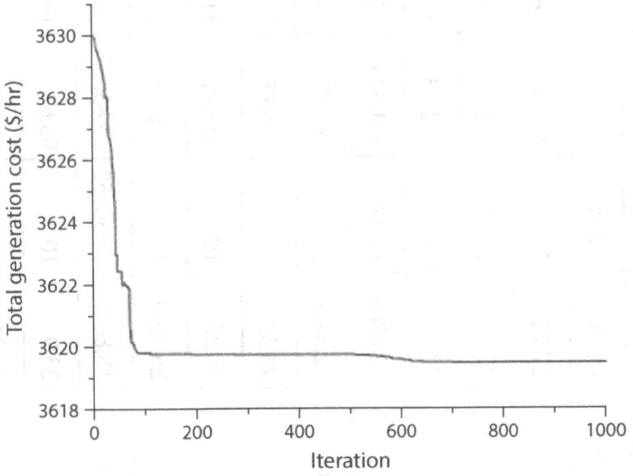

Figure 12.3 Convergence curve for three-unit system.

Table 12.2. The typical relationship between iterations and overall generating costs is seen in Figure 12.4.

Case Study 3: Consider a 10-unit system where there is transmission loss, restricted ramp rates, and no-go zones. In this analysis, we take into account load requirements of 2,700 W. After 500 iterations, the relative

Table 12.2 Power output for six-generator system including losses.

		P_{demand}	P1	P2	P3	P4	P5	P6	P_{out}	P_{loss}	Fuel cost
S. no.	**Method**	(MW)	(MW)	(MW)	(MW)	(MW)	(MW)	(MW)	(MW)	(MW)	(Rs/h)
1	**CSHO**	800	27.74	23.97	113.52	168.76	279.78	211.34	825.15	25.15	41,875.44
	Conventional		32.63	14.48	141.54	136.04	257.65	243	825.34	25.34	41,898.45
	PSO		32.67	14.45	141.73	136.56	257.37	242.54	825.32	25.3309	41,896.7
	ALO		32.6003	14.483	141.544	136.0413	257.6587	243.0033	825.3306	25.3307	41,896.6286
2	**CSHO**	700	32.8424	23.9518	163.0288	108.5607	185.1577	204.4979	718.0393	18.0393	36,911.34
	Conventional		28.33	10	118.95	118.67	230.75	212.8	719.5	19.5	36,914.01
	PSO		28.28	10	119.02	118.79	230.78	212.745	719.615	19.4319	36,912.2
	ALO		28.3031	10	118.955	118.6728	230.7596	212.7411	719.4316	19.4317	36,912.1444
3	**CSHO**	600	18.80595	25.56652	73.83081	107.0821	216.9169	171.9598	614.16	14.16	32,092.9335
	Conventional		23.9	10	95.63	100.7	202.82	182.02	615.07	15.07	32,096.58
	PSO		23.84	10	95.57	100.52	202.78	181.52	614.23	14.2374	32,094.7
	ALO		23.8713	10	95.6365	100.7064	202.8285	181.1945	614.2372	14.2372	32,094.6783

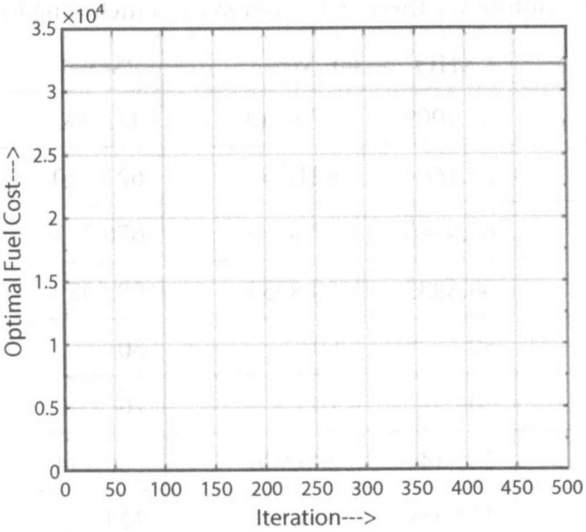

Figure 12.4 Convergence curve for six-unit system.

costs of various techniques of creation are compared in Table 12.3. The typical relationship between iterations and overall generating costs is seen in Figure 12.5.

Case Study 4: Consider a 13-unit system where there is transmission loss, restricted ramp rates, and no-go zones. In this analysis, we take into account load requirements of 1,800 W. After 1,000 iterations, the relative costs of various techniques of creation are compared in Tables 12.4 and 12.5. The typical relationship between iterations and overall generating costs is seen in Figure 12.6.

Case Study 5: Here, we assume a demand of 10,500 MW and a 40-unit system that takes into consideration the valve point loading impact. We do not bother with the transmission loss. Table 12.6 displays the cost of generating for many distinct approaches. Table 12.7 showcases the comparative analysis for different approaches. Figure 12.7 showcases the convergence curve for 40-unit system using CSHO technique.

Case Study 6: Here, we assume a demand of 49,342 MW and a 140-unit system that takes into consideration the valve point loading impact. We do not bother with the transmission loss. Table 12.8 displays the cost of generating for many distinct approaches. Table 12.9 showcases the comparison of the proposed technique with some already existing techniques. Figure 12.8 showcases the graphical representation between optimal fuel cost and number of iterations.

Table 12.3 Power output for three-generator system including losses.

Unit	CSHO	CM	PSO	ESCSA
P1	47.9009	47.8749	47.9492	47.9392
P2	62.1604	621,299	621,460	611,460
P3	67.9345	67.8696	678,343	678,343
P4	58.8933	58.8633	58.8639	58.9039
P5	50	50	50	52
P6	70	70	70	71
P7	206.444	206.806	206.794	206.7936
P8	224.948	224.591	224.613	224.7032
P9	377.317	377.534	377.482	377.4816
P10	383.681	383.476	383.431	383.4313
P Loss	49.2796	49.1165	49.1148	49.1148
Fuel cost (Rs/h)	81,108.9	81,115	81,115	81,113.35

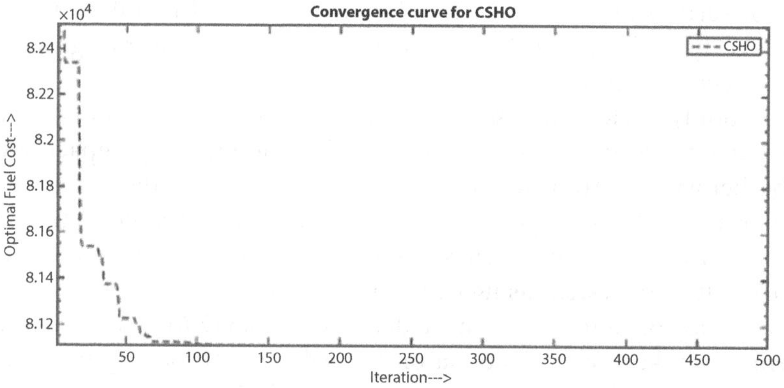

Figure 12.5 Convergence curve for 10 units.

Table 12.4 Power output for 13-generator system including losses.

Method	P1	P2	P3	P4	P5	P6	P7	P8	P9	P10	P11	P12	P13	Fuel cost
CSHO	538.559	149.8762	224.4966	109.9054	109.8903	109.8764	109.9194	109.8681	109.8885	77.42087	40.09216	55.17739	55.02971	17,969.14

Table 12.5 Comparison of fuel cost for 13 units (demand = 1,800 MW).

Method	HMAPSO	PPSO	MPPSO	DPSO	PSO	APPSO	CSHO
Fuel cost	17969.31	17971.01	17976.19	17976.31	18014.16	17978.89	17969.14

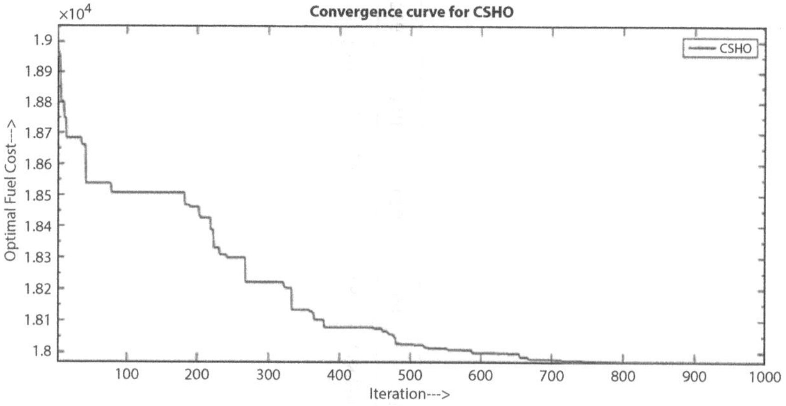

Figure 12.6 Convergence curve for 13 units.

Table 12.6 Power output for 40-generator system with valve point for demand = 10,500 MW without losses.

GEN	CSHO	PSO	PPSO	MPPSO	APPSO	DPPSO	HMAPSO
1	112.833	113.116	111.601	112.903	112.579	111.917	111.136
2	37.967	113.01	111.781	112.802	111.553	112.338	111.135
3	172.613	119.702	118.613	117.515	98.751	118.922	120
4	179.613	81.647	179.819	181.442	180.384	179.928	177.221
5	10.752	95.062	92.443	95.876	94.389	48.998	88.699
6	217.545	139.209	139.846	139.856	139.943	139.931	140
7	52.4938	299.127	296.703	299.452	298.937	299.61	260.157
8	138.102	287.491	284.566	298.277	285.827	298.206	284.723
9	359.303	292.316	285.164	299.043	298.381	285.372	285.523

(*Continued*)

Table 12.6 Power output for 40-generator system with valve point for demand = 10,500 MW without losses. (*Continued*)

GEN	CSHO	PSO	PPSO	MPPSO	APPSO	DPPSO	HMAPSO
10	55.3038	279.273	203.859	130.886	130.212	130.701	130
11	19.7283	169.766	94.283	243.53	94.385	94.849	168.805
12	20.8946	94.344	94.09	94.768	169.583	244.086	168.689
13	35.2175	214.871	304.83	215.033	214.617	214.739	304.123
14	215.695	304.79	304.173	304.739	304.886	304.504	304.678
15	394.448	304.563	304.467	304.694	304.547	304.744	304.317
16	214.967	304.302	304.177	215.146	304.584	304.501	304.317
17	489.199	489.173	489.544	497.407	489.452	489.515	489.187
18	578.939	491.336	489.773	489.459	497.472	489.534	489.455
19	511.186	510.88	511.28	511.867	512.816	511.567	512.097
20	601.001	511.474	510.904	548.4	548.992	511.374	511.349
21	615.772	524.814	524.092	523.396	524.652	525.246	523.247
22	785.874	524.775	523.121	525.206	523.399	523.979	523.515
23	523.309	525.563	523.242	524.971	548.895	548.599	523.454
24	792.587	522.712	524.26	523.66	525.871	523.314	523.453
25	702.817	503.211	523.283	523.624	523.814	523.259	523.492
26	792.599	524.199	523.074	527.932	523.565	524.36	523.307
27	7.81599	10.082	10.8	10.474	10.575	10.388	10
28	9.05675	10.663	10.742	11.074	11.177	10.552	10
29	9.77993	10.418	10.799	10.582	11.21	10.082	10
30	93.4637	94.244	94.475	96.403	96.178	96.422	88.691
31	211.716	189.377	189.245	189.338	189.999	189.692	190
32	210.032	189.796	189.995	189.849	189.924	189.82	190
33	222.916	189.813	188.081	189.739	189.714	189.954	190

(*Continued*)

Table 12.6 Power output for 40-generator system with valve point for demand = 10,500 MW without losses. (*Continued*)

GEN	CSHO	PSO	PPSO	MPPSO	APPSO	DPPSO	HMAPSO
34	17.3751	199.797	198.475	199.808	199.284	199.427	164.218
35	90.7142	199.284	197.528	199.994	199.599	199.905	200
36	17.3079	198.165	196.971	199.749	199.751	199.229	200
37	124.656	109.291	109.161	109.917	109.973	109.565	110
38	98.2264	109.087	109.9	109.41	109.506	109.741	110
39	153.628	109.909	109.855	109.728	109.363	109.575	110
40	602.552	512.348	510.984	512.053	511.261	511.554	511.009

Table 12.7 Comparison of simulation results for 40 units (load = 10,500 MW).

Method	PSO	PPSO	MPPSO	APPSO	DPSO	HMAPSO	CSHO
Fuel cost	122,324	121,788	122,225.7	122,044.63	122,160	121,587	120,772

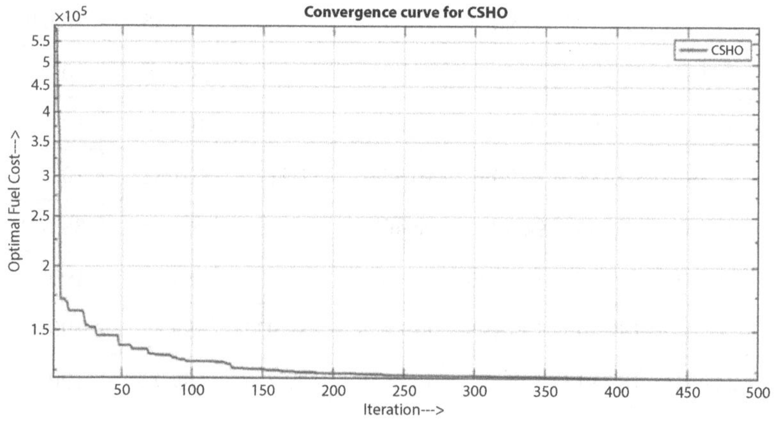

Figure 12.7 Convergence curve for 40-unit system.

Table 12.8 Power output for 140-generator system for demand = 49,342 MW without losses.

Unit	P_{out}	Unit	P_{out}	Unit	P_{out}	Unit	P_{out}	Unit	P_{out}	Unit	P_{out}	Unit	P_{out}
P1	112.9537	**P21**	505	**P41**	9.768932	**P61**	163.0002	**P81**	259.5392	**P101**	958	**P121**	176.3981
P2	189	**P22**	505	**P42**	5.526958	**P62**	95	**P82**	56.00934	**P102**	1007	**P122**	8.06709
P3	189.9633	**P23**	505	**P43**	238.365	**P63**	301.4734	**P83**	115.0004	**P103**	1006	**P123**	4.000052
P4	189.9998	**P24**	504.9924	**P44**	243.3615	**P64**	277.8478	**P84**	118.0037	**P104**	1012.999	**P124**	19.57877
P5	165.1845	**P25**	537	**P45**	206.9179	**P65**	453.1948	**P85**	115	**P105**	1019.991	**P125**	9.124038
P6	129.9098	**P26**	536.9429	**P46**	184.5184	**P66**	196	**P86**	207.0024	**P106**	953.9979	**P126**	18.92125
P7	490	**P27**	549	**P47**	203.5492	**P67**	411.4635	**P87**	207.0004	**P107**	951.9983	**P127**	10.91138
P8	490	**P28**	548.9963	**P48**	186.6405	**P68**	339.8344	**P88**	175	**P108**	1005.999	**P128**	112
P9	495.9999	**P29**	501	**P49**	198.5964	**P69**	131.3599	**P89**	200.8057	**P109**	1012.997	**P129**	4.175892
P10	496	**P30**	500.9878	**P50**	221.148	**P70**	279.3187	**P90**	210.278	**P110**	1020.999	**P130**	5.000186
P11	496	**P31**	505.9788	**P51**	367.5563	**P71**	165.2939	**P91**	196.9418	**P111**	1015	**P131**	7.475242
P12	495.9993	**P32**	506	**P52**	167.4609	**P72**	304.6648	**P92**	579.9925	**P112**	94.0057	**P132**	66.08453
P13	505.9663	**P33**	505.9992	**P53**	172.9659	**P73**	204.5445	**P93**	644.9904	**P113**	94.0019	**P133**	5.060165
P14	508.9998	**P34**	506	**P54**	165.4117	**P74**	187.6006	**P94**	983.9829	**P114**	94.09574	**P134**	51.84209
P15	506	**P35**	499.9998	**P55**	180.0002	**P75**	183.8519	**P95**	977.9975	**P115**	244	**P135**	47.51604
P16	505	**P36**	500	**P56**	180.0003	**P76**	207.4066	**P96**	681.9756	**P116**	273.5693	**P136**	41.0032
P17	505.8612	**P37**	240.9839	**P57**	149.4992	**P77**	175.8266	**P97**	720	**P117**	244.7998	**P137**	17.00015
P18	506	**P38**	241	**P58**	198.0181	**P78**	332.2586	**P98**	717.9253	**P118**	95.07854	**P138**	12.79178
P19	504.9992	**P39**	773.9994	**P59**	229.7698	**P79**	530.9241	**P99**	719.9941	**P119**	110.6841	**P139**	7.25453
P20	504.9971	**P40**	768.9547	**P60**	292.7676	**P80**	530.9994	**P100**	964	**P120**	116.0001	**P140**	29.76438

Table 12.9 Comparison of fuel cost for 140 units (load = 10,500 MW).

Method	Fuel cost (Rs/h)
CTPSO	1,657,962.73
CSPSO	1,657,962.73
COPSO	1,657,962.73
CCPSO	1,657,962.73
GSO	1,728,151.16
CQGSO	1,657,962.72
DEL	1,657,962.72
ACS	1,658,002.72
IACS	1,657,956.80
CSHO	**1,563,293.00**

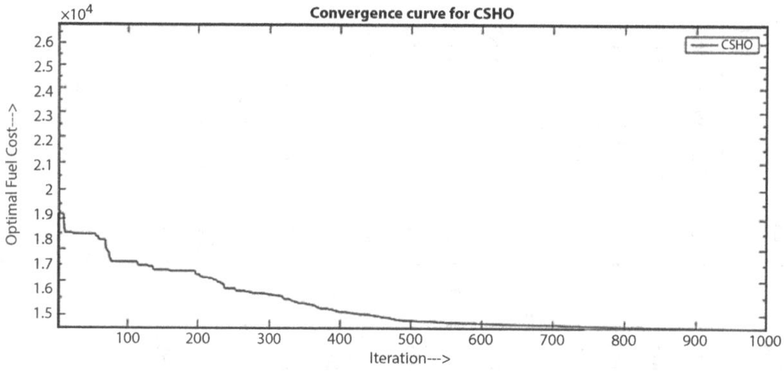

Figure 12.8 Convergence curve for 140-unit system.

12.10 Conclusion

In an electric power system, the goal of economic load dispatch is to optimize the generation levels of different power-generating units while considering factors such as fuel costs, operating constraints, and system stability.

The objective is to minimize the total cost of generation while meeting the demand for electricity and satisfying operational constraints such as generator limits, transmission line limits, and environmental constraints. Optimization problems are indeed complex and challenging due to the nonlinearities, constraints, and uncertainties involved in power system operation and planning [65]. Efficient solutions to these problems can help power system operators and planners make informed decisions to ensure reliable and economic operation of power systems while meeting energy demand and operational constraints. Of the optimization methods that show promise in this chapter, we suggest CSHO as a method for resolving both convex and non-convex ELD issues. The results of the suggested technique show promise in locating the best option to reduce generating costs and transmission losses to a minimum. After applying this strategy to six distinct test cases, its superiority over other optimizations techniques is shown. In conclusion, CSHO stands out as the best method for achieving low generating costs and low transmission losses.

References

[1] A. J. Conejo and L. Baringo, "Unit Commitment and Economic Dispatch," in *Power Systems Operations*, pp. 197–232, 2018, doi: 10.1007/978-3-319-69407-8_7.

[2] M. Dorigo, M. Birattari, and T. Stutzle, "Ant colony optimization," *IEEE Comput. Intell. Mag.*, vol. 1, no. 4, pp. 28–39, Nov. 2006, doi: 10.1109/MCI.2006.329691.

[3] E. Rashedi and H. N. Saeid, "BGSA: binary gravitational search algorithm," no. December 2009, pp. 727–745, 2010, doi: 10.1007/s11047-009-9175-3.

[4] X. Yang, "Bat algorithm for multi-objective optimisation," vol. 3, no. 5, p. 2011, 2011.

[5] H. C. Kuo and C. H. Lin, "Cultural Evolution Algorithm for Global Optimizations and its Applications," *J. Appl. Res. Technol.*, vol. 11, no. 4, pp. 510–522, 2013, doi: 10.1016/S1665-6423(13)71558-X.

[6] Arora, K.; Kumar, A.; Kamboj, V.K.; Prashar, D.; Shrestha, B.; Joshi, G.P. Impact of Renewable Energy Sources into Multi Area Multi-Source Load Frequency Control of Interrelated Power System. *Mathematics*, 2021, *9*, 186.

[7] S. Mirjalili and A. Lewis, "Adaptive gbest-guided gravitational search algorithm," 2014, doi: 10.1007/s00521-014-1640-y.

[8] X. Yang, "Firefly Algorithms for Multimodal Optimization," pp. 169–178, 2009.

[9] H. Faris, I. Aljarah, M. A. Al-betar, and S. Mirjalili, "Grey wolf optimizer: a review of recent variants and applications," *Neural Comput. Appl.*, no. November, 2017, doi: 10.1007/s00521-017-3272-5.
[10] S. Ali, M. Pradhan, P. Kumar, and T. Pal, "Opposition-based krill herd algorithm applied to economic load dispatch problem," *Ain Shams Eng. J.*, 2016, doi: 10.1016/j.asej.2016.02.003.
[11] S. Mirjalili, "Moth-Flame Optimization Algorithm: A Novel Nature-inspired Heuristic Paradigm," *Knowl.-Based Syst.*, no. July, 2015, doi: 10.1016/j.knosys.2015.07.006.
[12] M. Naik, M. R. Nath, and S. Sahany, "A New Adaptive Cuckoo Search Algorithm," pp. 1–5, 2015.
[13] F. Merrikh-Bayat, "The runner-root algorithm: A metaheuristic for solving unimodal and multimodal optimization problems inspired by runners and roots of plants in nature," *Appl. Soft Comput. J.*, vol. 33, pp. 292–303, 2015, doi: 10.1016/j.asoc.2015.04.048.
[14] Arora, K.; Kumar, A.; Kamboj, V.K.; Prashar, D.; Jha, S.; Shrestha, B.; Joshi, G.P. Optimization Methodologies and Testing on Standard Benchmark Functions of Load Frequency Control for Interconnected Multi Area Power System in Smart Grids. *Mathematics,* 2020, *8*, 980.
[15] M. D. Li, H. Zhao, X. W. Weng, and T. Han, "Advances in Engineering Software A novel nature-inspired algorithm for optimization: Virus colony search," *Adv. Eng. Software*, vol. 92, pp. 65–88, 2016, doi: 10.1016/j.advengsoft.2015.11.004.
[16] G. Wang, L. Guo, A. H. Gandomi, G. Hao, and H. Wang, "Chaotic Krill Herd algorithm," *Inf. Sci. (N. Y.)*, 2014, doi: 10.1016/j.ins.2014.02.123.
[17] S. Mirjalili, "The Ant Lion Optimizer," *Adv. Eng. Software*, vol. 83, pp. 80–98, 2015, doi: 10.1016/j.advengsoft.2015.01.010.
[18] S. Mirjalili, "Dragonfly algorithm: a new meta-heuristic optimization technique for solving single-objective, discrete, and multi-objective problems," *Neural Comput. Appl.*, vol. 27, no. 4, pp. 1053–1073, 2016, doi: 10.1007/s00521-015-1920-1.
[19] H. Abedinpourshotorban, S. Mariyam, and Z. Beheshti, "Electromagnetic field optimization: A physics-inspired metaheuristic optimization algorithm," *Swarm Evol. Comput.*, pp. 1–14, 2015, doi: 10.1016/j.swevo.2015.07.002.
[20] J. Li, Y. L S. T. J. Xia, and J. Li, *An Improved Cuckoo Search Algorithm with Self-Adaptive Knowledge Learning*, vol. 0123456789. Springer London, 2019. doi: 10.1007/s00521-019-04178-w.
[21] Y. Zheng, "Computers & Operations Research Water wave optimization: A new nature-inspired metaheuristic," *Comput. Oper. Res.*, vol. 55, pp. 1–11, 2015, doi: 10.1016/j.cor.2014.10.008.
[22] E. Rashedi, H. Nezamabadi-pour, and S. Saryazdi, "GSA: A Gravitational Search Algorithm," vol. 179, pp. 2232–2248, 2009, doi: 10.1016/j.ins.2009.03.004.

[23] A. H. Kashan, "Computers & Operations Research A new metaheuristic for optimization: Optics inspired optimization (OIO)," vol. 55, pp. 99–125, 2015, doi: 10.1016/j.cor.2014.10.011.

[24] X. Li, J. Zhang, and M. Yin, "Animal migration optimization: an optimization algorithm inspired by animal migration behavior," 2013, doi: 10.1007/s00521-013-1433-8.

[25] G. Wang, S. Deb, G. Wang, and S. Deb, "Earthworm optimization algorithm: a bio-inspired metaheuristic algorithm for global optimization problems," no. November, 2015, doi: 10.1504/IJBIC.2015.10004283.

[26] D. Simon and S. Member, "Biogeography-Based Optimization," vol. 12, no. 6, pp. 702–713, 2008.

[27] M. Ghaemi, "Expert Systems with Applications Forest Optimization Algorithm," *Expert Syst. Appl.*, no. May, 2014, doi: 10.1016/j.eswa.2014.05.009.

[28] S. Saremi, S. Mirjalili, and A. Lewis, "Advances in Engineering Software Grasshopper Optimisation Algorithm: Theory and application," vol. 105, pp. 30–47, 2017, doi: 10.1016/j.advengsoft.2017.01.004.

[29] A. S. Computing and N. Ghorbani, "Exchange market algorithm," no. April, 2016, doi: 10.1016/j.asoc.2014.02.006.

[30] X. Yang, "Flower Pollination Algorithm for Global Optimization," pp. 240–249, 2012.

[31] R. Y. M. Nakamura, L. A. M. Pereira, K. A. Costa, D. Rodrigues, and J. P. Papa, "BBA: A Binary Bat Algorithm for Feature Selection," 2012, doi: 10.1109/SIBGRAPI.2012.47.

[32] S. Mirjalili, S. M. Mirjalili, and A. Hatamlou, "Multi-Verse Optimizer: a nature-inspired algorithm for global optimization," no. 30, 2015, doi: 10.1007/s00521-015-1870-7.

[33] Y. Tan and Y. Zhu, "Fireworks Algorithm for Optimization," pp. 355–364, 2010.

[34] S. Karimkashi, S. Member, and A. A. Kishk, "Invasive Weed Optimization and Its Features in Electromagnetics," vol. 58, no. 4, pp. 1269–1278, 2010.

[35] S. Das, A. Biswas, S. Dasgupta, and A. Abraham, "Bacterial Foraging Optimization Algorithm: Theoretical Foundations, Analysis, and, Applications," vol. 3, pp. 23–55, 2009.

[36] A. H. Kashan, "League Championship Algorithm (LCA): An algorithm for global optimization inspired by sport championships," *Appl. Soft Comput. J.*, vol. 16, pp. 171–200, 2014, doi: 10.1016/j.asoc.2013.12.005.

[37] C. B. Optimization, *Colliding Bodies Optimization*. 2014. doi: 10.1007/978-3-319-05549-7.

[38] H. Shareef, A. A. Ibrahim, and A. H. Mutlag, "Lightning search algorithm," *Appl. Soft Comput. J.*, 2015, doi: 10.1016/j.asoc.2015.07.028.

[39] E. Cuevas, M. Gonz, D. Zaldivar, P. Marco, and G. Garc, "An Algorithm for Global Optimization Inspired by Collective Animal Behavior," vol. 2012, 2012, doi: 10.1155/2012/638275.

[40] E. Mechanics, "Shuffled frog-leaping algorithm: a memetic meta-heuristic for discrete optimization," vol. 38, no. 2, pp. 129–154, 2006.

[41] M. Cheng and D. Prayogo, "Symbiotic Organisms Search: A new metaheuristic optimization algorithm," *Comput. Struct.*, 2014, doi: 10.1016/j.compstruc.2014.03.007.

[42] A. I. Hafez, H. M. Zawbaa, E. Emary, and A. E. Hassanien, "Sine Cosine Optimization Algorithm for Feature Selection," pp. 1–5, 2016.

[43] I. Aljarah, H. Faris, and S. Mirjalili, "Optimizing connection weights in neural networks using the whale optimization algorithm," *Soft Comput.*, 2016, doi: 10.1007/s00500-016-2442-1.

[44] X. Meng, X. Z. Gao, L. Lu, and Y. Liu, "Journal of Experimental & Theoretical A new bio-inspired optimisation algorithm: Bird Swarm Algorithm," no. June, pp. 37–41, 2015, doi: 10.1080/0952813X.2015.1042530.

[45] H. Salimi, "Knowledge-Based Systems Stochastic Fractal Search: A powerful metaheuristic algorithm," *Knowl.-Based Syst.*, 2014, doi: 10.1016/j.knosys.2014.07.025.

[46] C. Dai, Y. Zhu, W. Chen, and C. Engineering, "Seeker Optimization Algorithm," pp. 167–176, 2007.

[47] S. Mirjalili, A. H. Gandomi, S. Zahra, and S. Saremi, "Salp Swarm Algorithm: A bio-inspired optimizer for engineering design problems," *Adv. Eng. Software*, vol. 0, pp. 1–29, 2017, doi: 10.1016/j.advengsoft.2017.07.002.

[48] H. Eskandar, A. Sadollah, A. Bahreininejad, and M. Hamdi, "Water cycle algorithm – A novel metaheuristic optimization method for solving constrained engineering optimization problems," *Comput. Struct.*, vol. 110–111, pp. 151–166, 2012, doi: 10.1016/j.compstruc.2012.07.010.

[49] P. Civicioglu, "Backtracking Search Optimization Algorithm for numerical optimization problems," *Appl. Math. Comput.*, vol. 219, no. 15, pp. 8121–8144, 2013, doi: 10.1016/j.amc.2013.02.017.

[50] A. Sadollah, A. Bahreininejad, H. Eskandar, and M. Hamdi, "Mine blast algorithm: A new population based algorithm for solving constrained engineering optimization problems," *Appl. Soft Comput. J.*, 2012, doi: 10.1016/j.asoc.2012.11.026.

[51] G. Wang, S. Deb, and Z. Cui, "Monarch butterfly optimization," *Neural Comput. Appl.*, 2015, doi: 10.1007/s00521-015-1923-y.

[52] Z. Bayraktar, M. Komurcu, and D. H. Werner, "Wind Driven Optimization (WDO): A Novel Nature-Inspired Optimization Algorithm and its Application to Electromagnetics (1)," no. 1, pp. 0–3, 2010.

[53] M. Shahrouzi and A. Salehi, "Imperialist Competitive Learner-Based Optimization: A Hybrid Method to Solve Engineering Problems," *Iran University of Science & Technology*, vol. 10, no. 1, pp. 155–180, 2020, Accessed: Mar. 07, 2023. [Online]. Available: http://ijoce.iust.ac.ir/article-1-427-en.html

[54] C. Dai, W. Chen, L. Ran, Y. Zhang, and Y. Du, "Human Group Optimizer with Local Search," pp. 310–320, 2011.

[55] A. H. Gandomi, "Interior search algorithm (ISA): A novel approach for global optimization," *ISA Trans.*, pp. 1–16, 2014, doi: 10.1016/j.isatra.2014.03.018.

[56] S. C. Satapathy, A. Naik, and K. Parvathi, "A teaching learning based optimization based on orthogonal design for solving global optimization problems," pp. 1–12, 2013.

[57] G. Model, "Weighted Superposition Attraction (WSA): A swarm intelligence," 2015, doi: 10.1016/j.asoc.2015.10.036.

[58] A. J. Lockett, "No free lunch theorems," *Nat. Comput. Series*, vol. 1, no. 1, pp. 287–322, 2020, doi: 10.1007/978-3-662-62007-6_12.

[59] D. H. Wolpert and W. G. Macready, "No free lunch theorems for optimization," *IEEE Trans. Evol. Comput.*, vol. 1, no. 1, pp. 67–82, 1997, doi: 10.1109/4235.585893.

[60] T. Mishra, A. K. Singh, and V. K. Kamboj, "A cost-effective solution to security constrained dispatch problem of Islanded microgrid using chaotic spotted Hyena optimization," https://doi.org/10.1080/23080477.2022.2092670, pp. 1–11, Jun. 2022, doi: 10.1080/23080477.2022.2092670.

[61] G. Dhiman and V. Kumar, "Advances in Engineering Software Spotted hyena optimizer: A novel bio-inspired based metaheuristic technique for engineering applications," *Adv. Eng. Software*, vol. 114, pp. 48–70, 2017, doi: 10.1016/j.advengsoft.2017.05.014.

[62] D. C. Walters and G. B. Sheble, "Genetic algorithm solution of economic dispatch with valve point loading," *IEEE Trans. Power Syst.*, vol. 8, no. 3, pp. 1325–1332, 1993, doi: 10.1109/59.260861.

[63] N. I. Nwulu and X. Xia, "Optimal dispatch for a microgrid incorporating renewables and demand response," *Renew. Energy*, vol. 101, pp. 16–28, 2017, doi: 10.1016/j.renene.2016.08.026.

[64] A. Y. Abdelaziz, E. S. Ali, and S. M. A. Elazim, "Electrical Power and Energy Systems Combined economic and emission dispatch solution using Flower Pollination Algorithm," *Int. J. Electr. Power Energy Syst.*, vol. 80, pp. 264–274, 2016, doi: 10.1016/j.ijepes.2015.11.093.

[65] Kumari, C.L., Kamboj, V.K., Bath, S.K. *et al.* A boosted chimp optimizer for numerical and engineering design optimization challenges. *Eng. Comput.*, (2022). https://doi.org/10.1007/s00366-021-01591-5.

13

Simulation of Automatic Search of Charging Station for Electric Bikes

Saroja S.[1*], Haseena S.[2], Hariharan M.[2] and Raghul Priyadharshan M.[2]

[1]*Department of Computer Applications, National Institute of Technology, Trichy, Tamil Nadu, India*

[2]*Department of Information Technology, Mepco Schlenk Engineering College, Sivakasi, Tamil Nadu, India*

Abstract

Less reliance on fossil fuels is one of the main goals behind the increasing number of electric bikes that are currently on the road. If there are not enough charging stations, however, it will be difficult for the electric bike rider. Electric bike riders may have to wait longer for their turns, which will be uncomfortable for them, if there are only a few charging stations in the city and a lot of charging requests are being made at once. If the charging station is devoted and competent in its charging, then the stability of the Intelligent Grid may also be in jeopardy. Two algorithms—Electric Bike Supply Equipment (EBSE) and Load Balancing of Smart Grid (LBSR)—can be used by the proposed model to resolve these issues. LBSR works to maintain grid stability by attempting to balance the load. EBSE is responsible for selecting the charging location and organizing the charging and unloading processes. The workings of the proposed two algorithms, LBSR and EBSE, are verified by simulating a sample electric vehicular network using SUMO 1.11.0. Simulation results demonstrate the capability of the two methods in reducing waiting times.

Keywords: Load balancing of smart grid, smart grid, electric bike supply equipment

**Corresponding author*: saroja@nitt.edu

Krishan Arora, Suman Lata Tripathi and Himanshu Sharma (eds.) Electric Vehicle Design: Design, Simulation and Applications, (295–312) © 2024 Scrivener Publishing LLC

13.1 Introduction

A motorcycle is a two- or three-wheeled motor vehicle that is also called a motorbike, bike, cycle, or trike (if three-wheeled). Motorcycles are designed for a variety of uses, including long-distance travel, commuting, cruising, sports (including racing), and off-road riding. Riding a motorbike and participating in other related social activities, such as joining a motorcycle club and attending motorcycle rallies, are all examples of motorcycling.

The first internal combustion, petroleum-fueled motorbike was the Daimler Reitwagen, which was built in Germany by Gottlieb Daimler and Wilhelm Maybach in 1885. In 1894, Hildebrand and Wolfmüller produced the first series of bikes. Today, the majority of two-wheeled vehicles are made in India. Motorcycle and scooter segments make up the majority of the two-wheeler market. The production of two-wheelers for the domestic market increased from 23.2 million units in 2017–2018 to 24.5 million units in 2018–2019.

Currently, two-wheeler production makes up 80% of all Indian automobile production. More than 25 million two-wheelers can be produced annually by Indian manufacturers. Two-wheeler sales in India will be further boosted by the rise in per capita income, high fuel efficiency, simple financing options, final mile delivery for the e-commerce sector, government policy for farmers, and preference for two-wheelers in eMobility. By 2026, the annual production of two-wheelers is anticipated to reach 50.6 million units. Motorcycles are considered utilitarian in poor countries due to reduced pricing and better fuel economy. Except for car-centric Japan, the Asia-Pacific and Southern and Eastern Asia regions account for 58% of all motorbikes in the world.

Motorcycles had 37 times the number of fatalities per vehicle mile traveled than cars, according to the US Department of Transportation. Compared to traditional motorbikes, electric bikes (EBs) require significantly less maintenance. There is no oil or cooling system to maintain, and there are fewer parts overall, resulting in less maintenance. In addition, compared to petrol and electric motorcycles, maintenance costs are minimal. You do not have to buy any petrol, and you also do not have to charge the bike. Electric bikes are always ready to ride—no planning or foresight is required. You can get on and go whenever the mood strikes you!

The proposed work's main goal is to lessen the load demand for EBs while coordinating charging and discharging processes to cut costs. In order to accomplish this, the chapter suggests two algorithms, namely, the Load Balancing of Smart Grid (LBSR) and Electric Bike Supply Equipment

(EBSE), respectively. Using information like the EB profile and price profile, the load balancing algorithm of the smart grid is in charge of scheduling EBs for charging or discharging services. The best available charging or discharging station for the EB will be selected by the EBSE selection model. The discharging services are provided by the bidirectional power flow between the EB and the smart grid, which enables the EB to work as a distributed energy storage system.

Take the example of a man who owns an EB and plans to charge it before leaving for work. Many EBs are already awaiting their turn to charge, so consumers will have to wait until that person's turn arrives, which will negatively affect that person and make it more difficult to locate another charging station (CS) nearby. In this situation, the proposed model can be utilized to locate a CS nearby and to decide whether to charge or discharge the EB.

Most of the currently available models lack both management for the high heat produced while charging a large number of EBs at once and management for the stability of the smart grid. A voltage issue could develop if more power is needed during peak load hours to charge the increasing number of EBs. All of these problems can be resolved by the two algorithms.

An electrical network with two-way communication is known as a smart grid. Electricity for EB riders is provided solely by the Smart Grid. The Smart Grid can communicate with the EB, helping to cut costs and energy use. As a result, in our strategy, we only use smart grid technologies to communicate with EBs.

The Smart Grid and EBs may be able to communicate with one another. Before the plug-in phase, these smart grid facilities can be informed about the status of EBs and the equipment that supplies them.

The Smart Grid and the EB might exchange data in the following forms:

- Electricity purchase costs (which vary depending on the amount of load), which are moderately high.
- The EBs receive the most recent EBSE status from Smart Grid.
- The EBSE status shows the closest available EBSE and the number of vehicles that must be supplied on each EBSE.

The number of vehicles needed to charge can be calculated on the basis of the line by locating the closest available CS and obtaining the EBSE status on one EBSE. This decreases the amount of time users must wait. The data are all received by the EB. The advantages listed above are brought by the two algorithms. They are in charge of the balancing load in smart grids as well as coordinating charging and discharging process for EBs.

13.2 Related Works

In "Stochastic modelling of EB charging at charging station" [1], the authors concentrated primarily on charging the EB, by omitting the discharging process. The charging procedure and the discharging process are critical for EBs. If we simply examine the charging process for all time, that is, if we ignore the discharging process, then we will experience problems like energy leakage from the EB and inefficient use of electrical energy. If discharging mechanism is also taken for consideration, then the surplus energy can be used by other users for the charging process of the EB. However, the author put forth efficient strategies for charging the EB. The most essential factor that needs considerable priority is the total time required for EBs to charge to their full capacity. This factor helps in balancing the number of EBs on the grid, which the author does not highlight.

In "Smart grid augmented by HetNet" [2], the authors wanted to speed up the process of charging and discharging an EB. It was a great idea to shorten the charging time for the EB. The technique requires changing the battery in the EB. Although it can be finished quickly, there are certain downsides. Because you only need one alternative battery and it is also somewhat expensive, you should ideally maintain some alternative batteries that are somewhat less flexible than the standard charging and discharging technique. The battery is quite pricey when compared to the price of a typical charging method.

Regarding technology for batteries and energy [3], users who ride EBs may have "range anxiety," which is the perception that the battery is running low and that they are hampered by the lack of a CS. New battery and energy technologies were introduced to prevent range anxiety. Thus, hybrids and plug-in hybrids in this technological spectrum reduce anxiety. Energy conservation is the main objective of battery and energy technology for EBs. Energy can be saved by placing more energy-efficient batteries in the same area. The fact that an increasing number of individuals are using EBs is another application of this technology. Charge anxiety is currently plaguing users. The three different charger levels are put in to use to reduce the worry that EBs bring when it comes to charging. Here, the charging technology changes the grid's AC electricity to DC power. With the help of a dynamic resource sharing model, queueing models and greedy control rules are used to implement quick charging mechanisms [4].

A multi-objective optimization problem is formulated with the objectives of maximizing the voltage security margin while minimizing total power losses, total electrical energy expenditures, and total power source emissions [5]. Faced with rising energy costs, pollutant emissions, and security concerns for plug-in electric vehicles (PHEVs), modern electrical networks and energy management systems must develop an ideal system of operation and management for PEVs. Qin and Zhang [6] proposed charging scheduling with minimal waiting in a network of electric vehicles (EVs) and CSs with the goal of reducing waiting time incurred in the charging process. Intention-aware routing for EV CSs is proposed in [7, 8] and focuses on eliminating CS delays by intelligently organizing charging activities in terms of space and time.

The frequency, time, and location of charging and discharging all have an impact on the quality of experience (QoE) of driving an EV. Scheduling techniques for EV charging and discharging should consider both the QoE of using EVs and the QoE of not using EVs, as well as the load capacity of the electricity grid. The Stable Matching Algorithm (SMA) is presented for assigning EVs to CSs as a many-to-one matching game for traveling plan-aware scheduling [9]. Then, the power exchange between charging and discharging EVs in the same parking lot as a many-to-many matching model with linkages is characterized, and the Pareto Optimal Matching Algorithm (POMA) for cooperative EV charging and discharging in parking patterns is proposed. According to simulation studies, the SMA may significantly improve the average system utility for EV charging in driving patterns, whereas the POMA can increase the amount of electricity offloaded from the grid, hence improving power grid stability.

A data mining model [10] was built to investigate the idiosyncrasies of EB charging demand in a geographic area. This work's technique is divided into two steps. Scheduling EV charging lessens the pressure on distribution and transmission networks while guaranteeing that all vehicles are charged at the same time. The vehicles have been charged. In the first step, the number of vehicles is counted. Using day-ahead data, the amount to be charged each hour is optimized. In the second stage, the maximum demands for a refund is determined. The number of vehicles that can be charged based on operating conditions for the next hour must be tracked in order to maintain distribution dependability.

13.3 Methodology

Long-distance travel, commuting, cruising, sports (including racing), and off-road riding are just a few of the applications for motorcycles.

Motorcycling includes riding a motorcycle and various related social activities like joining a motorcycle club and attending motorcycle rallies. Electric bikes require far less maintenance than conventional motorcycles. There is no need to maintain an oil or cooling system, and there are fewer parts in general, requiring less care. Furthermore, maintenance costs are low as compared to gasoline and electric motorcycles. You do not need to buy gas or charge the bike. Electric bikes are constantly ready to ride, so there is no need to plan ahead. Riders are free to come and go whenever the mood strikes. Using these algorithms to discover alternative CSs in the vicinity, EB waiting times may be significantly reduced, and EB balance and Smart Grid stability can also be maintained.

The proposed model's main goal is to reduce load demand while also synchronizing charging and discharging activities in order to save money. The Smart Grid's load balancing is in charge of planning EBs for charging or discharging services based on data such the EB profile and price profile. The EBSE selection model will guide the EB to the most appropriate charging or discharging station. The discharging services are provided by the bidirectional power flow between the EB and the smart grid, allowing the EB to serve as a distributed energy storage system.

It is primarily based on a load balancer control version mixed with an EBSE choice scheme, and it is aimed at reducing load balancers caused by a small number of EBs charging at EBSEs by utilizing the unused conserved electricity in EB batteries within the context of Time of Use Pricing (TOUP). Second, it seeks to synchronize charging and discharging processes to improve grid stability while also maximizing charging and discharging time and price. In the suggested design, managing the EBs is as simple as coordinating and managing the charging and discharging activities.

13.3.1 Functional Requirements

- A load balancer algorithm should be used by the rider to charge the EBs.
- A load balancer algorithm should be used by the rider to discharge the EBs.
- The riders can detect the nearest CS.
- The riders are scheduled for the charging services by using the EBSE algorithm.

- The riders are scheduled for the discharging services by using the EBSE algorithm.
- The riders are charging or discharging the EBs with the minimal time.

Figure 13.1 depicts the architecture of the proposed model. It is used to find a nearby CS and determine whether the EB should be charged or discharged. It is principally based on a load balancer control version combined with an EBSE choice scheme and is targeted first at decreasing load balancers due to a handful of EBs charging at EBSEs by utilizing the unused saved electricity in EB batteries within the context of TOUP. Second, it tries to synchronize charging and discharging operations to maximize value (charging/discharging time and price) while also improving grid stability.

The modules that make up the work flow are as follows and are depicted in Figure 13.2:

- Data Collection
- Load Balancing of Smart Grid (LBSR)
- Electric Bike Supply Equipment (EBSE)
- Routing Module

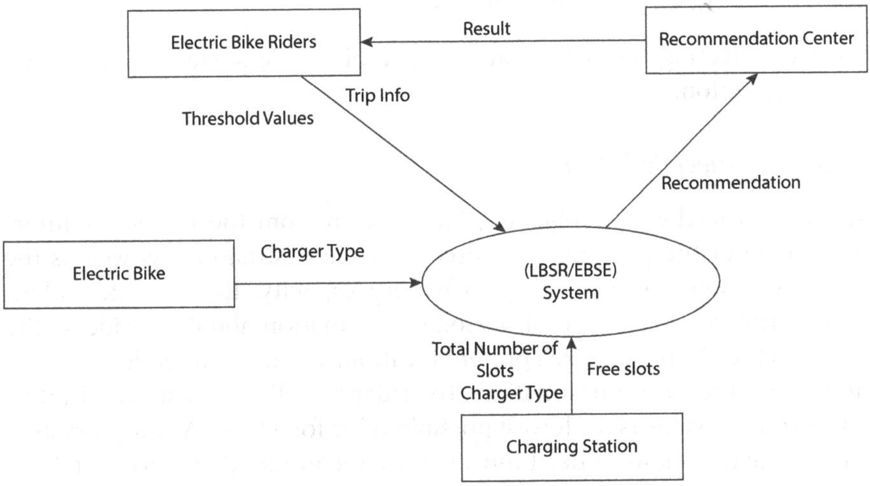

Figure 13.1 System architecture.

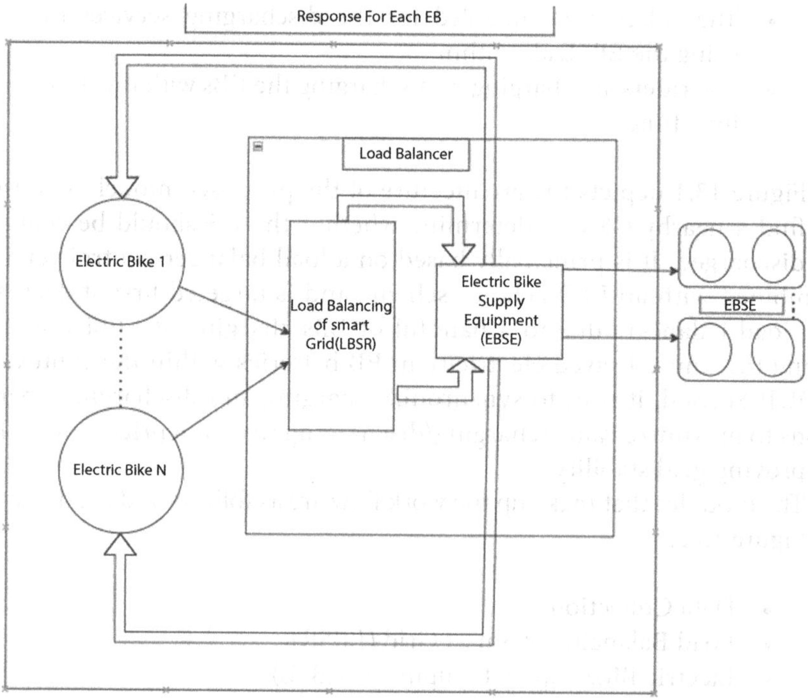

Figure 13.2 Modules of the proposed work.

13.3.2 Data Collection Process

We are collecting the data from: electric bike riders, electric bikes, and charging station.

13.3.2.1 Electric Bike Riders

The system fetches the following information from the EB riders: information about the person, the source, and the destination, as well as the threshold value corresponding to battery capacity. Electric bike riders' information is a collection of personal information about the riders. The source is the electronic bike's present location. The user's intended destination is referred to as a destination. The rider establishes a threshold value. The threshold value is the lowest possible value for a bike. When the system crosses the threshold value, it must make a recommendation to the bike.

13.3.2.2 Electric Bikes

The system fetches the following information from the EBs: driving efficiency, plug type, max capacity, maximum trip distance, charging curves, and state of charge are all factors to consider. Because driving economy is dependent on the road, it will vary. It may be 0.9 on the highway. In rural areas, it is 0.6, whereas, in cities, it is 0.8. Every electronic bike has a number of different sorts of charging and discharging facilities. At least one plug type is present in an electronic bike, and it may have a variety of plug kinds. Charging is accomplished with this plug type.

13.3.2.3 Charging Station

The system fetches the following information from the CS: number of slots, number of vacant slots, plug type, longitude, and latitude. There is a plug type for every electrical bike. These plugs are used to charge electric automobiles. There are number of slots in this. The entire number of slot machines is represented by this number. There are no openings available. It is done in real time to display the number of available slots at a certain CS.

13.4 Load Balancing of Smart Grid

There is communication between the Smart Grid and the EB in the load balancer algorithm control. The EB profiles, such as the position of the EB, the current status of the charge of the EB, the trip required by the EB before the charging process, and the maximum waiting time tolerated by the EB, are shared before the start of the fall season. Based on the EB profile and vehicle requirements, determine which EB participates in which procedure, such as charging and discharging. After determining which procedure to use for the EB, such as charging or discharging, and carrying out that procedure. The demand for electricity is the core notion of load balancer control. The LBSR algorithm is primarily motivated by the rising cost of electricity.

EB travel plans are communicated with Smart Grid at the outset, and the LBSR algorithm assesses how much energy is required for the EB to go each time a journey is booked. That is, to determine whether a suitable amount of battery power is required and whether the current state of charge in the EB is adequate. If the current charge level is insufficient, then the LBSR model can be used to schedule the charging of the EB. The EB is charged up to one threshold level or the desired level of the present charge condition.

If EB does not have any plans to travel, then he or she will be discharged in the second circumstance. There is a limit to the draining process, just as there is to the charging process (desired level of current state of the charge). That is, until the battery depletion threshold is met, the EB continues to charge.

The profile of the EB is updated with the Smart Grid in the LBSR model, and this operation is repeated every T seconds. It is also possible to adjust the EB's programmed plan at any moment, although this must be done before to the plug-in phase. Load balancer control produces two types of results. The output that we get from an EB can be used to request charging or discharging services.

13.5 Electric Bike Supply Equipment (EBSE) Algorithm

The LBSR algorithm produces the outcome of scheduling the EB for charging or discharging service. The EBSE selection model algorithm's primary goal is to direct the EB to plug-in to maintain smart grid stability during the day. One of the most essential aspects of the EBSE selection model is that it reduces the user's waiting time during the day. Smart Grid transmits or receives real-time data on EBSE statuses (which can be busy or active), electrical energy prices, load period prices, and average daily prices.

We can compute the EB's arrival time at the EBSE by computing two parameters: traffic conditions and the EB's present position. In the fundamentals of first come first serve (FCFS) queueing strategy, we solely need a queueing theory to represent the charging or discharging paradigm. According to the following discipline, the cut-off level for entry to the charging process for the coming EB is as follows:

1) Serve any EB that arrives when the number of busy plugin sockets is less than the cut-off level.
2) Whenever one of the cut-off level busy plugin sockets becomes free, serve an EB scheduled for charging service (EB Ch).
3) An EB planned for discharging service (EB Disch) enters service immediately when a plugin socket becomes free; otherwise, it can be deferred.

We successfully achieved the goal of these projects by implementing the LBSR model and the EBSE model, which was to reduce waiting time and improve the Smart Grid's stability. Introduce the mechanism for calculating alternative supply stations, which should aid users in locating the closest supply station.

13.6 Simulation

To replicate our project, we are employing the SUMO idea. Simulation Urban Mobility (SUMO) is a free, open-source, highly portable network simulation program. It is designed to handle enormous numbers of networks. Based on our LBSR and EBSE algorithms, as well as the time and distance, the bikes are recommended to be charged or discharged at the nearest CS.

- OpenStreetMap: This project is a free and open-source initiative. It is used to obtain a map of any location on the planet. The file downloaded is an OSM file. The acronym OSM stands for Open Street Map.
- Netconvert: This is a utility that imports a road network from many sources and builds a road network that other tools in the imported packages can exploit. We use the OpenStreetMap import file and the net.xml export file in our project. To convert an OSM file to a net.xml format, run the following command: netconvert --osm-files osmfilename.osm -o xmlfilename.net.xml, where --osm-file – denotes input of netconvert and it is an osm file format, -o – denotes output, net.xml – output file extension.
- Polyconvert: This imports geometrical shapes (polygons or points of interest) from different sources and converts them to a representation that may be visualized using sumogui: As input files, we use the osm and type file formats: Polyconvert –osm-files filename.osm –net-file filename.net.xml – type-file filename.typ.xml -o filename.poly.xml, The output of both the command above is a poly.xml file.
- RandomTrips: These are used to generate a vehicle's random journey. We have not given the vehicle any direction. Each vehicle takes a different route. It is a Python program. The sumo directory contains the random trip python file. The version that we are dealing with is Sumo 1.11.0. Python randomTrips.py -n filename.net.xml -r filename.rou. file -e 50 -l, -n denotes that it is a network file, net.xml, and it contains the information about the network; -r denotes a random file, rou. file, and it contains the information about the random routes; -e denotes vehicles, and 50 vehicles are mentioned in the configuration.

Information about the network is stored in a netfile. A roufile is used to hold information about each vehicle route. The supplementary file contains information about extraneous items like reroute, bus stop, and CS, and the file name is saved as .sumocfg or .sumo.cfgy.

13.6.1 Running Simulation

The following command is used to run the simulation:

sumo-gui filename. sumocfg

The simulation will start, and, for Sumo compilation, the following command is used:

sumo -c filename. Sumocfg

13.6.2 Simulation Screenshots

Figures 13.3–13.5 shows the screenshots and the command prompt output from SUMO relevant to the charging or discharging procedure. We can infer the number of EBs that are in charging and discharging lines, as well as how much capacity each EB has.

Figure 13.3 Initial road, charging station, and electric bike movement configuration.

Figure 13.4 A queueing model depicts electric bikes in the charging or discharging process.

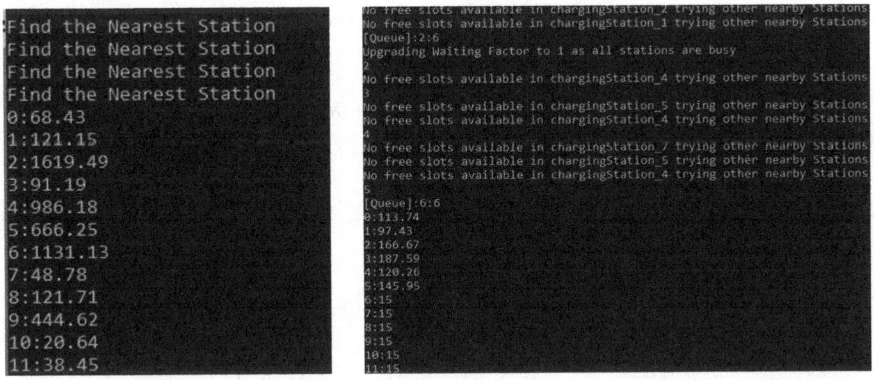

Figure 13.5 The SUMO tool displays a command prompt for the peak level of the electric bike charging and discharging queue.

13.7 Performance Metrics

The comparison of the existing model with our suggested model for various periods of time to check the average grid power required by the EB is shown in Figures 13.6–13.8. Different periods are as follows: [0–25] hours, [0–15] hours, and [15–25] hours. When compared to the existing models, our suggested model demonstrates a reduction in average grid power, as shown in Figures 13.6–13.8.

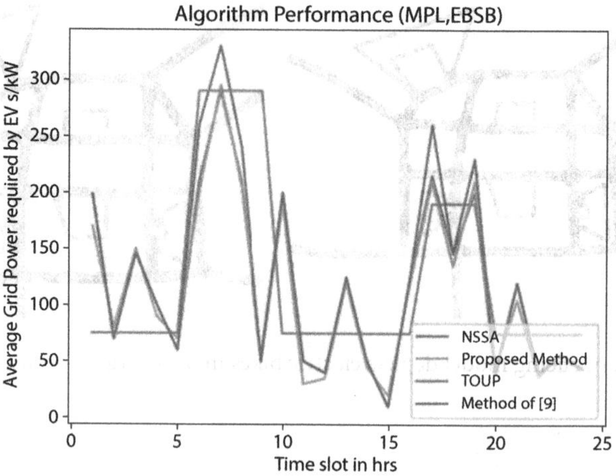

Figure 13.6 Time slot (hours) vs. average grid power required by EV in 24 h.

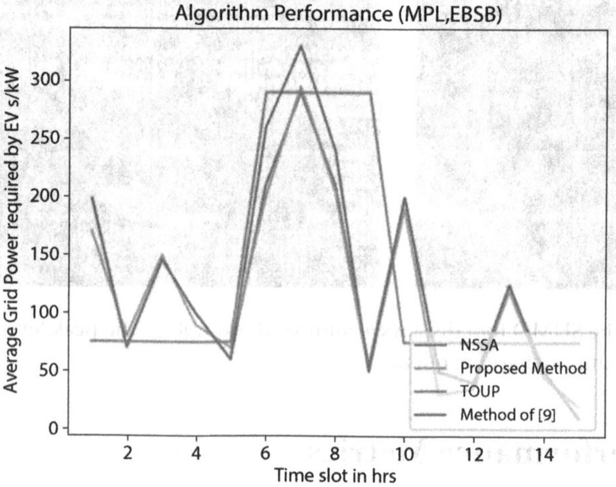

Figure 13.7 Time slot (hours) vs. average grid power required by EV in [0-15] h.

13.8 Method and Existing Method

Figure 13.9 shows how the existing approach reduces the time it takes for EBs to charge at the CS and how our suggested model reduces the time it takes for EBs to charge at the CS. Our proposed model curve is lower at all locations along the average waiting time axis when compared to the existing model.

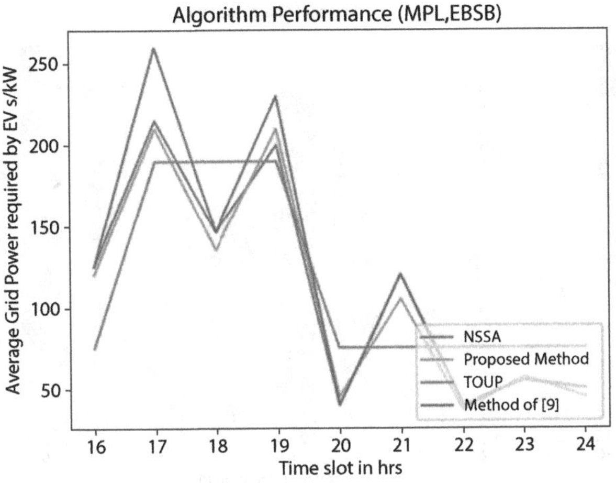

Figure 13.8 Time slot (hours) vs. average grid power required by EV in [15–25] h.

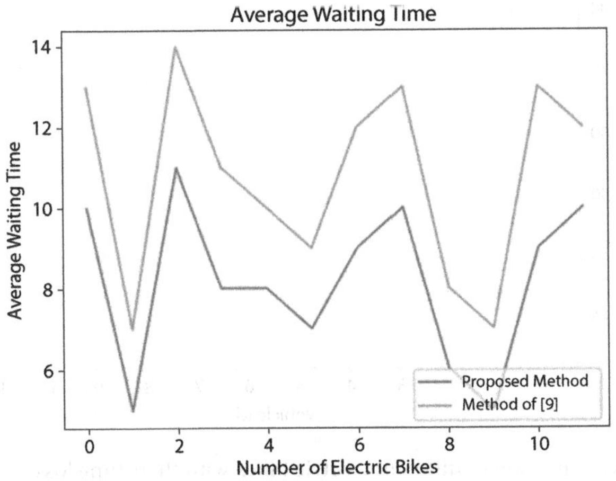

Figure 13.9 Number of electric bikes vs. average waiting time with our proposed model.

The power capacity of various CSs is depicted in Figure 13.10. The different CSs have varying power capacities. In the graph above, there is an up and own variety.

Figure 13.11 shows how different EBs compare in terms of time loss. The amount of time lost by various vehicles varies. The time loss of the eleventh bike in this graph is significant, indicating that more time is required.

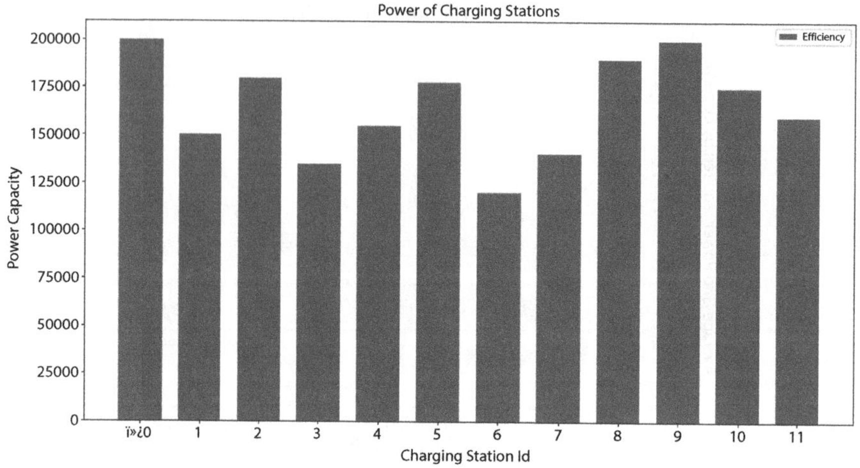

Figure 13.10 Power capacity of different charging stations.

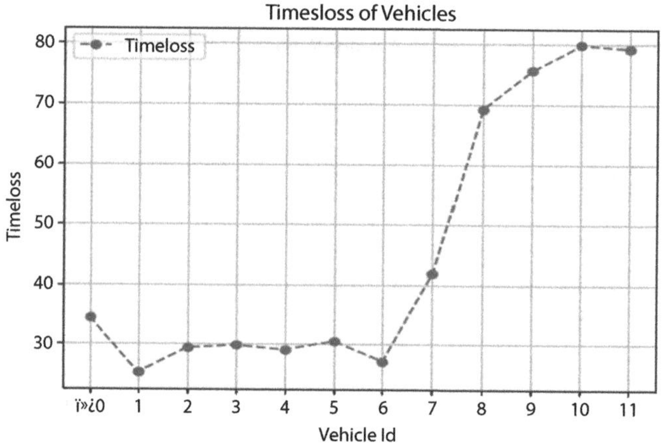

Figure 13.11 Comparison of different electric bikes with their time loss.

Conclusion and Future Work

In this work, we provide the EBs' charging/discharging services based on the First-In–First-Serve Model; the ultimate aim of our proposed model is to coordinate the charging and discharging process. In our proposed model, two algorithms were used: the LBSR and the EBSE algorithms. LBSR is used to decide whether the EB needs to be in the charging process or in discharging process, and our algorithm uses the unused energy available from the EB that is unused energy taken from the EB and conserved via the Smart Grid and supply to the EBs, in which bikes need to be charging [11]. The second algorithm is to EBSE, and the main responsibility of this algorithm is to schedule the bike for the charging or discharging process. These two algorithms are tested through the SUMO Simulation, which coordinates the EB charging and the EB discharging process in an efficient way.

We plan to expand our suggested EB charging/discharging scheme into a global guiding system for EBs in the future, which will be viewed as a Smart Grid enabler. This suggestion is beneficial. The methodology might be more difficult, particularly when considering the energy transaction features in a microgrid idea. Moreover, EBs can also be regarded as a major player in the public electricity sector, notably in markets with many billing service providers, and in the framework of smart cities with an additional limitation of a free market, for instance, possible limitations could be included in the constraints. Incentives for EB owners to allow them to charge their bikes at load times when the price and demand are both high, which is a good moment to buy, help to maintain grid stability.

References

[1] C. Farkas and L. Prikler. (2012). Stochastic modeling of EB charging at charging stations. *ICREPQ*, Spain, 1046–1051.

[2] Wang, M., Ismail, M., Zhang, R., Shen, X., Serpedin, E. and Qaraqe, K. (2018). Spatio-Temporal Coordinated V2V Energy Swapping Strategy for Mobile PEVs. *IEEE Trans. Smart Grid*, 9(3), pp. 1566–1579. doi: https://doi.org/10.1109/tsg.2016.2593667.

[3] Arora, K.; Kumar, A.; Kamboj, V.K.; Prashar, D.; Jha, S.; Shrestha, B.; Joshi, G.P. Optimization Methodologies and Testing on Standard Benchmark Functions of Load Frequency Control for Interconnected Multi Area Power System in Smart Grids. *Mathematics*, 2020, 8, 980.

[4] Aveklouris, A., Vlasiou, M. and Zwart, B. (2019). A Stochastic Resource-Sharing Network for Electric Vehicle Charging. *IEEE Trans. Control Netw.*, 6(3), pp. 1050–1061. doi: https://doi.org/10.1109/tcns.2019.2915651.

[5] Khorramdel, B., Khorramdel, H., Aghaei, J., Heidari, A. and Agelidis, V.G. (2015). Voltage Security Considerations in Optimal Operation of BEVs/PHEVs Integrated Microgrids. *IEEE Trans. Smart Grid*, 6(4), pp. 1575–1587. doi: https://doi.org/10.1109/tsg.2015.2394499.

[6] Qin H. and Zhang W. (2011) Charging scheduling with minimal waiting in a network of electric vehicles and charging stations. *8th ACM Int. Workshop VANET*, 51–60.

[7] Zeng, M., Leng, S., Zhang, Y. and He, J. (2018). QoE-Aware Power Management in Vehicle-to-Grid Networks: A Matching-Theoretic Approach. *IEEE Trans. Smart Grid*, 9(4), pp. 2468–2477. doi: https://doi.org/10.1109/tsg.2016.2613546.

[8] Aravinthan, V. and Jewell, W. (2015). Controlled Electric Vehicle Charging for Mitigating Impacts on Distribution Assets. *IEEE Trans. Smart Grid*, 6(2), pp.999–1009. doi: https://doi.org/10.1109/tsg.2015.2389875.

[9] De Weerdt, M. M., Gerding, E H., Stein S., Robu V. and Jennings N. R. (2013) Intention-aware routing to minimize delays at electric vehicle charging stations. *23rd Int. Joint Conf. Artif. Intell.*, 83–89.

[10] Xydas, E., Marmaras, C., Cipcigan, L.M., Jenkins, N., Carroll, S. and Barker, M. (2016). A data-driven approach for characterising the charging demand of electric vehicles: A UK case study. *Appl. Energy*, [online] 162, pp. 763–771. doi: https://doi.org/10.1016/j.apenergy.2015.10.151.

[11] R. Sahani, M. Daga, I. Jamatia, P. Kumar and S. L. Tripathi, "Design of a self-charging electric vehicle with embedded multiple natural energy sources," *2023 IEEE Devices for Integrated Circuit (DevIC)*, Kalyani, India, 2023, pp. 296-301, doi: 10.1109/DevIC57758.2023.10134774.

14

Self-Charging Electrical Vehicle Design and Analysis with MATLAB

Ramchandra Sahani, Ishak Jamatia, Manas Daga, Pavan Kumar and Suman Lata Tripathi*

SEEE, Lovely Professional University, Jalandhar, Punjab, India

Abstract
Limited petroleum resources and the high cost of transportation are major reasons that pull researchers to work in the area of making hybrid electric vehicles. Also, the requirement to control air pollution and noise pollution makes hybrid electric vehicles a suitable choice and useful for a sustainable environment. This chapter describes the integration of electric vehicles with natural energy resources instead of conventional fuel. Solar and wind energy are taken for integration with battery charging systems through the vehicle itself. Solar panels are fitted in the roof and sides of the vehicle while a wind turbine is proposed to be fitted on the wheels. Design parameters are analyzed using MATLAB tool for minimum speed, required torque, and supply voltage. The overall system includes different blocks such as motor control, battery management system, charging system, solar panel, and wind turbine. In this chapter, a prototype model is demonstrated for electric vehicles by integrating multiple natural energy resources.

Keywords: Natural energy resources, electric vehicle, MATLAB, smart charging, battery management system

14.1 Introduction

Non-renewable resources such as coal and petroleum typically power electric vehicles (EVs), but recent research has explored the use of hybrid or battery EVs, as well as the charging of Li-ion batteries with renewable

**Corresponding author*: tri.suman78@gmail.com

Krishan Arora, Suman Lata Tripathi and Himanshu Sharma (eds.) Electric Vehicle Design: Design, Simulation and Applications, (313–334) © 2024 Scrivener Publishing LLC

energy sources and smart charging concepts. Solar energy has shown promise in this regard, with advances in battery charging and panel optimization algorithms and techniques. This study proposes a modification to an electric vehicle that runs solely on renewable energy sources, specifically solar and wind energy [1–5]. The vehicle is equipped with solar panels on the roof and a wind turbine installed on the bonnet, which powers the inner circuits and digital boards. Solar panels and wind turbines collect energy that is used to charge the battery, which powers the vehicle. The battery is charged in the presence of solar panels and a wind turbine, using a specific charging mechanism.

14.2 Natural Energy Sources for Self-Charging Electrical Vehicles

The use of renewable energy sources such as solar, wind, and hydropower is essential in the development of self-charging EVs. These vehicles have built-in technology that enables them to collect and generate electricity from renewable energy sources, without relying on traditional charging infrastructure [6–8]. Solar energy is one of the most promising renewable energy sources for self-charging EVs. The solar panels mounted on the vehicle can absorb sunlight and convert it into electricity, which can be used to charge the battery. Additionally, these solar panels can also be used to power the vehicle's air conditioning and other electronic components, reducing the reliance on the battery [9, 10]. Wind energy is another promising renewable energy source for self-charging EVs. Wind turbines can be installed on the vehicle to generate electricity from the wind, which can be used to power the vehicle's inner circuits and digital boards [11–17]. Hydropower is also a potential source of renewable energy for self-charging EVs. The energy generated from water movement can be harnessed using hydroelectric turbines and used to charge emissions, which can only be achieved through carbon capture technologies. Figure 14.1 describes the power controller system of electric vehicle.

The use of these natural energy sources in self-charging EVs has several advantages. It reduces the dependence on traditional charging infrastructure, provides greater independence and convenience to vehicle owners, and reduces carbon emissions by using renewable energy sources. In addition, the cost of using renewable energy sources is decreasing rapidly, making it a more attractive option for self-charging EVs [18–22].

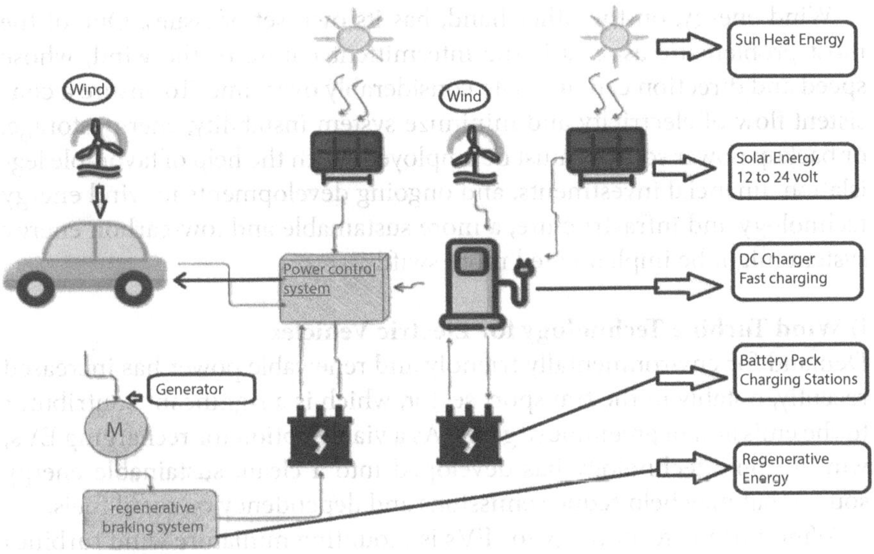

Figure 14.1 Manual power controller system in EV.

14.2.1 Wind Energy

Wind energy is the process of converting the kinetic energy of the wind into a usable source of energy, most commonly electricity. Although it has been a dependable and sustainable source of energy for millennia, interest in it has recently increased due to its ability to meet rising energy demand while reducing greenhouse gas emissions and reliance on fossil fuels. Wind turbines employ large blades that revolve when the wind blows over them to generate electricity. The whirling blades move a shaft connected to a generator to create power. Wind turbines can be built on land or offshore, depending on local legislation and the availability of wind resources.

Wind energy has seen considerable growth in utilization over the last 10 years. With an installed capacity of around 700 GW, wind energy generated 7.2% of world energy consumption in 2020. The top five wind energy producers are China, the United States, Germany, India, and Spain. Wind energy offers several advantages. It is a clean energy source because it is a free and abundant resource that produces no emissions or pollution. Furthermore, because wind energy is a renewable resource, it can always be produced in greater quantities. Local wind energy generation may reduce dependency on imported energy sources while increasing job possibilities in the renewable energy sector.

Wind energy, on the other hand, has its own set of issues. One of the most problematic aspects is the intermittent nature of the wind, whose speed and direction can fluctuate considerably over time. To ensure a consistent flow of electricity and minimize system instability, energy storage, or backup power sources must be employed. With the help of favorable legislation, financial investments, and ongoing developments in wind energy technology and infrastructure, a more sustainable and low-carbon energy system might be implemented more swiftly.

i) Wind Turbine Technology for Electric Vehicles

Demand for environmentally friendly and renewable power has increased recently, notably in the transport sector, which is a significant contributor to the emission of greenhouse gases. As a viable option for recharging EVs, wind turbine technology has developed into a clean, sustainable energy source that may help reduce emissions and dependency on fossil fuels.

Wind turbine technology for EVs is mounting miniature wind turbines on the vehicle that can create power by spinning the blades while the vehicle is in motion. The created electrical may subsequently be collected in a battery and employed to power the electric power plant of the vehicle. Wind turbines can be used to power vehicles in a few different ways. The most common way is to use the energy from wind turbines to charge the battery of an EV. This allows the EV to run for longer distances. Wind turbines can also be used to directly power a vehicle, such as in a wind-powered sailboat. Additionally, wind turbines can be used to supplement the power of a hybrid vehicle, allowing it to run more efficiently. Figures 14.2a and b present the views of outer and inner wind turbine, respectively.

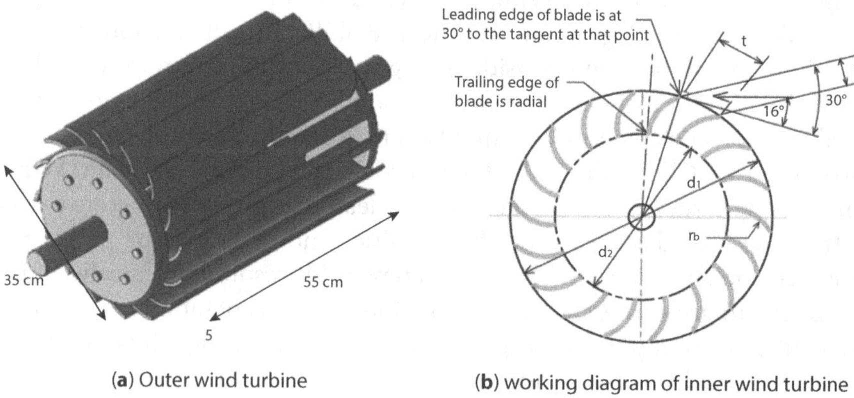

Figure 14.2 (a) Outer wind turbine. (b) Working diagram of inner wind turbine.

Here, d_1 = 18 cm, d_2 = 35 cm, and L = 55 cm.
According to Betz's law, the power from a wind turbine is max.
Pmax = $1/2\rho Cpd_1d_2V^3$, Cp = 16/25, Thus, for the theoretical ideal rotor, its experimental value is a maximum of 0.4 for Savonius turbines.
P = Power output, watts;
Cp = 0.30 experimental value, (theoretical maximum = 0.59);
A = Rotor swept area, 0.1925 m^2;
V = (Wind speed + Vehicle speed), 12.50 m/s;
Applying force = 3.60 N;
$\mathbf{P_{max}}$ = 36.0-watt induced energy with only 10 m/s speed of the vehicle.

A detailed description of vehicle speed and generated energy is described in Table 14.1.

ii) Mechanical Gear Control

Mechanical control systems for wind turbines use mechanical components such as gears, pulleys, and levers to control the blade pitch and rotational speed of the turbine. These types of systems are typically used on small wind turbines and are often integrated with digital control systems to control and tune the turbine's performance.

iii) Alternators

An alternator is an electrical generator that converts mechanical energy into electrical energy in the form of alternating current. Alternators are commonly used in automobiles to charge the battery but here are connected

Table 14.1 Vehicle speed and generated energy at different applied force.

(Vehicle speed + Air speed)	Applying force	Generate energy
10 m/s	3.6 N	36.0 watts
20 m/s	14.4 N	288 watts
30 m/s	32.4 N	972 watts
40 m/s	57.6 N	2304 watts
50 m/s	90 N	4500 watts
60 m/s	129.6 N	7776 watts

to the wind turbine and produce to power the electrical system when the vehicle is running.

Alternators can also be used as a source of renewable energy in vehicles, where they are used to generate electricity from sources such as wind or solar power. With the best NdFeB Permanent Magnet and high-grade pure copper winding, the generator can produce power stably and highly efficiently. Designed with a specialized rotor and rotator, featuring low resistance during startup and efficient cooling. Utilizing a gearless, direct-drive system with a Permanent Magnet Generator, it ensures easy and safe operation and maintenance. Table 14.2 provides the specification of low-rpm alternator.

The basic components of a wind turbine include the rotor blades, a hub that connects the blades to the main shaft, a gearbox that increases the rotational speed of the blade, a generator that converts the rotational energy into electrical energy, and a tower that supports the entire structure, which help the alternator produce a lot of energy. The capacity to generate power while the vehicle is in motion, which removes the need for outside charging infrastructure, is one of the main benefits of using wind turbines for electric cars. This can considerably extend the range of the vehicle and eliminate the need for frequent recharge, making EVs more practical and pleasant for daily use.

Several companies and researchers have investigated the application of wind turbine technology in EVs. Light Year, a Dutch startup, has created a solar-powered electric car with an inbuilt wind turbine that generates additional energy while driving. Similarly, researchers at the University of

Table 14.2 Specification of low-rpm alternator, 600 r/m 100-watt alternator.

Wattage	**100 watts**
Item weight	3 kilograms
Voltage	24 volts
Output wattage	100 watts
Color	Copper
Start torque	0.12 Nm
Output current	AC
Frequency	60 Hz

Texas in Arlington created a prototype EV with a retractable wind turbine capable of producing up to 600 watts of power. However, incorporating wind turbine technology into EVs has some drawbacks. One of the key issues is that small wind turbines only produce a limited quantity of electricity, which may not be sufficient to fully power the vehicle. Furthermore, wind turbines can add weight and complexity to vehicles, reducing efficiency and increasing manufacturing costs. Despite these challenges, the establishment of environmentally friendly modes of transport could benefit substantially from the use of wind turbines for electric cars. The adoption of this groundbreaking technology may be accelerated with the aid of supporting policies and investments, paving the way for a more sustainable future.

iv) Advantages and Disadvantages of Wind Energy

In recent years, interest in wind energy has grown as a practical way to satisfy the increasing energy needs of the globe while decreasing emissions of greenhouse gases and reducing dependence on petroleum and coal. Wind power is a fresh, eco-friendly, and abundant source of energy. However, just like other energy sources, wind energy has benefits and drawbacks.

One of wind energy's main benefits is that it does not emit atmospheric carbon dioxide or other pollutants while producing electricity, making it a sustainable and environmentally friendly source of energy. In contrast to fossil fuels, which are limited and diminishing, wind energy is recyclable, indicating that it may be renewed and is always available.

A further benefit of wind energy is its reasonable price, which has increasingly improved as it positively compares with conventional sources of electricity that depend on petroleum and coal. Thanks to improvements in technological advances, the economics of production, and favorable policies and expenditures, the price of wind energy has dropped substantially. However, there are also certain disadvantages to using wind power. The intermittent nature of the wind, whose velocity and direction can change dramatically over time, is one of the key difficulties. This might lead to grid unpredictability, necessitating the adoption of storage for energy or additional energy sources to guarantee a steady flow of electricity.

Additionally, especially when set up in residential regions, wind farms can have an effect on the local community in terms of sight and sound. The possible negative effects of wind energy on animals and the natural world, especially when blades are erected in sensitive areas or routes of migration are still another drawback. There are worries about the possible effects of offshore turbines on aquatic creatures as well as the possibility of collision between bats and birds and wind turbine blades. Despite these difficulties, wind energy has a glowing future as an important element of the global power mixture. It is crucial to keep improving wind energy infrastructures and technology in order to address problems and maximize the benefits of this sustainable and ecologically benign source of electricity.

14.2.2 Solar Energy

Solar energy has become one of the human greatest potential renewable energy sources right now. To collect sunlight and convert it into power, many technical techniques are applied. Because it is less expensive, abundant, and protective of the environment, solar energy has the potential to replace traditional energy sources such as natural gas, petroleum, and charcoal. The following article will go into further detail on solar power and all its parts.

The sun is a massive gaseous orb that emits energy in the form of heat and light. This natural resource can be produced via photovoltaic, concentrated solar power, or solar cooling and heating systems. The most common method is photovoltaic solar power generation. They are composed of solar cells which use the photovoltaic effect in order to transform sunlight into electricity. Concentrated solar power plants employ mirrors or lenses to focus light into a tiny area, causing temperatures to rise that may be used to generate energy. One of the most significant benefits of sunlight is that it is completely renewable. Solar energy is a type of finite source of energy because the sun is going to continue to shine for billions of years. Solar energy is also environmentally friendly because it emits no carbon dioxide or pollution that causes global warming. As a result, it provides a very appealing alternative for reducing global warming and minimizing our carbon footprint relying on fossil fuels.

However, sunlight can present some challenges. The main impediment is the unpredictable characteristics of solar energy. Because it can only be utilized during the day while is affected by the weather, sunlight is not a constant supplier of power. This indicates that energy storage and grid integration technologies have to be established in order to make solar power a reliable source of electricity. Finally, solar power is a highly promising alternative to existing energy sources. Because it is plentiful and clean, and renewable, it is an attractive option for preventing global warming and reducing reliance on fossil fuels. Solar power has an opportunity to supplant other forms of energy due to advances in technology, the development of energy storage devices, and grid-planned integration in the years to come. Figure 14.3 shows a power controller system of EVs.

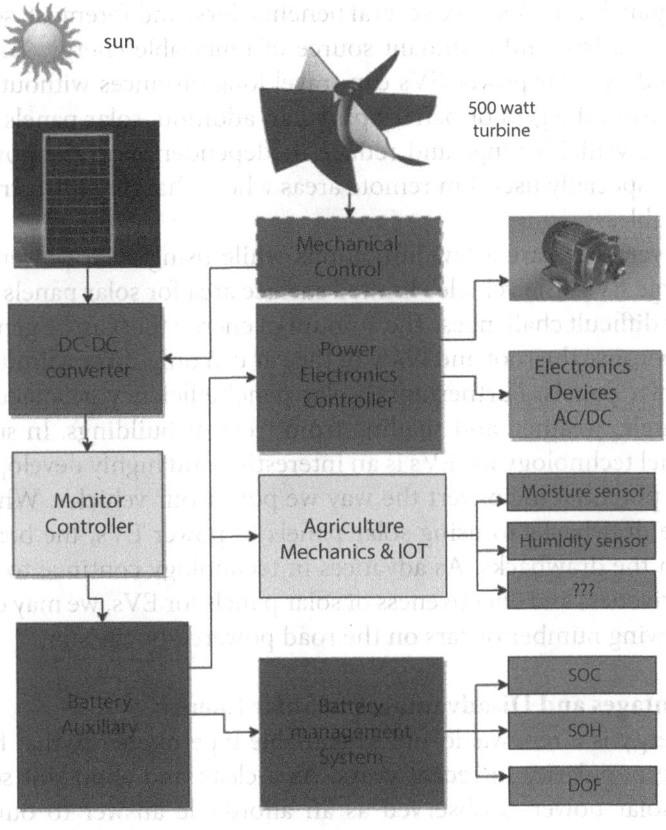

Figure 14.3 Working process of power controller system in EV.

i) Solar Panel Technology for Electric Vehicles

Solar panel technology for EVs is rapidly changing with the potential to modify the way we power our vehicles. The idea behind solar panels for electric cars is simple: instead of fully depending on electricity from the grid or battery packs to power the vehicle, solar panels on the roof or body of the vehicle can assemble energy from the sun and harmonize with the power source. In this essay, we will observe the technology behind solar panels for EVs and evaluate the benefits and drawbacks of using them. Solar panels for EVs apply the same basic technology as solar panels in general. Photovoltaic cells, which convert sunlight into electricity, are used in solar panels' effect. Depending on the size and use of the solar panels, the electricity generated by panels can be used to amplify the power source of the EV or even power it completely.

Solar panels for EVs have several benefits. First and foremost, solar panels provide a free and abundant source of renewable energy. This means that fueled by solar power EVs can travel long distances without needing to refuel from the grid or battery packs. In addition, solar panels can help extend the vehicle's range and reduce its dependence on the power grid, which is especially useful in remote areas where charging infrastructure is not available.

However, there are a few limitations while using solar power stations that charge EVs. The vehicle's limited surface area for solar panels is one of the most difficult challenges. The amount of energy that can be generated is limited because the roof and body of the car can only carry a limited number of solar panels. Furthermore, solar panel efficiency gets factors such as sun angle, weather, and shading from trees or buildings. In summary, solar panel technology for EVs is an interesting and highly developing area with the potential to convert the way we power our vehicles. While there are some drawbacks to using solar panels to power EVs, the benefits far outweigh the drawbacks. As advances in technology continue to improve the effectiveness and effectiveness of solar panels for EVs, we may expect to see a growing number of cars on the road powered by the sun.

ii) Advantages and Disadvantages of Solar Energy

Solar energy is a renewable and sustainable type of energy that has been acquiring popularity in recent years. As a clean and abundant source of energy, solar power is observed as an affordable answer to our energy needs while reducing carbon emissions. Solar energy, like other technologies, offers advantages and disadvantages. In this section, we will observe the advantages and disadvantages of solar energy.

Advantages of Solar Energy

1. Renewable and Sustainable: Solar energy is a renewable and sustainable energy source, so it can be used based as a daily basis without decreasing the source.
2. Clean Energy: Solar energy is a non-greenhouse fuel-emitting, ecologically friendly energy source that does not contribute to climate change.
3. Reduces Electricity Expenses: Using solar energy to power homes and businesses can significantly reduce or eliminate electricity expenses.
4. Low Maintenance: Solar panels require less maintenance, making them a low-cost long-term investment.

Figure 14.4 Manual power controller system in EV.

5. Energy Independence: Solar energy lets households and companies generate their own electricity and decrease their reliance on the grid.

Disadvantages of Solar Energy

1. High Initial Investment: The cost of installing solar panels has been certified as too expensive for particular households and companies.
2. Intermittency: As solar energy is only obtainable during the day and is affected by weather conditions, it is considered an intermittent source of energy.
3. Energy Storage: Without an efficient method of conserving extra energy, it cannot be used at night if it is used during the day, making it less practical for certain purposes.
4. Land Use: Large-scale solar farms require an extensive amount of land, which has the capacity for disrupting ecosystems and habitats.
5. Manufacturing and Recycling: The manufacturing and recycling of solar panels can have an impact on the environment and demand the use of resources. Figure 14.4 presents a prototype model of the EV design [22].

14.3 Arduino-Based Control Systems in Electric Vehicles

Hardware Design and Selection: Designing and selecting hardware for an Arduino-based control system in an EV requires careful consideration of various factors. Here are some key points to keep in mind:

Voltage and Current Ratings: EVs typically operate at high voltages and currents, which can be dangerous if not handled properly. Make sure to select components that meet the voltage and current requirements of your EV and that are designed to handle the high power levels involved.
Processor Speed and Memory: The Arduino platform comes in various models with differing processing power and memory capacity. Choose a model that is suitable for the specific control tasks you have in mind, and that has ample memory to store the required software and data.
Sensors and Actuators: EV control systems rely heavily on sensors to monitor various parameters and actuators to control various subsystems, such as the motor, battery, and charging system. Make sure to select sensors and actuators that are compatible with the Arduino platform and that meet the accuracy and reliability requirements of your EV.
Communication Interfaces: EV control systems often require communication with external devices, such as displays, chargers, and other vehicles. Choose communication interfaces that are compatible with the Arduino platform and that meet the communication requirements of your EV.
Environmental Considerations: EVs operate in harsh environments, with high temperatures, vibrations, and electromagnetic interference. Make sure to select components that are designed to withstand these conditions, and that are properly insulated and shielded.
Safety Considerations: EVs pose unique safety challenges, such as high voltages, high currents, and the potential for thermal runaway in the battery. Make sure to design and select components that meet the safety standards and regulations applicable to your EV and that incorporate appropriate safety features, such as fuses, overcurrent protection, and thermal management.
Software Design and Programming: Designing and programming control systems for EVs using an Arduino platform can be a complex task, but there are several key steps you can follow to ensure success.
Define the System Requirements: Begin by defining the functional requirements of the control system. This includes the specific tasks the system needs to perform, such as monitoring battery voltage, controlling motor speed, and managing regenerative braking.
Select the Components: Once you have defined the system requirements, select the appropriate hardware components such as sensors, actuators, and communication modules. Ensure that these components are compatible with the Arduino platform and can be easily integrated into the system.
Develop the Software: Develop the software to control the system components. This may involve writing code to read sensor data, control the motor, or manage the battery charging system. The code should be written

in a structured and modular way to ensure ease of maintenance and future updates.
Test the System: Test the system to ensure that it meets the requirements and performs as expected. This may involve both hardware and software testing, such as checking sensor readings, motor performance, and communication between components.
Optimize the System: Once the system is tested and functional, optimize the code to improve system performance and efficiency. This may involve optimizing code for faster execution, reducing power consumption, or improving the accuracy of sensor readings.
Document the System: Lastly, document the system design and software code for future reference. This will help to ensure that the system can be maintained and updated by other developers in the future.

14.4 MATLAB-Based Simulation and Modeling for Self-Charging Electric Vehicles

The battery model is a critical component of any simulation and modeling software used for self-charging EVs. It models the behavior and characteristics of the battery used in the EV, such as its capacity, voltage, and current output. Here are some key considerations for developing a battery model in MATLAB:

i) Battery Model: This component models the behavior and characteristics of the battery used in the EV. Figure 14.5 presents a battery management system model on MATLAB.
ii) Battery Chemistry: The first step in developing a battery model is to understand the chemistry of the battery. Different chemistries have different characteristics, such as the rate of charge and discharge, the number of cycles that they can perform, and their temperature sensitivity. Identify the specific chemistry of the battery used in the EV and use this information to develop the battery model. Figure 14.6 presents a battery power calculation model.
iii) Electrical Characteristics: The battery model should also incorporate the electrical characteristics of the battery, such as its voltage and current output. This information can be obtained through testing and experimentation and can be used to develop a mathematical model that accurately represents the battery's behavior.
iv) Capacity: The battery model should include the capacity of the battery, which is the amount of energy that can be stored in the battery.

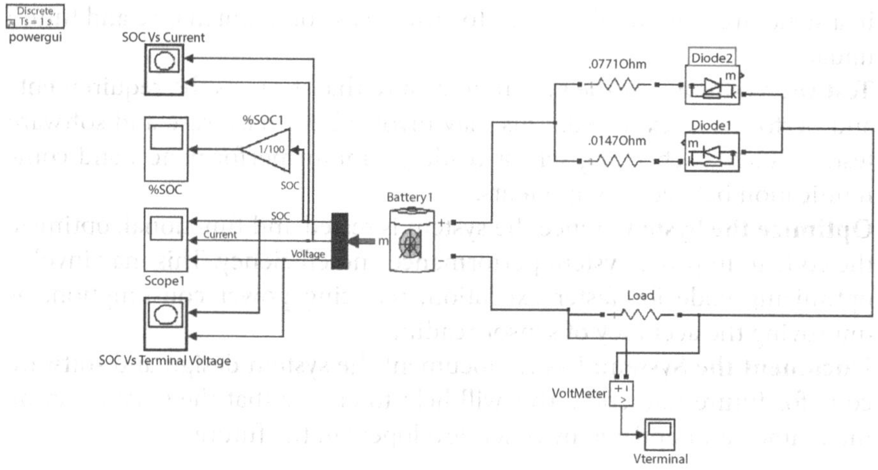

Figure 14.5 Battery management system.

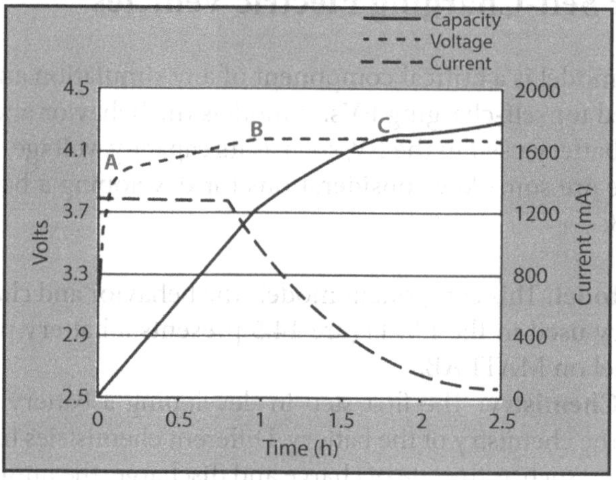

Figure 14.6 Battery power calculations.

This information can be determined through testing and can be used to calculate the range of the EV.

v) Temperature Sensitivity: The battery model should also incorporate the temperature sensitivity of the battery, as this can affect the battery's performance and life span. This information can be obtained through experimentation and can be incorporated into the battery model using temperature-dependent equations. Figure 14.7 shows battery charging capacity at different temperature.

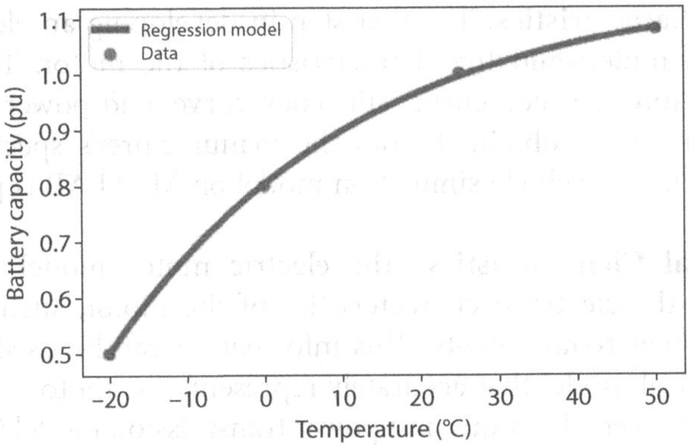

Figure 14.7 Battery charging and temperature.

14.5 Electric Motor Model

This component models the behavior and characteristics of the electric motor used to drive the vehicle. MATLAB-based simulation and modeling is used for self-charging EVs. The electric motor model simulates the behavior and characteristics of the motor used in the EV, such as torque output and efficiency. Here are some key considerations for developing an electric motor model in MATLAB as shown in Figure 14.8.

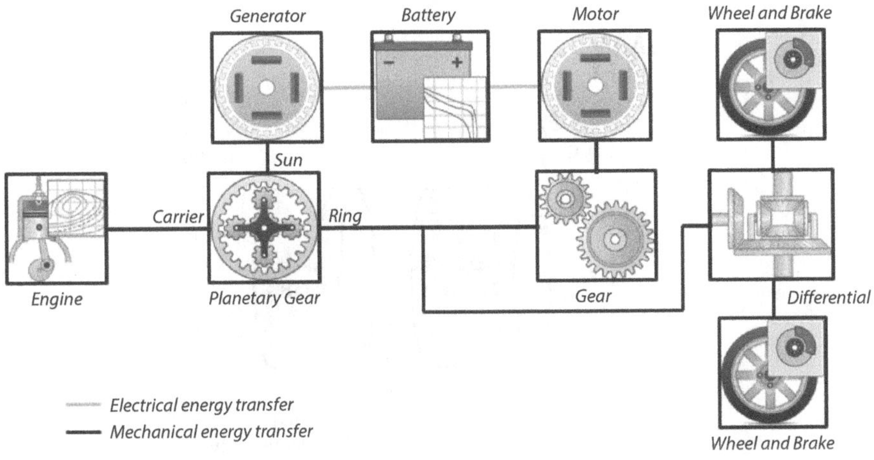

Figure 14.8 Basic EV simulation: block diagram.

i) Motor Characteristics: The first step in developing an electric motor model is to understand the characteristics of the motor. This includes the motor's torque-speed curve, efficiency curve, and power rating. This information can be obtained from the manufacturer's specifications or through testing. A vehicle simulation model on MATLAB is presented in Figure 14.9.

ii) Electrical Characteristics: The electric motor model should also incorporate the electrical characteristics of the motor, such as its voltage and current requirements. This information can be used to develop a mathematical model that accurately represents the motor's behavior. In Figure 14.10, a vehicle torque and power transmission model is shown.

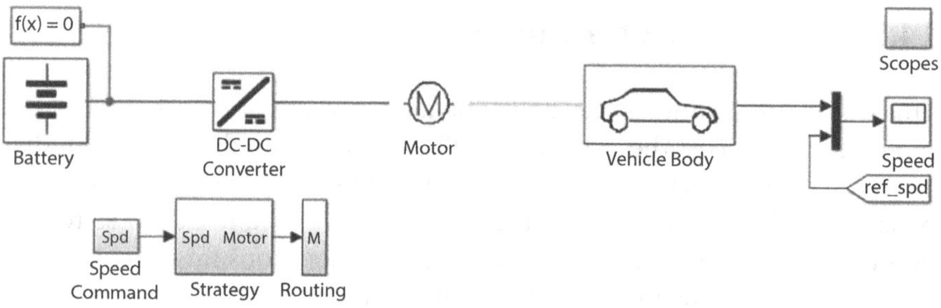

Figure 14.9 MATLAB EV simulation model.

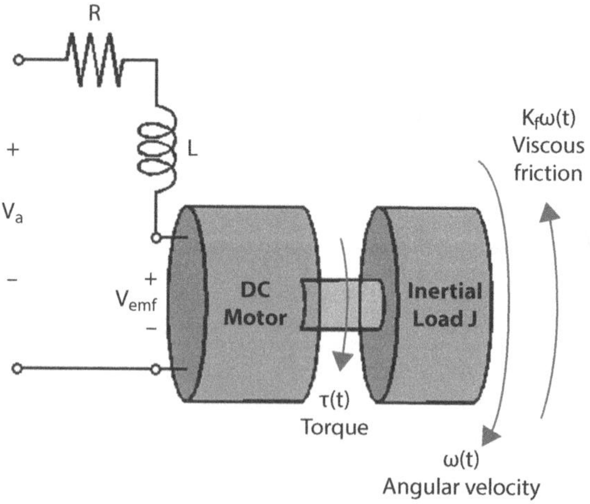

Figure 14.10 Torque and power transmission.

14.6 Results for Vehicle Performance

The electric motor model should also include the control system used to operate the motor, such as a pulse width modulation controller. The control system can be modeled using MATLAB's simulation software, which allows for the simulation of the control algorithm and the motor's response to different inputs. Figure 14.11 shows the performance of the motor in the EV model.

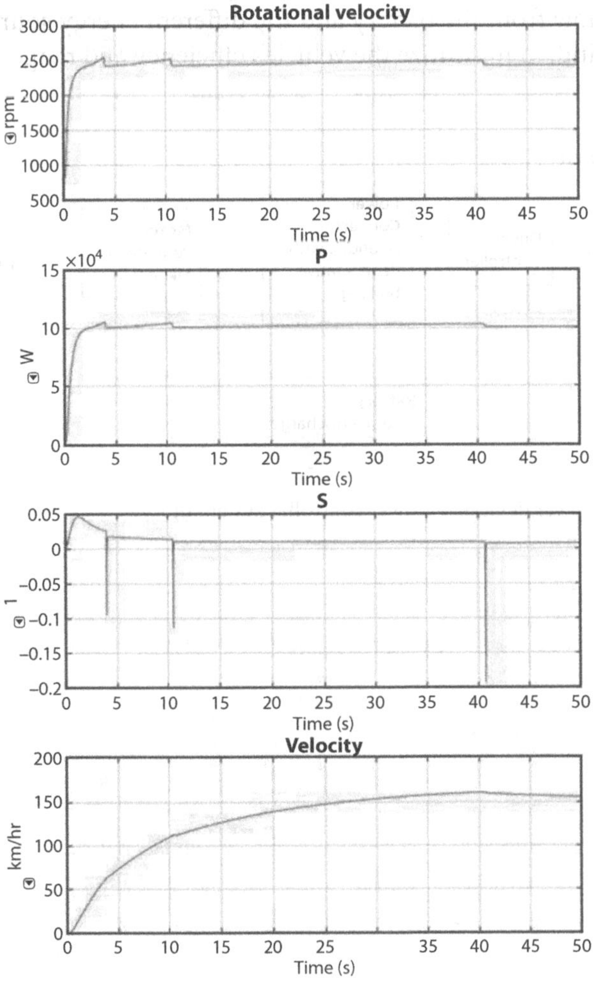

Figure 14.11 Performance of motor system in EV model.

14.7 Power Electronics Model

This component models the power electronics used to control the power flow from the battery to the motor. Figure 14.12 shows use of different power devices in the power controller of EV model. Figure 14.13 shows an elaborated simulation model of power controller of EV in MATLAB.

14.8 Energy Management System

This component models the energy management system used to manage the power flow from the battery and the different energy sources, such as solar and wind, to maximize the vehicle's efficiency and range.

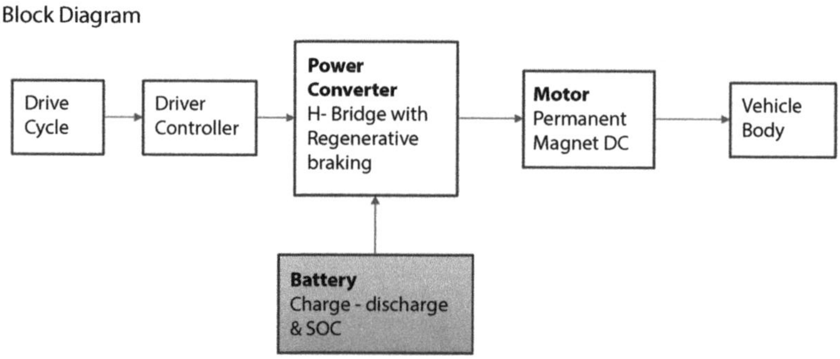

Figure 14.12 Power devices in power controller system of EV.

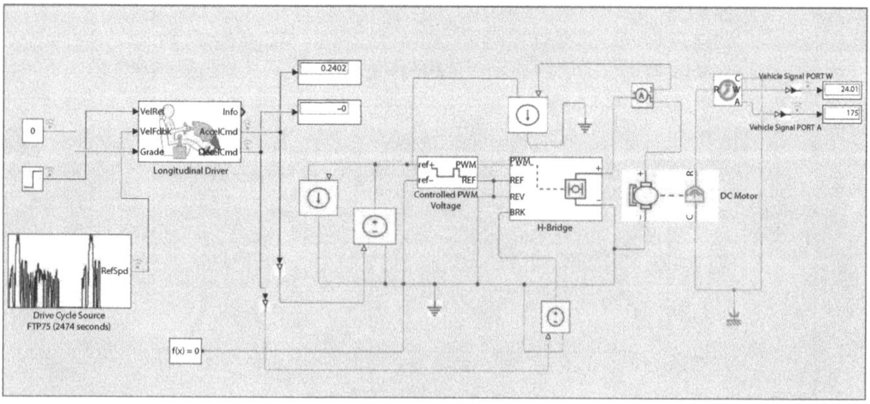

Figure 14.13 Manual power controller system in EV.

i) Back EMF: Electric Vehicle (EV) is a new type of vehicle, which has been introduced in the Indian market recently. These vehicles are propelled by electric motors, which are powered by rechargeable battery packs. The electric motor that powers the wheels of the car plays an important role during braking. When the brake pedal is pushed, the regenerative braking circuit switches the motor on such that it now operates in reverse to offset the direction of the wheels. Electric vehicles have many advantages over non-electric vehicles, one of which is the braking system. The generated voltage must be increased through voltage boosters that increase voltage level during regenerative braking of BLDC motors through a coil, diode, and semiconductor MOSFET switch. A DC-DC boost converter is a suitable choice for voltage amplifications or boosting. Figure 14.14 shows the regenerative power system designed in MATLAB.

ii) Solar energy: A MATLAB-based solar power simulation model is shown in Figure 14.15. When the temperature of the solar panels continues to rise to 144°F, which is about 62°C, the efficiency will drop slightly. However, the world's countries rarely get hotter than 62°C, including the Sahara Desert, which is close to having the highest temperature in the world. In most countries of the world, the temperature remains between 15°C and 35°C throughout the year or except in winter.

iii) Vehicle Model Wheel Transmission: This component models the vehicle itself, including its physical characteristics and dynamics, such as weight, drag, and tire friction. This electric car is designed that will be charged from a natural source to make this vehicle attractive enough by using wind energy, solar energy, and regenerative energy in it. In this, a

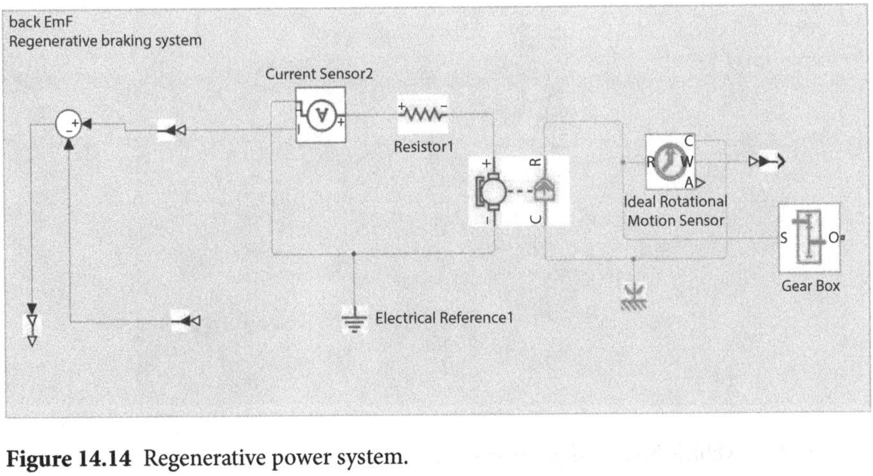

Figure 14.14 Regenerative power system.

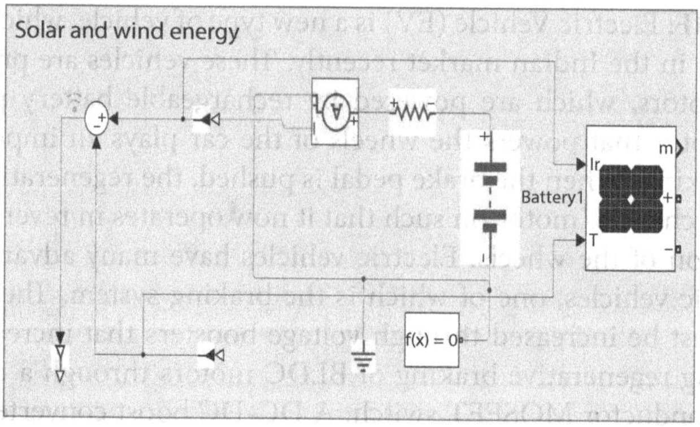

Figure 14.15 Solar power controller.

charging station will be installed for backup power of the vehicle, which will be fully dependent on the energy-powered battery. The presented EV will be useful in many areas like agriculture and transport. Apart from transportation systems, farmers can use these added features of transport in farms without using other costly power resources [22]. Incorporating various sensors into electric vehicle (EV) systems to enhance the efficiency and resourcefulness of our current EV technology. Figure 14.16 shows vehicle body and wheel setup model on MATLAB.

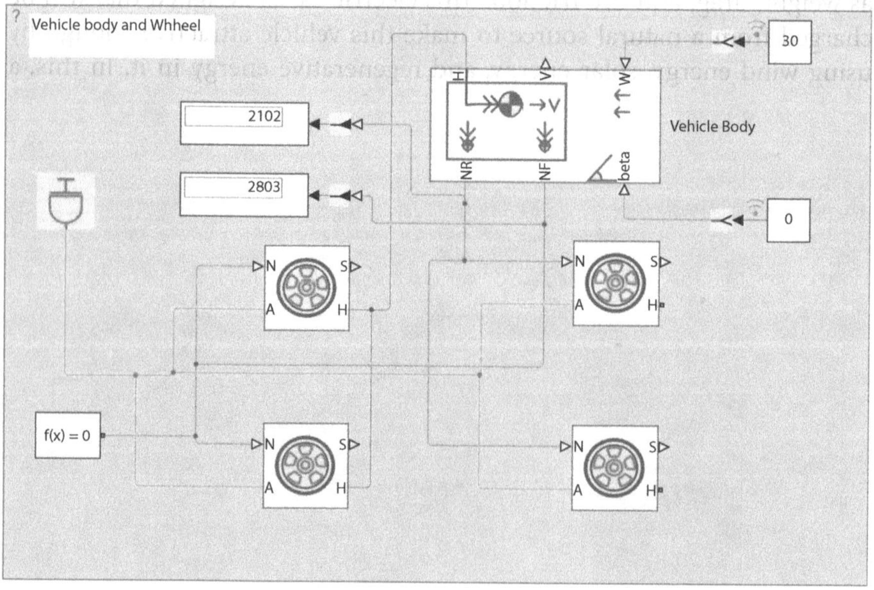

Figure 14.16 Vehicle body with a wheel setup.

14.9 Conclusion

This study mainly provides a glimpse of the design and simulation of an EV in MATLAB to evaluate for design parameters and performance of an EV. The vehicle shows the integration of a solar and wind energy management system for battery charging, and an EV model is developed for specific application at lower cost specifically for agriculture transport with an eco-friendly sustainable energy system.

References

[1] K. W. E. Cheng, "Recent development on electric vehicles," *2009 3rd International Conference on Power Electronics Systems and Applications (PESA)*, Hong Kong, China, 2009, pp. 1-5.

[2] B. Frieske, M. Kloetzke and F. Mauser, "Trends in vehicle concept and key technology development for hybrid and battery electric vehicles," *2013 World Electric Vehicle Symposium and Exhibition (EVS27)*, Barcelona, Spain, 2013, pp. 1-12.

[3] A. Gaurav and A. Gaur, "Modelling of Hybrid Electric Vehicle Charger and Study the Simulation Results," *2020 International Conference on Emerging Frontiers in Electrical and Electronic Technologies (ICEFEET)*, Patna, India, 2020, pp. 1-6.

[4] R. B. Singh and N. A. Shrivastava, "Viability analysis of Self-governing Hybrid Electric Vehicle Charging System," *2021 International Conference on Control, Automation, Power and Signal Processing (CAPS)*, Jabalpur, India, 2021, pp. 1-6.

[5] Simran, S. L. Tripathi, "Development of Solar Cell," Nova Publishers, USA, Feb. 2020.

[6] S. L. Tripathi, S. Ganguli, A. Kumar, T. Magradze *Intelligent Green Technologies for Sustainable Smart Cities*, Scrivener Publishing & Wiley, USA 2022.

[7] S. L. Tripathi, S. Padmanabhan (eds). *Green Energy: Fundamentals, Concepts, and Applications*, Scrivener Publishing, Wiley 2020.

[8] S. L. Tripathi, M. K. Dubey, V. Rishiwal, S. Padmanabhan, *Introduction to AI techniques for Renewable Energy System*, Boca Raton, FL: CRC Press, 2021.

[9] Kumari, C.L., Kamboj, V.K., Bath, S.K. *et al.* A boosted chimp optimizer for numerical and engineering design optimization challenges. *Engineering with Computers* (2022). https://doi.org/10.1007/s00366-021-01591-5.

[10] S. L. Tripathi, P. A. Alvi, U. Subramaniam (eds.). *Electrical and Electronic Devices, Circuits and Materials: Technological challenges and Solutions*, Scrivener Publishing, Wiley, 2021.

[11] J. Porter, "Lightyear One is a prototype 'solar car', with 450 miles of range", June 25, 2019.

[12] Tripathi S. L., Sinha S. K. (eds.). *Recent Advancement in Electronic Devices, Circuit and Materials*, New York: Nova Science Publishers, 2010.

[13] Goyal, N., Sharma, S., Kumar Rana, A., & Tripathi, S.L. (eds.). *Internet of Things: Robotic and Drone Technology*, 2022 (1st ed.), CRC Press.

[14] Tripathi S. L., Alvi P. A., Subramaniam U. (eds.). *Electrical and Electronic Devices, Circuits and Materials: Technological Challenges and Solutions*, Scrivener Publishing & Wiley, USA 2021.

[15] Tripathi S. L., Padmanabhan S., Kumar D., Raja P. (Eds.) *Design and Development of Energy Efficient Systems*, Scrivener Publishing & Wiley, USA 2021.

[16] Thillaiarasu, N., Lata Tripathi, S., & Dhinakaran, V. (eds.). (2022). *Artificial Intelligence for Internet of Things: Design Principle, Modernization, and Techniques* (1st ed.). CRC Press Boca Raton, UK 2022.

[17] Tripathi S. L. "Low Power High Performance Tunnel FET: Analysis for IOT applications" Handbook of Research on the Internet of Things Applications in Robotics and Automation, IGI global publisher, 2020.

[18] Lata Tripathi, S., & Dwivedi, S. (eds.). "Electronic Devices and Circuit Design: Challenges and Applications in the Internet of Things" 2022 (1st ed.). Apple Academic Press.

[19] Arora, K., Tripathi, S.L., & Padmanaban, S. *Smart Electrical Grid System: Design Principle, Modernization, and Techniques* (1st ed.). CRC Press, Boca Raton, UK 2022.

[20] B. S. Ram, S. L. Tripathi, "Operations of Doubly Fed Induction Generators Applied in Green Energy Systems" in *Green Energy: Solar Energy, Photovoltaics, and Smart Cities*, Scrivener Publishing & Wiley, USA 2021.

[21] S. L. Tripathi, S. Padmanabhan, D. Kumar, P. Raja (eds.). *Design and Development of Energy Efficient Systems*, Scrivener Publishing & Wiley, USA 2021.

[22] R. Sahani, M. Daga, I. Jamatia, P. Kumar and S. L. Tripathi, "Design of a self-charging electric vehicle with embedded multiple natural energy sources," *2023 IEEE Devices for Integrated Circuit (DevIC)*, Kalyani, India, 2023, pp. 296-301, doi: 10.1109/DevIC57758.2023.10134774.

Index

Also of Interest

By the same editors

INDUSTRIAL CONTROL SYSTEMS, Edited by Vipin Chandra Pal, Suman Lata Tripathi, and Souvik Ganguli, ISBN: 9781119829256. This volume serves as a comprehensive guide in the journey of industrial control systems with a multidisciplinary approach to the key engineering problems in the 21st century.

MODERN AUTOMOTIVE ELECTRICAL SYSTEMS: Theory and Applications, Edited by Pedram Asef, Sanjeevikumar Padmanaban, and Andrew Lapthorn, ISBN: 1119801047. Presenting the concepts and advances of modern automotive electrical systems, this volume, written and edited by a global team of experts, also goes into the practical applications for the engineer, student, and other industry professionals.

NANODEVICES FOR INTEGRATED CIRCUIT DESIGN, Edited by Suman Lata Tripathi, Abhishek Kumar, K. Srinivasa Rao, and Prasantha R. Mudimela, ISBN: 9781394185788. Written and edited by a team of experts in the field, this important new volume broadly covers the design of nano-devices and their integrated applications in digital and analog integrated circuits (IC) design.

EXPLAINABLE MACHINE LEARNING MODELS AND ARCHITECTURES: Real-Time System Implementation, Edited by Suman Lata Tripathi and Mufti Mahmud, ISBN: 9781394185849. This cutting-edge new volume covers the hardware architecture implementation, the software implementation approach, and the efficient hardware of machine learning applications.

MACHINE LEARNING TECHNIQUES FOR VLSI CHIP DESIGN, Edited by Abhishek Kumar, Suman Lata Tripathi, and K. Srinivasa Rao, ISBN: 9781119910398. This cutting-edge new volume covers the hardware architecture implementation, the software implementation approach, and the efficient hardware of machine learning applications with FPGA or CMOS circuits, and many other aspects and applications of machine learning techniques for VLSI chip design.

INTELLIGENT GREEN TECHNOLOGIES FOR SMART CITIES, Edited by Suman Lata Tripathi, Souvik Ganguli, Abhishek Kumar, and Tengiz Magradze, ISBN: 9781119816065. Presenting the concepts and fundamentals of smart cities and developing "green" technologies, this volume, written and edited by a global team of experts, also goes into the practical applications that can be utilized across multiple disciplines and industries, for both the engineer and the student.

DESIGN AND DEVELOPMENT OF EFFICIENT ENERGY SYSTEMS, edited by Suman Lata Tripathi, Dushyant Kumar Singh, Sanjeevikumar Padmanaban, and P. Raja, ISBN 9781119761631. Covering the concepts and fundamentals of efficient energy systems, this volume, written and edited by a global team of experts, also goes into the practical applications that can be utilized across multiple industries, for both the engineer and the student.

Electrical and Electronic Devices, Circuits, and Materials: Technical Challenges and Solutions, edited by Suman Lata Tripathi, Parvej Ahmad Alvi, and Umashankar Subramaniam, ISBN 9781119750369. Covering every aspect of the design and improvement needed for solid-state electronic devices and circuit and their reliability issues, this new volume also includes overall system design for all kinds of analog and digital applications and developments in power systems.

Green Energy: Solar Energy, Photovoltaics, and Smart Cities, edited by Suman Lata Tripathi and Sanjeevikumar Padmanaban, ISBN 9781119760764. Covering the concepts and fundamentals of green energy, this volume, written and edited by a global team of experts, also goes into the practical applications that can be utilized across multiple industries, for both the engineer and the student.

Check out these other related titles from Scrivener Publishing

HYDROGEN ELECTRICAL VEHICLES, Edited by Mehmet Sankir and Nurdan Demirci Sankir, ISBN: 9781394166381. Hydrogen electrical vehicles are an essential component of the "Green New Deal" and this book covers cutting-edge technologies designed for fuel-cell-powered cars.

PAVEMENT ASSET MANAGEMENT, By Ralph Haas, W.R. Hudson, and Lynne Cowe Falls, ISBN: 9781119038702. Whether you are new to managing pavement assets or a seasoned practitioner, this book provides the most up-to-date information in the field on the history, technology, practice and look ahead.

Printed in the USA/Agawam, MA
September 16, 2026

910661.109